Student Lecture Notebook for

LIFE ON EARTH
and
LIFE ON EARTH WITH PHYSIOLOGY

Eighth Edition

Teresa Audesirk
University of Colorado at Denver and Health Science Center

Gerald Audesirk
University of Colorado at Denver and Health Science Center

Bruce E. Byers
University of Massachusetts, Amherst

PEARSON
Benjamin Cummings

San Francisco Boston New York
Cape Town Hong Kong London Madrid Mexico City
Montreal Munich Paris Singapore Sydney Tokyo Toronto

Editor-in-Chief: Beth Wilbur
Senior Acquisitions Editor: Chalon Bridges
Executive Director of Development: Deborah Gale
Associate Editor: Leata Holloway
Assistant Editor: Rebecca Johnson
Executive Marketing Manager: Lauren Harp
Executive Managing Editor: Erin Gregg
Managing Editor: Mike Early
Production Supervisor: Shannon Tozier
Project Management: Michael Krapovicky/Pine Tree Composition, Inc.
Composition: Laserwords Private Limited
Cover Design: Maureen Eide and John Christiana
Cover Production: Seventeenth Street Studios
Manufacturing Manager: Pam Augspurger
Text and Cover Printer: QuebecorWorld, Dubuque

Cover Image: Rockhopper Penguins; The Neck, Saunders Island, Falkland Islands, by Laura Crawford Williams

ISBN 10-digit: 0-13-195780-5
13-digit: 978-0-13-195780-0

1 2 3 4 5 6 7 8 9 10—QWD—11 10 09 08 07
www.aw-bc.com

Table of Contents

To The Student

The Student Lecture Notebook is designed to assist you during your biology course. It includes art from the textbook with space available for your note taking needs. Since you will not have to redraw the art in class, you can focus your attention on the lecture while you mark up the pages.

Visit www.aw-bc.com/audesirk and log in to the Web site selected by your instructor to view additional resources, including self-grading quizzes, animations, and Internet links.

Community	Two or more populations of different species living and interacting in the same area	 antelope, hawk, grass
Population	Members of one species inhabiting the same area	 herd of pronghorn antelope
Multicellular Organism	An individual living thing composed of many cells	 pronghorn antelope
Organ System	Two or more organs working together in the execution of a specific bodily function	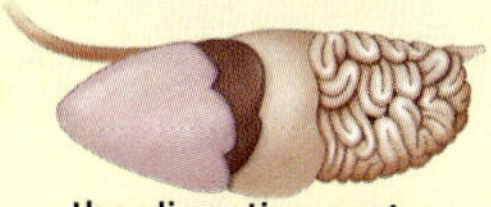 the digestive system
Organ	A structure usually composed of several tissue types that form a functional unit	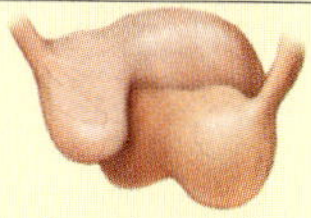 the stomach
Tissue	A group of similar cells that perform a specific function	 epithelial tissue
Cell	The smallest unit of life	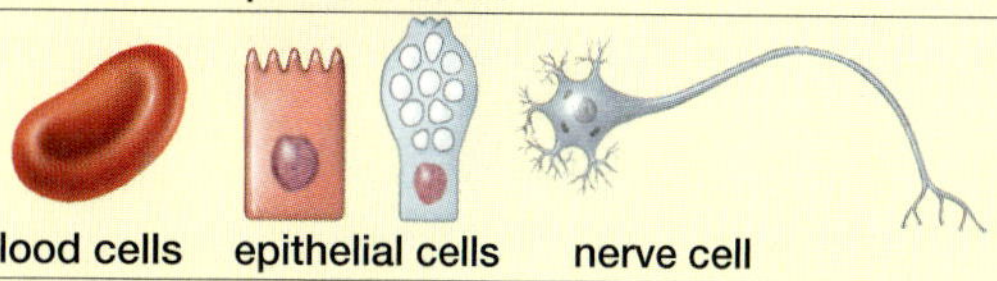 red blood cells; epithelial cells; nerve cell
Molecule	A combination of atoms	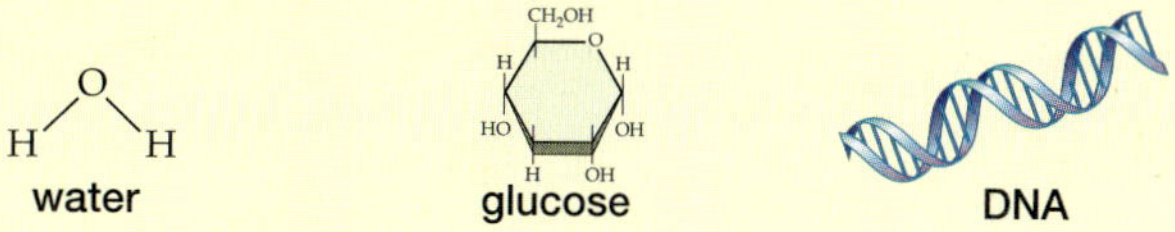 water (H_2O); glucose; DNA
Atom	The smallest particle of an element that retains the properties of that element	 hydrogen; carbon; nitrogen; oxygen

Figure 1-1 Levels of organization

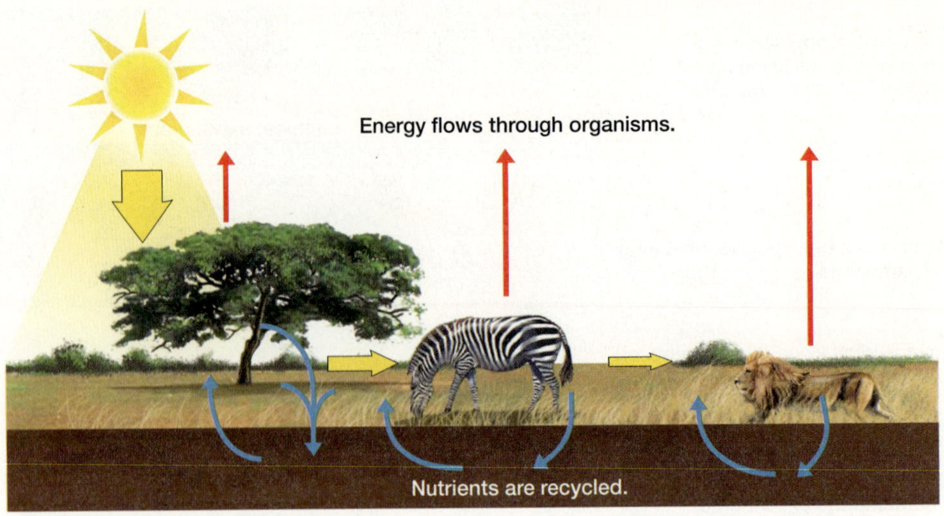

Figure 1-10 Energy flow and nutrient cycles

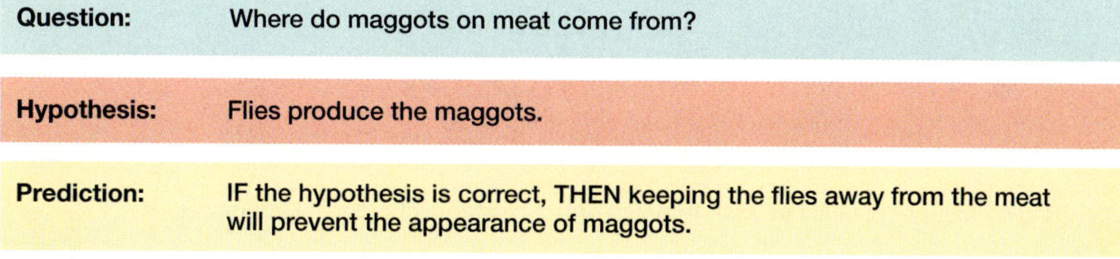

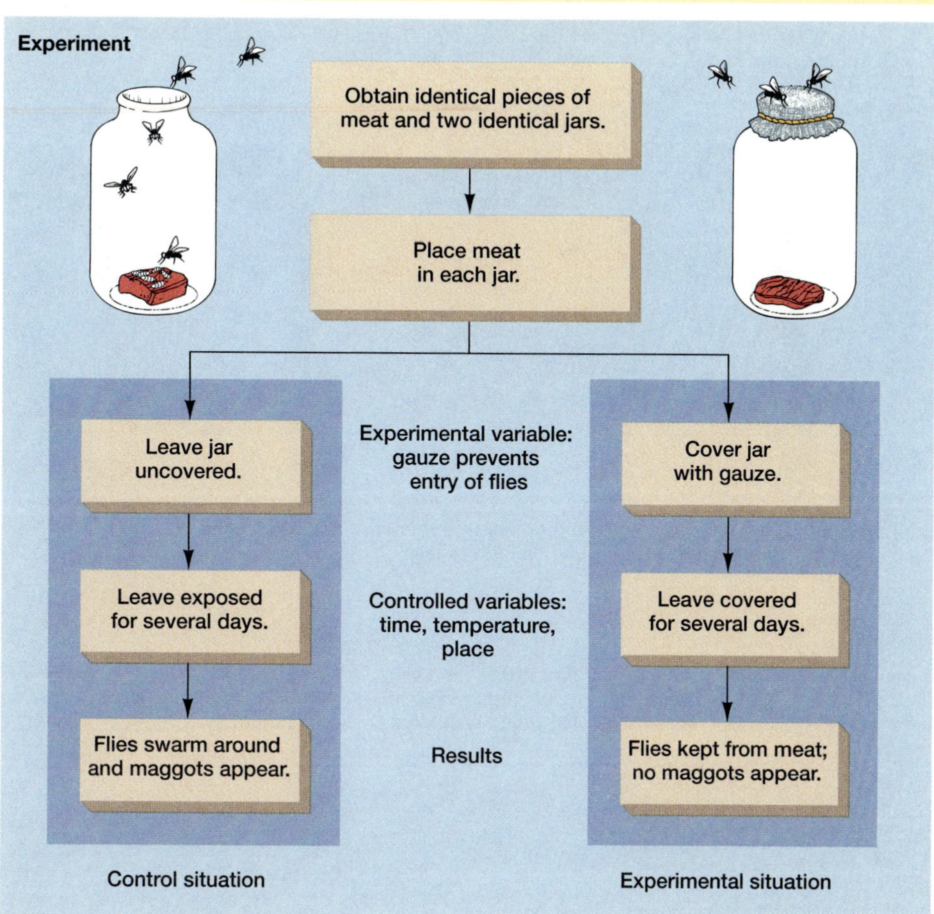

Figure E1-1 Experiments of Redi

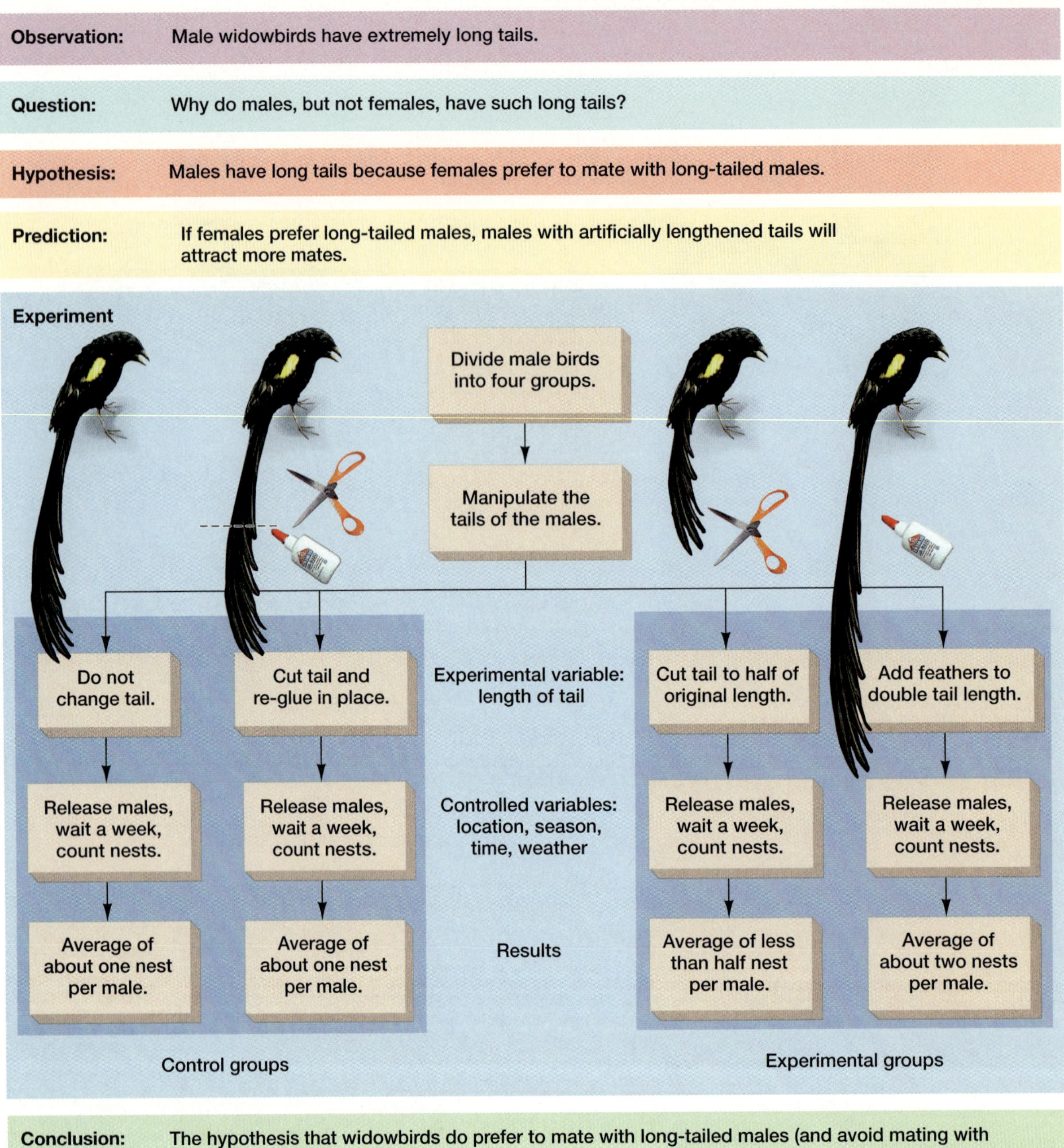

Figure E1-3 Experiments of Malte Andersson

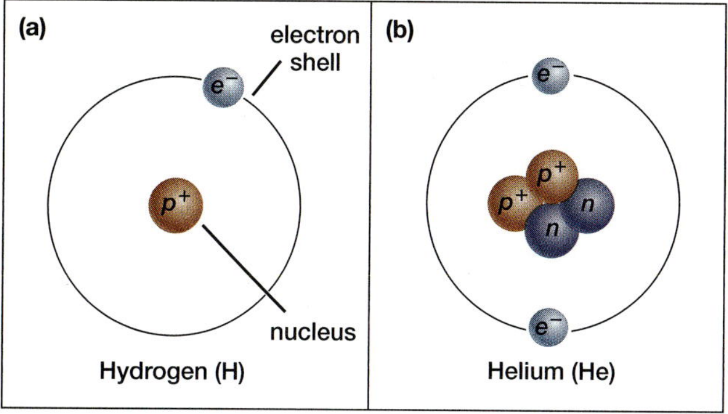

Figure 2.1 Hydrogen and helium

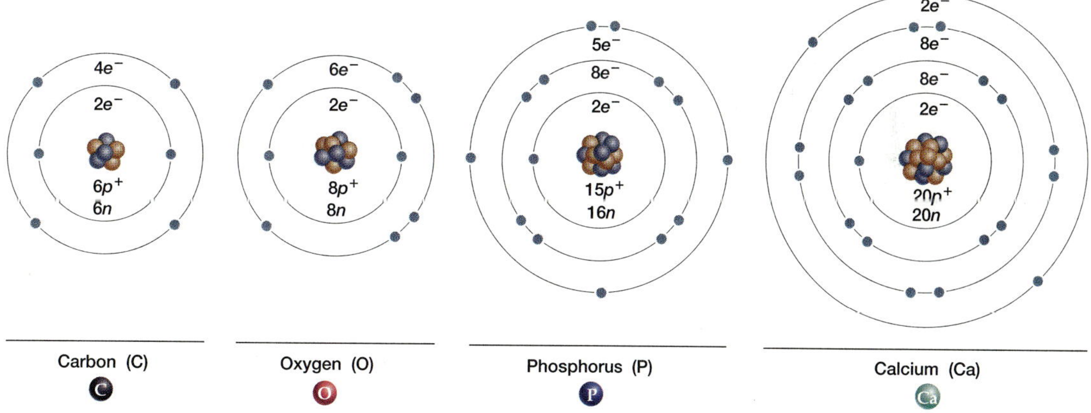

Figure 2-2 Electron shells

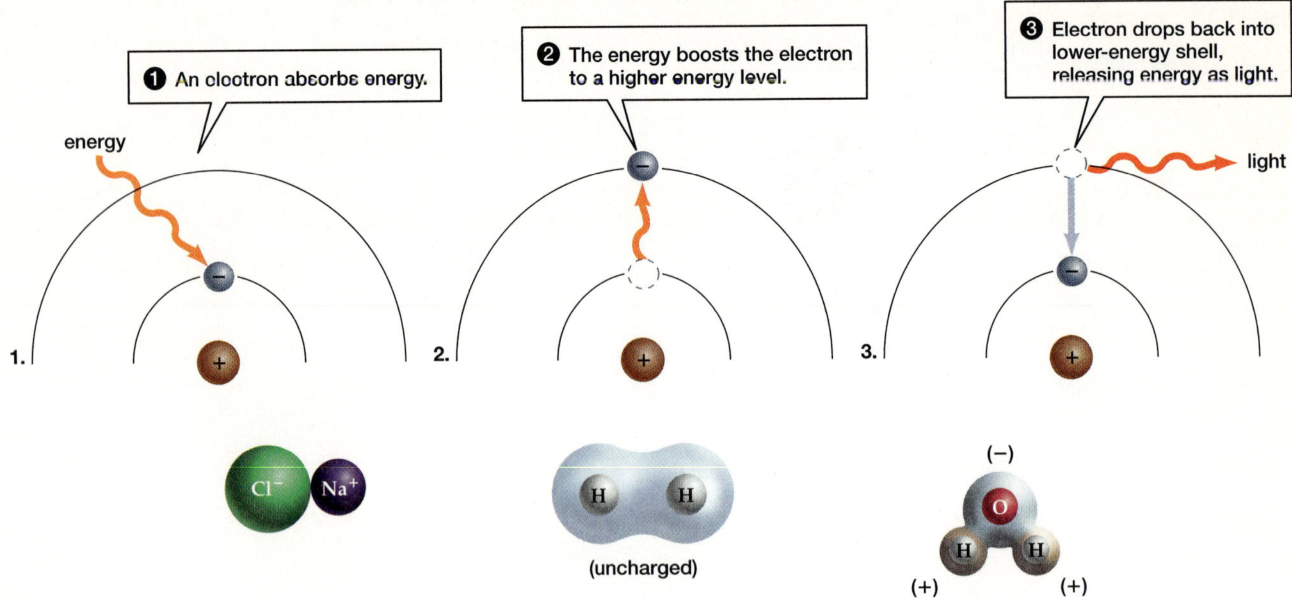

Figures 2-3, 2-5, 2-7, 2-8 Energy capture and release, Ionic bonds, Nonpolar covalent bond, Polar covalent bonds in water

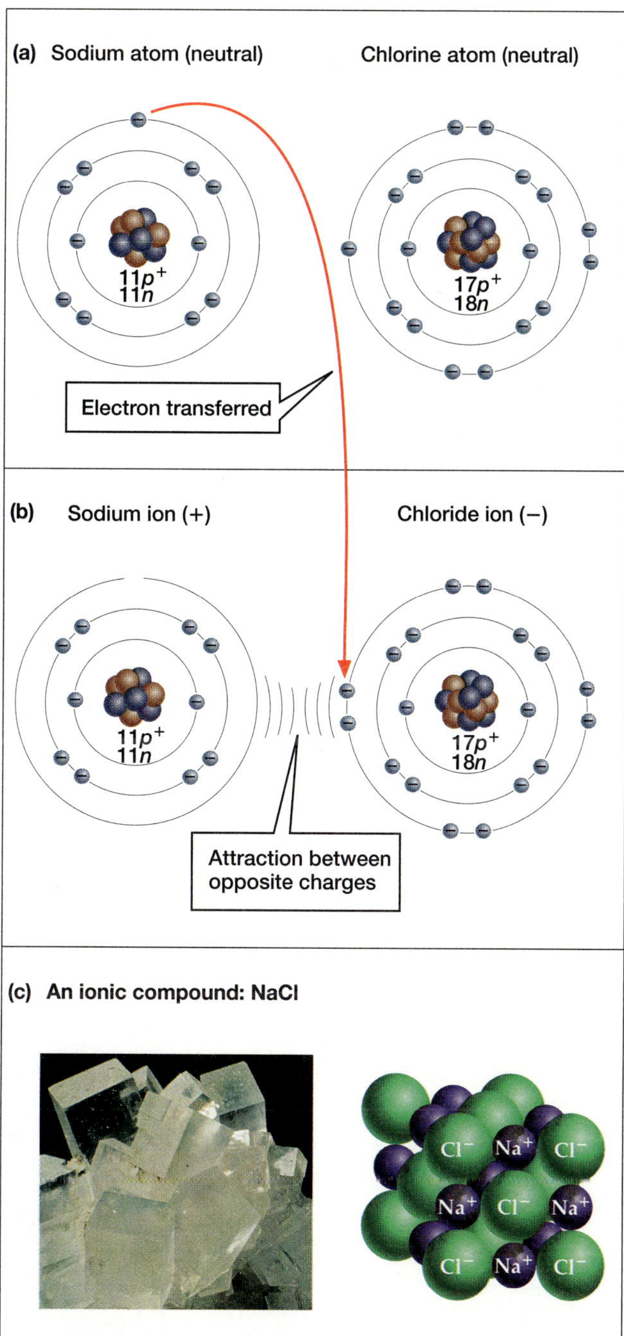

Figure 2-4 Ions and ionic bonds

Nonpolar covalent bonding

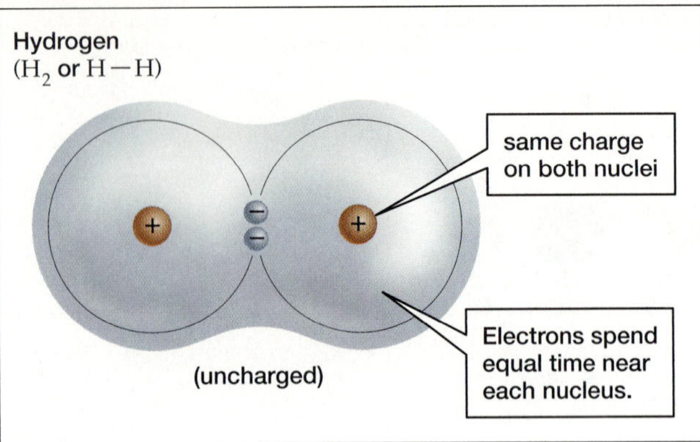

Figure 2-6a Nonpolar covalent

Figure 2-10 Hydrogen bonds

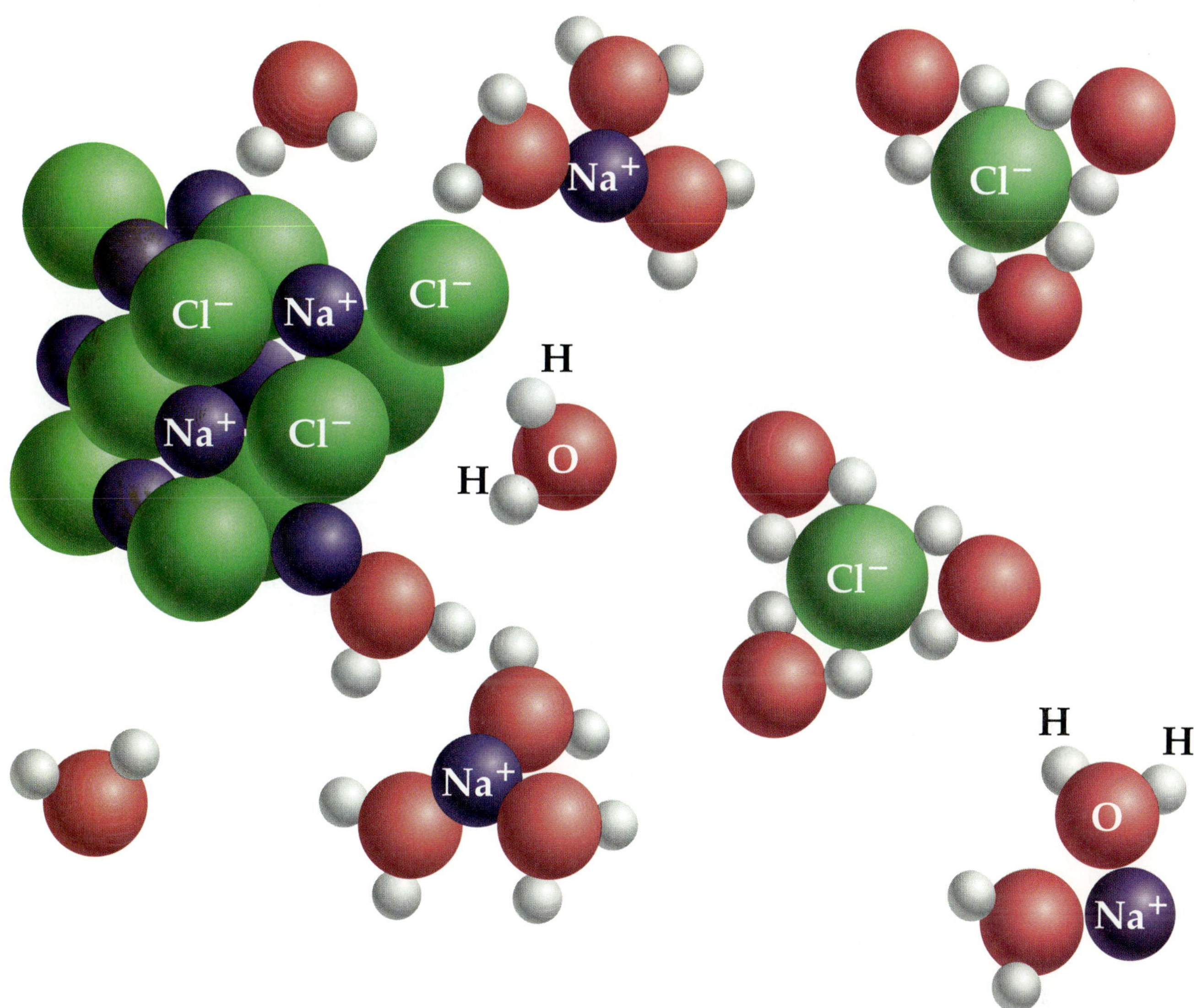

Figure 2-11 Water as a solvent

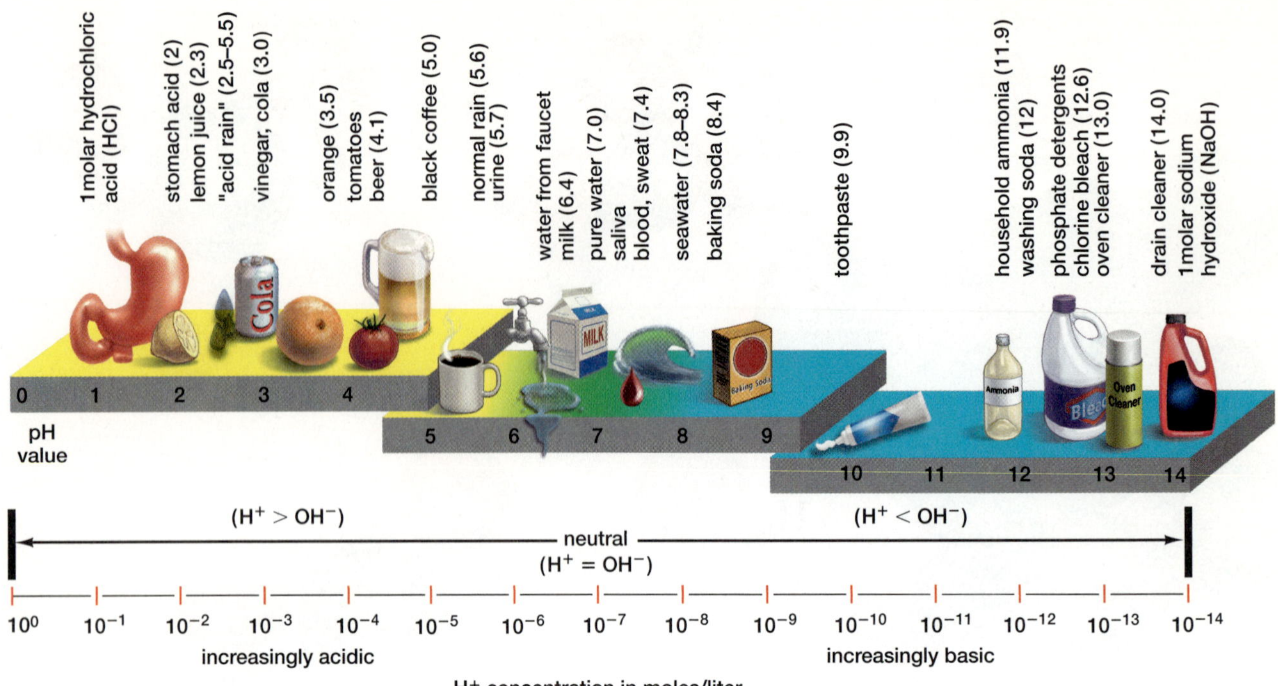

Figure 2-15 pH

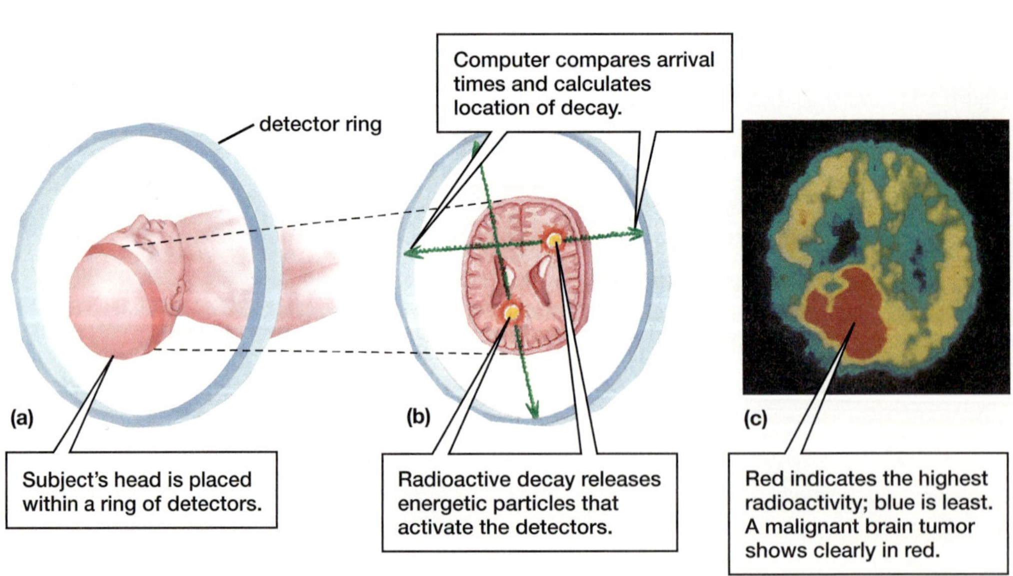

Figure E2-1 How positron emission topography works

Atom	Capacity of Outer Electron Shell	Electrons in Outer Shell	Number of Covalent Bonds Usually Formed	Common Bonding Patterns
Hydrogen	2	1	1	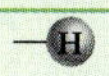 —H
Carbon	8	4	4	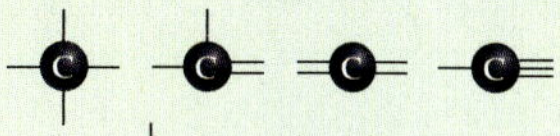 C (four single bonds), C (two single + one double), C (two double), C (one single + one triple)
Nitrogen	8	5	3	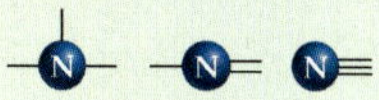 N (three single bonds), N (one single + one double), N (triple)
Oxygen	8	6	2	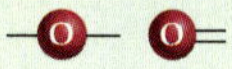 O (two single bonds), O (double)
Phosphorus	8	5	5	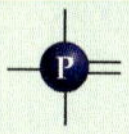 P (three single + one double)
Sulfur	8	6	2	 S (two single bonds)

Table 2-3 Bonding patterns

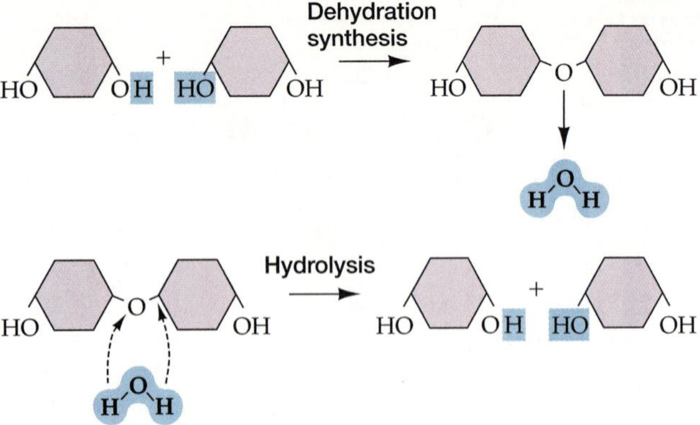

Figures 3-1, 3-2 Dehydration synthesis, hydrolysis

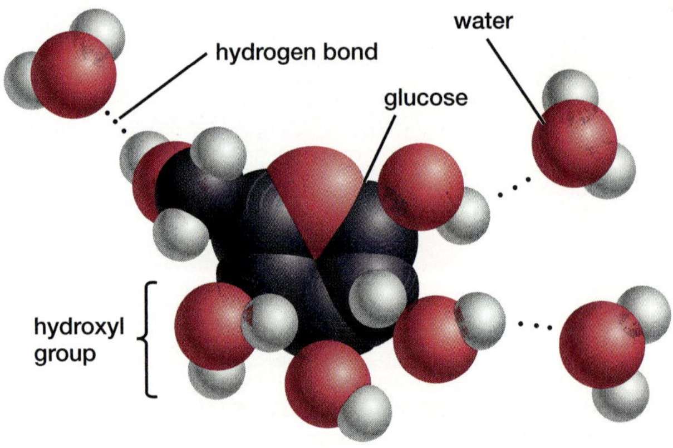

Figure 3-3 Sugar dissolving

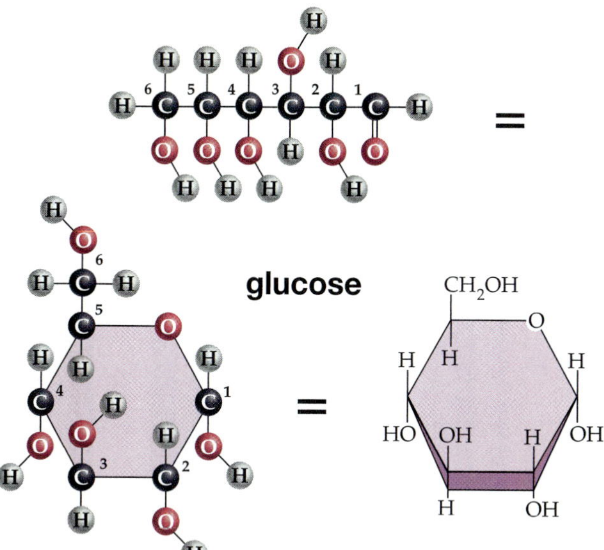

Figure 3-4 Glucose

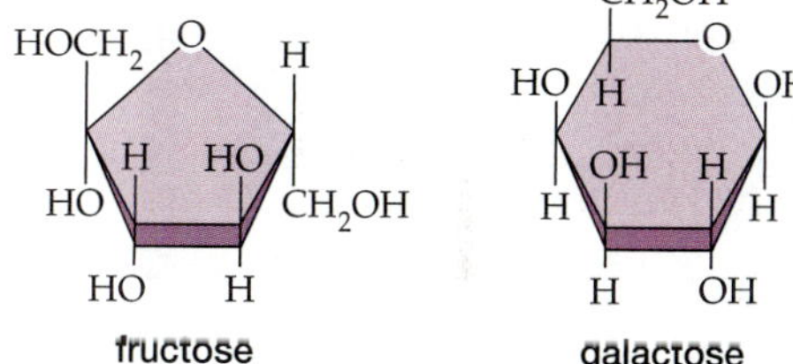

Figure 3-5 Monosaccharides

ribose
deoxyribose

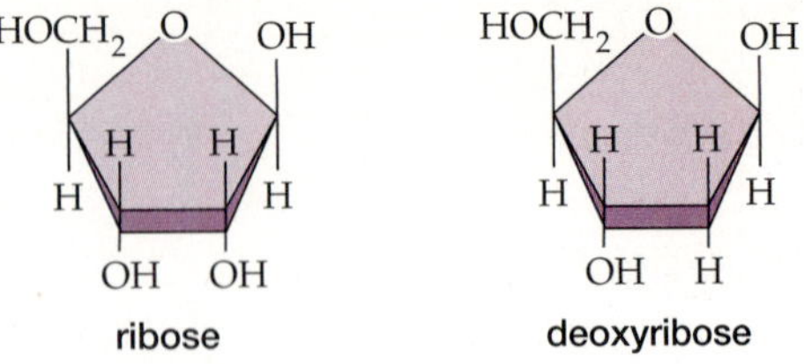

Figure 3-6 Ribose sugars

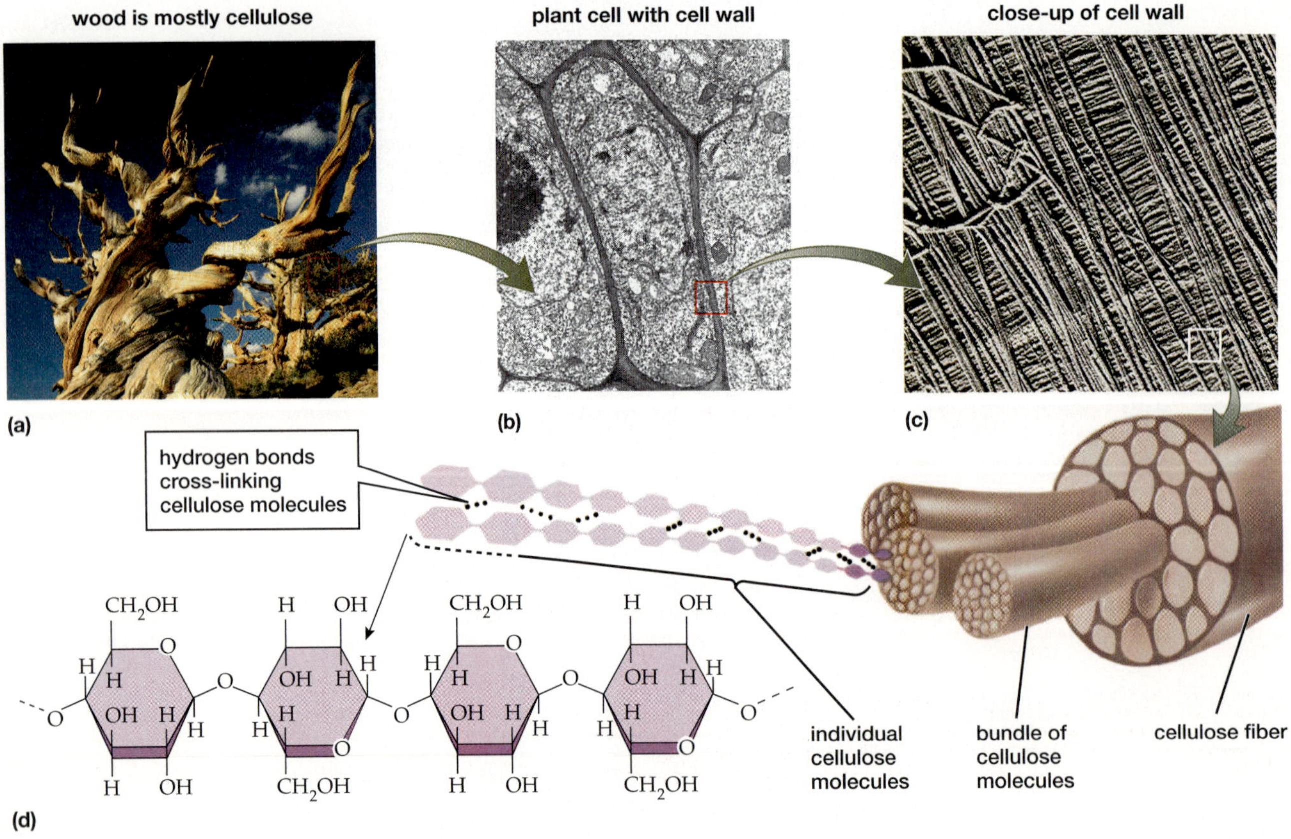

wood is mostly cellulose
plant cell with cell wall
close-up of cell wall
(a)
(b)
(c)
hydrogen bonds cross-linking cellulose molecules
individual cellulose molecules
bundle of cellulose molecules
cellulose fiber
(d)

Figure 3-9 Cellulose

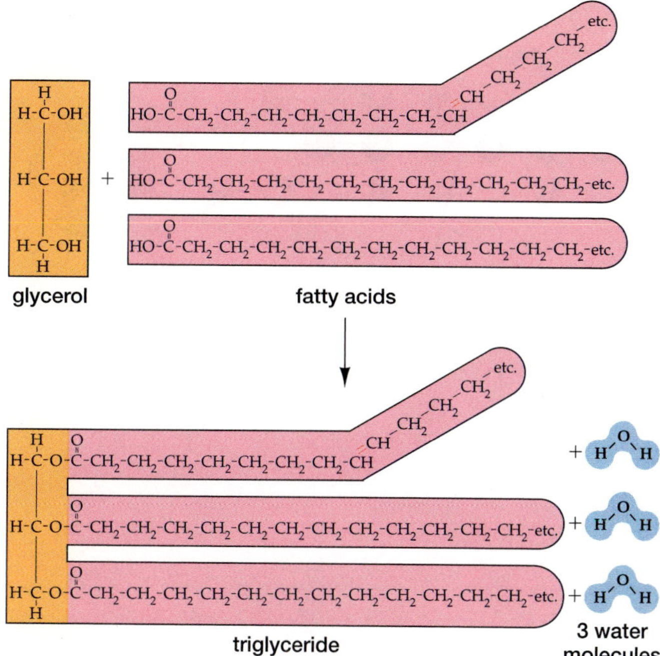

Figure 3-11 Glycerol, fatty acids

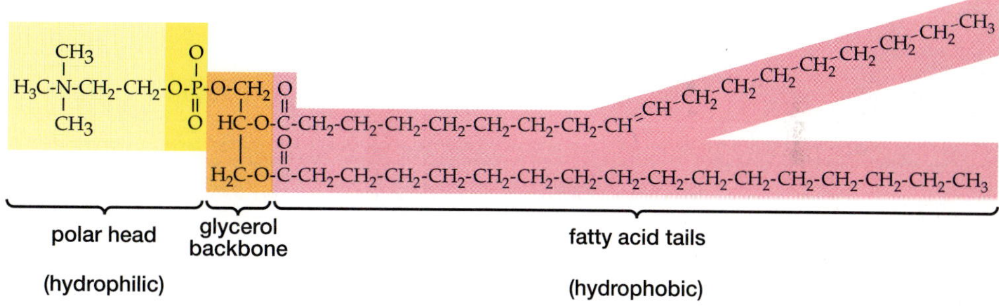

Figure 3-15 Phospholipids

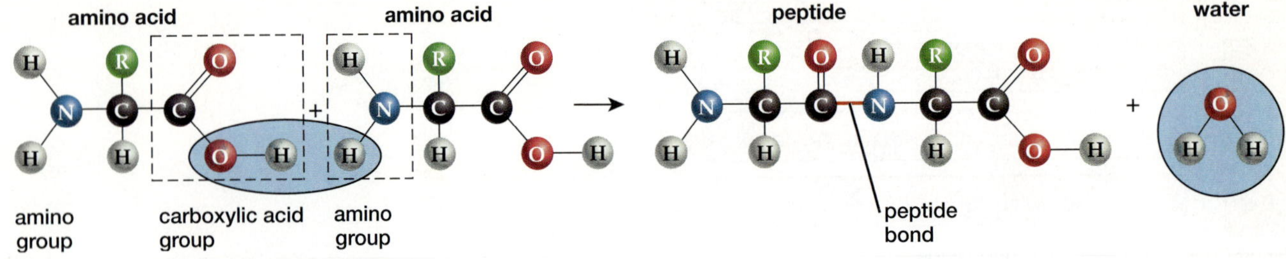

Figure 3-20 Protein synthesis

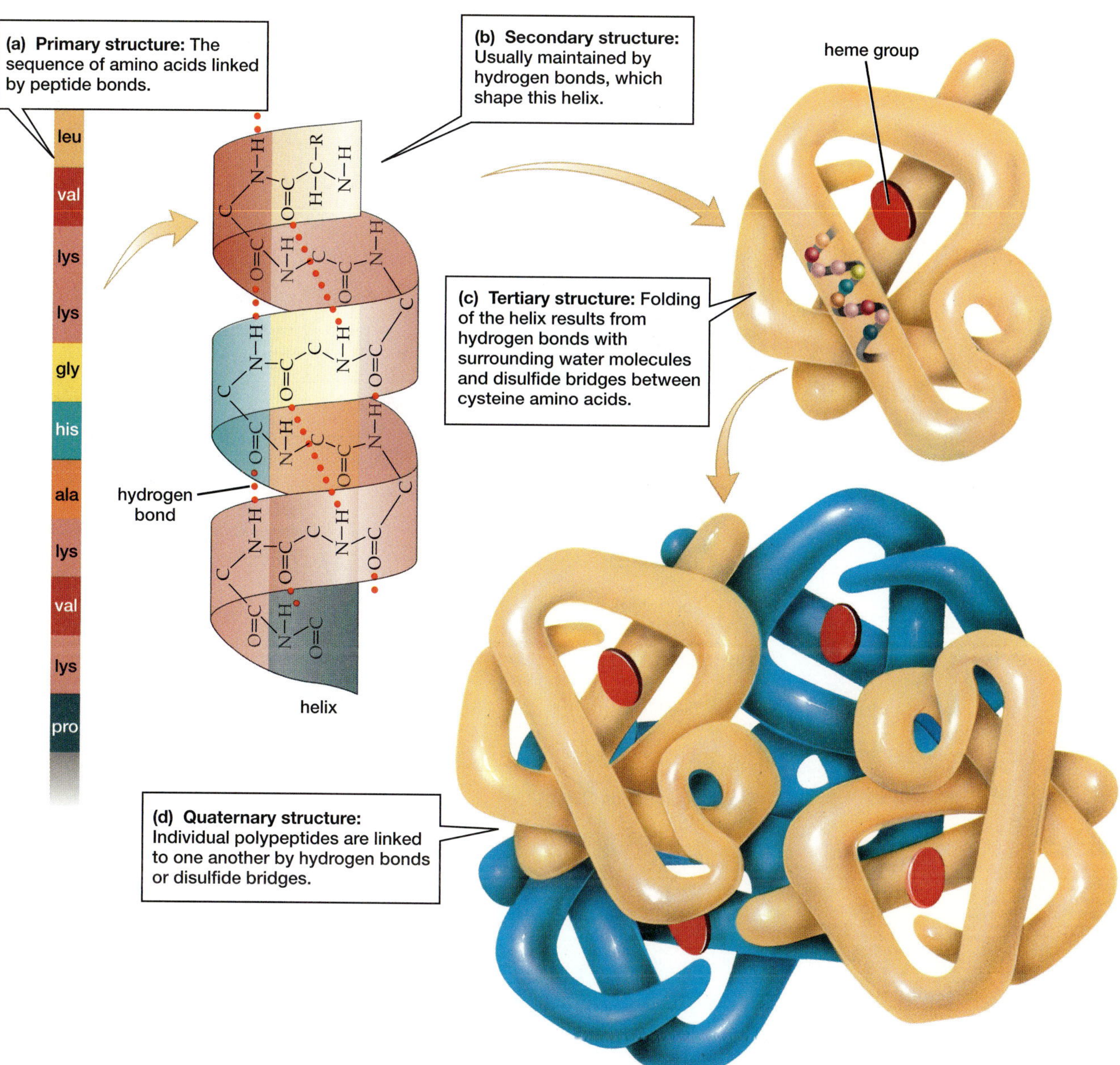

Figure 3-21 Levels of protein structure

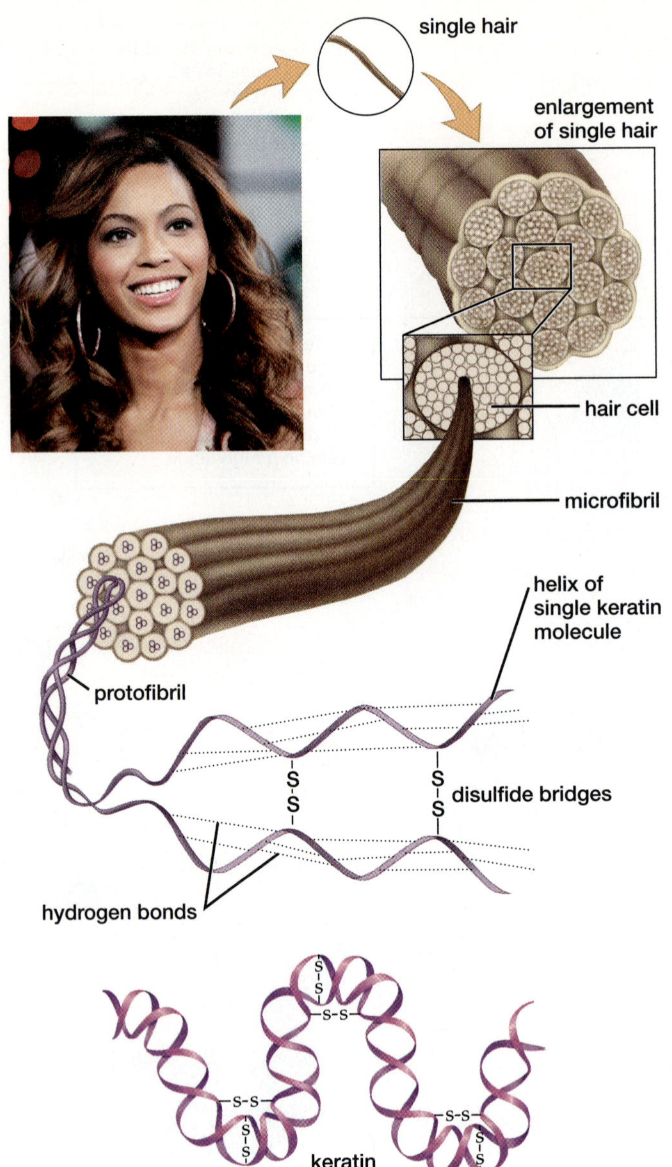

Figures E3-4, 3-23 Hair

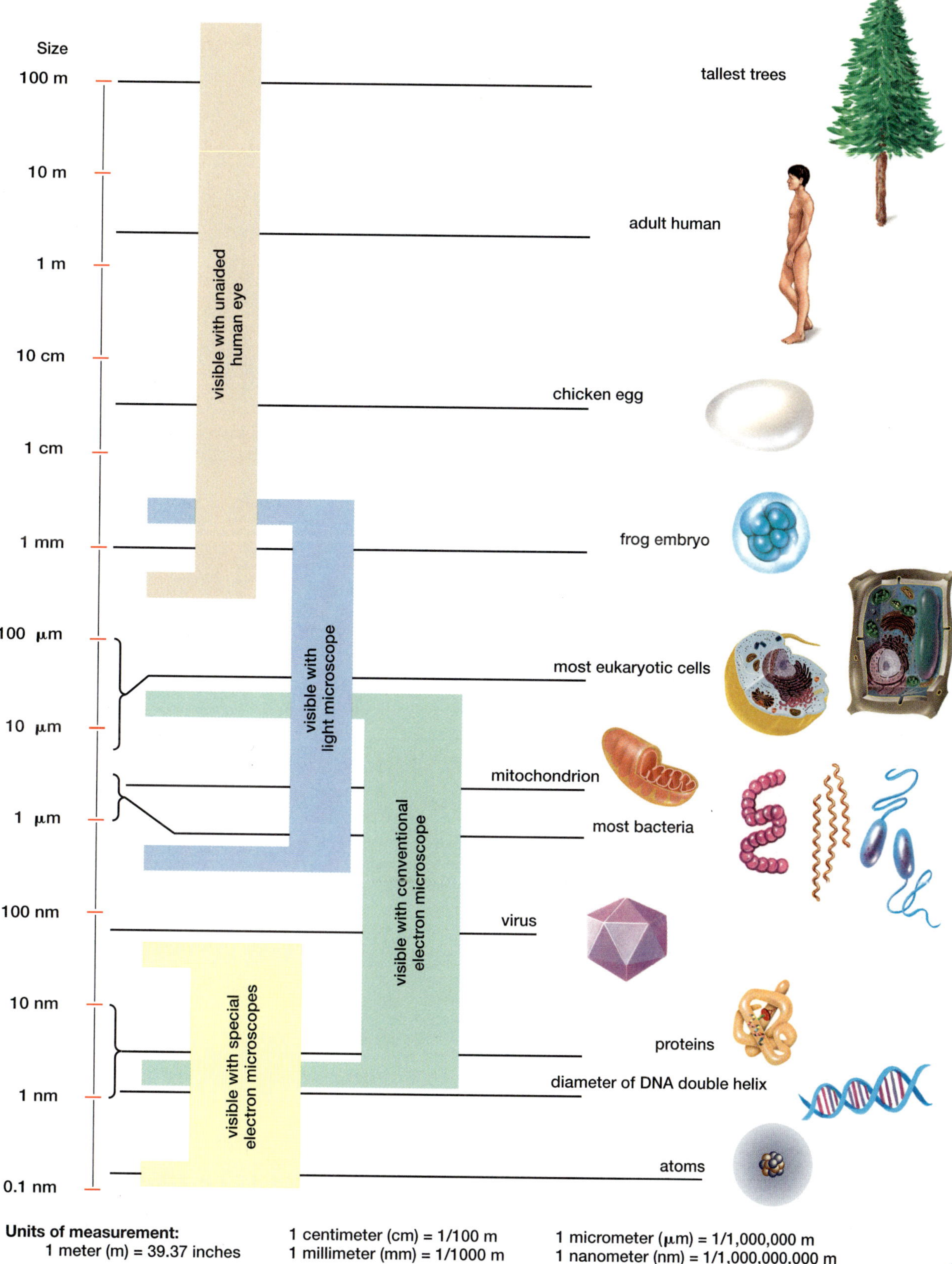

Units of measurement:
1 meter (m) = 39.37 inches
1 centimeter (cm) = 1/100 m
1 millimeter (mm) = 1/1000 m
1 micrometer (µm) = 1/1,000,000 m
1 nanometer (nm) = 1/1,000,000,000 m

Figure 4-1 Size scale

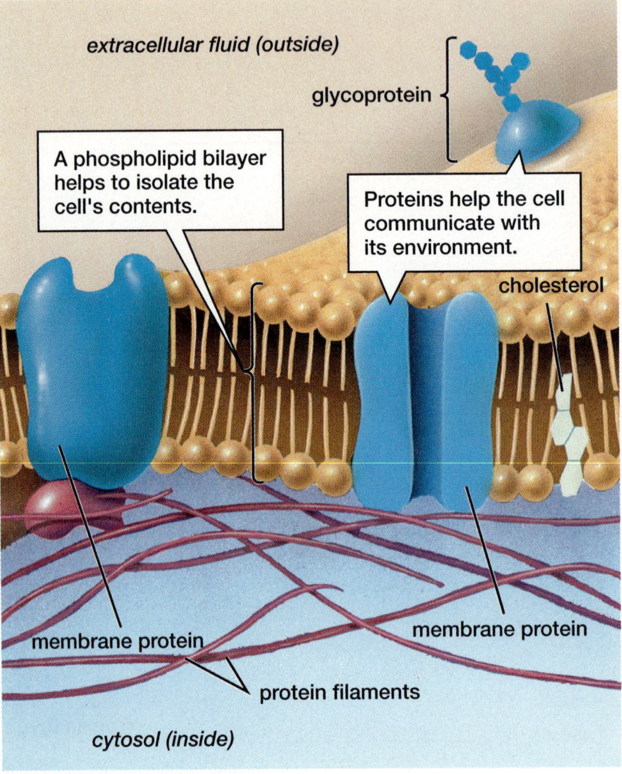

Figure 4-2 Plasma membrane

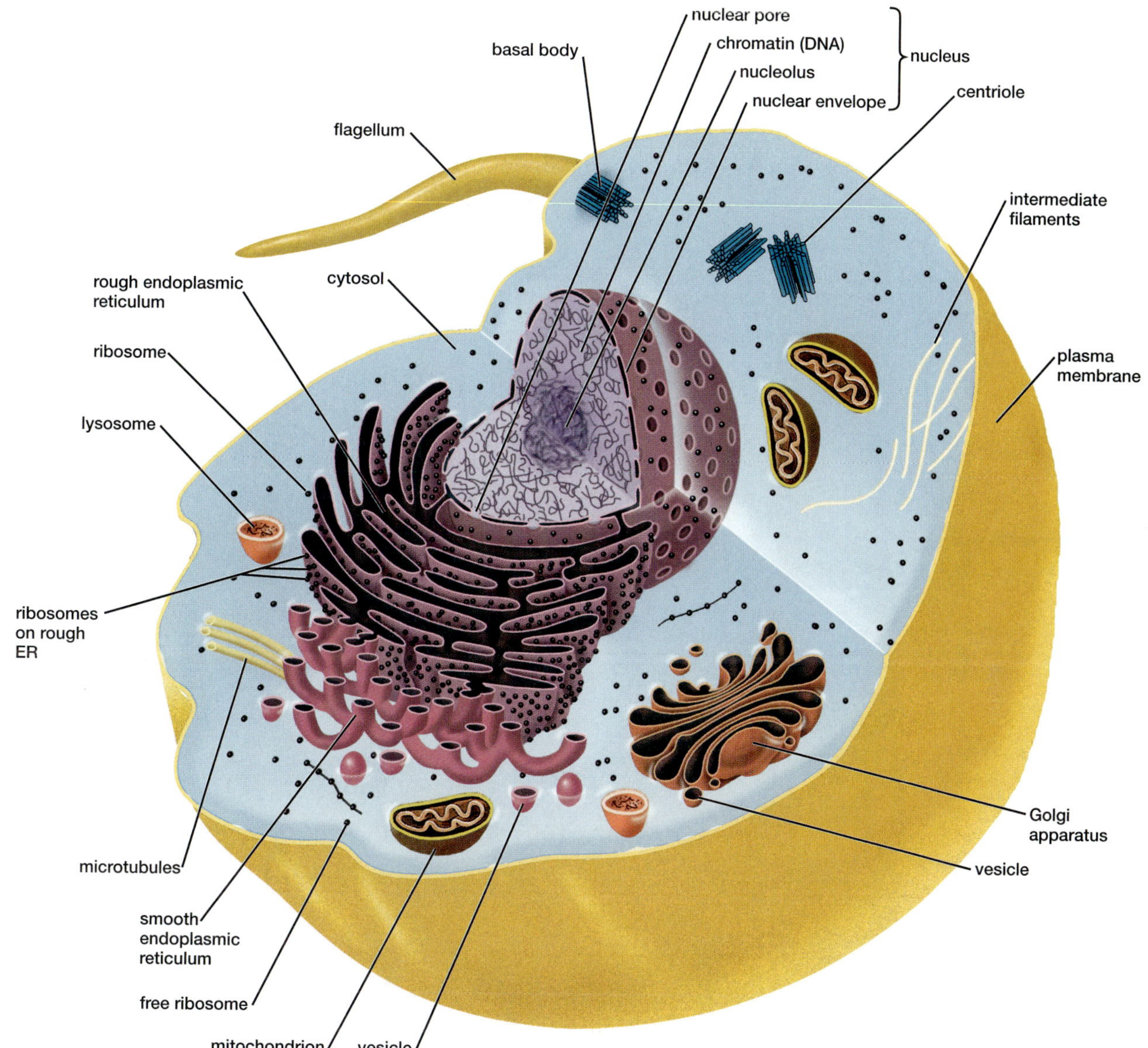

Figure 4-3 Animal cell

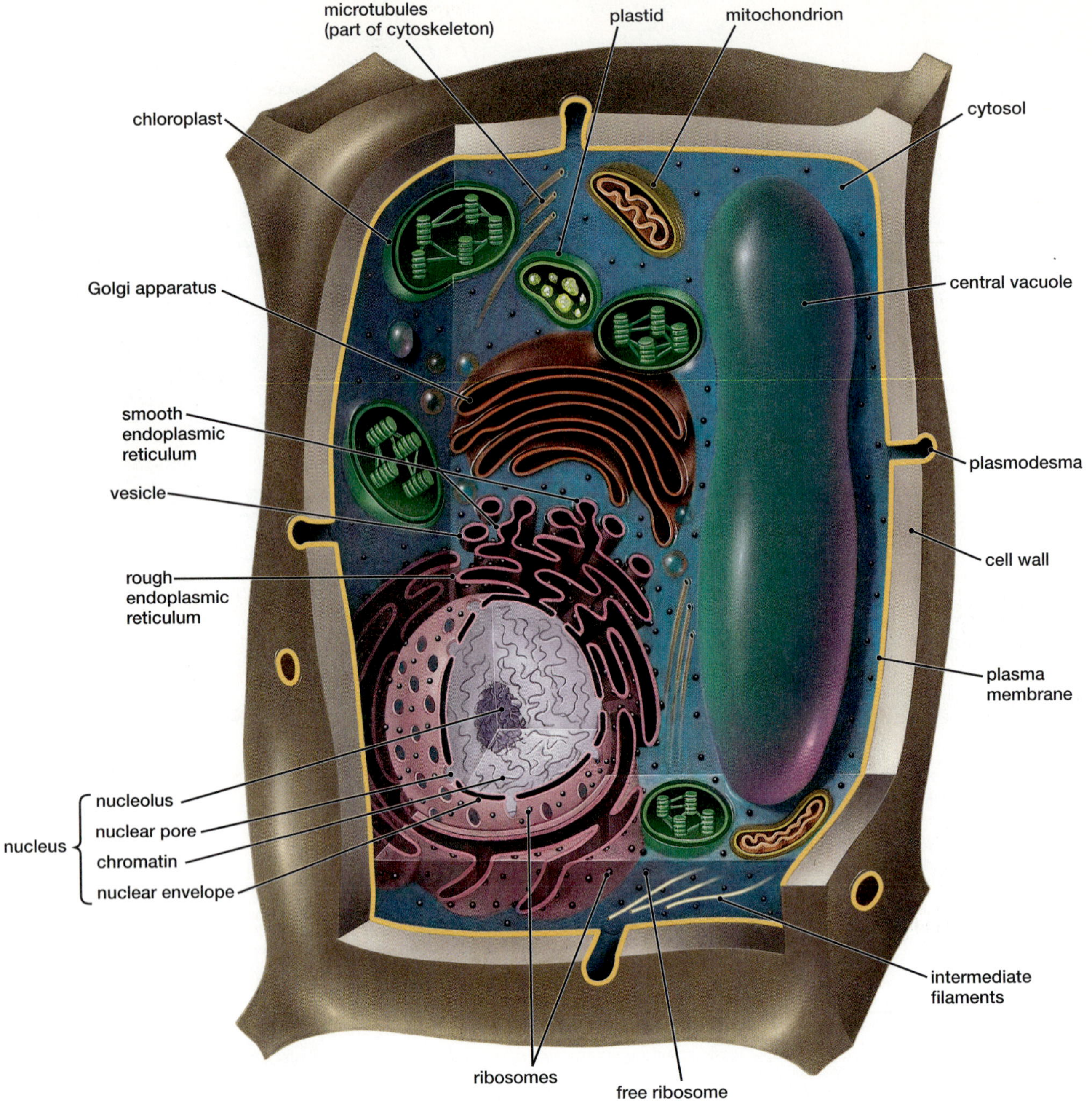

Figure 4-4 Plant cell

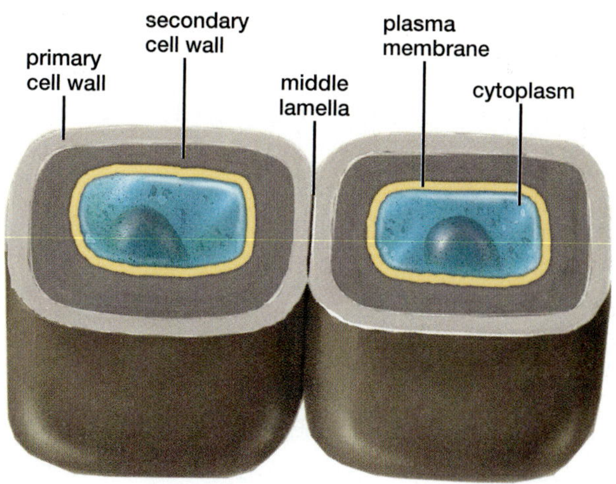

Figure 4-5 Plant cell wall

(a)

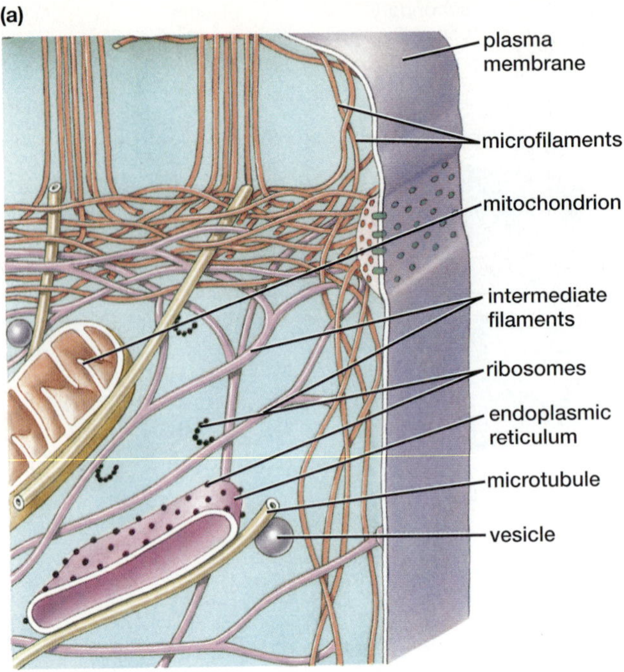

(b)

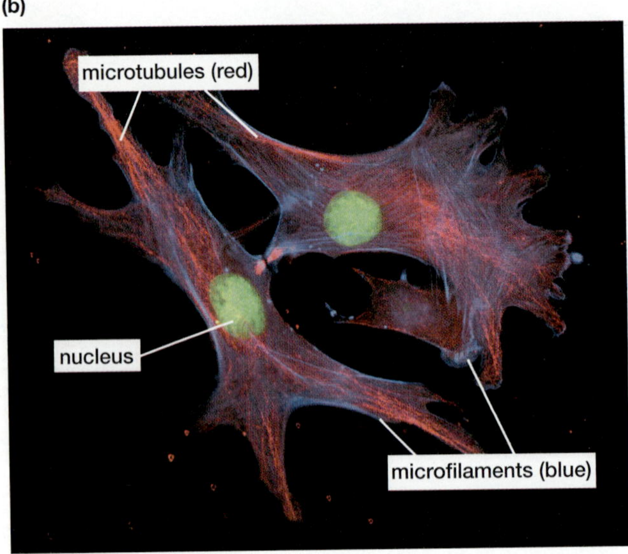

Figure 4-6 Cytoskeleton

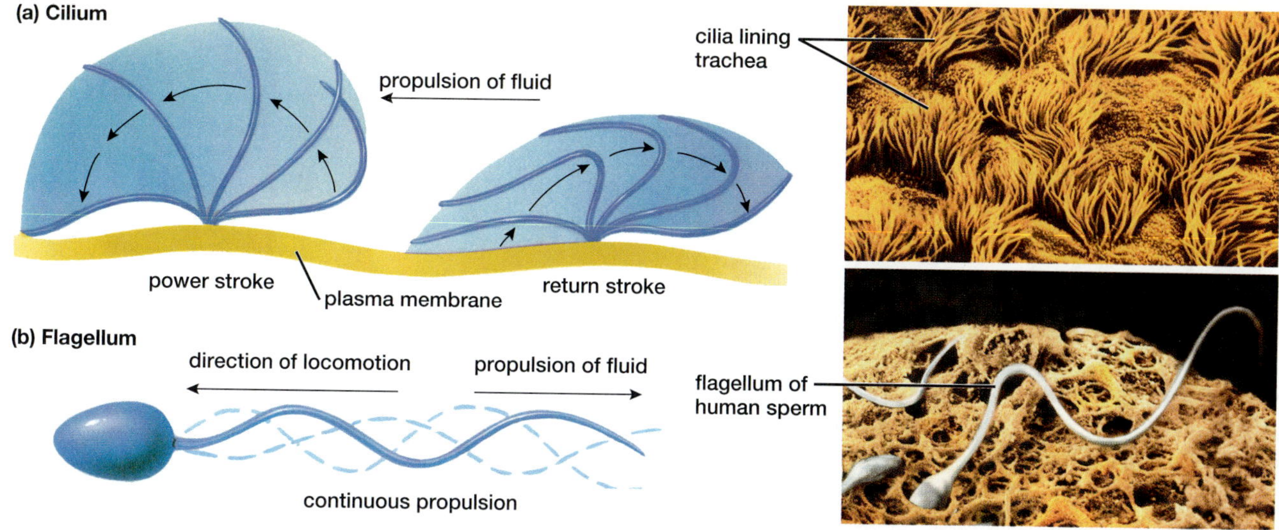

Figure 4-8 Cilia/flagella movement

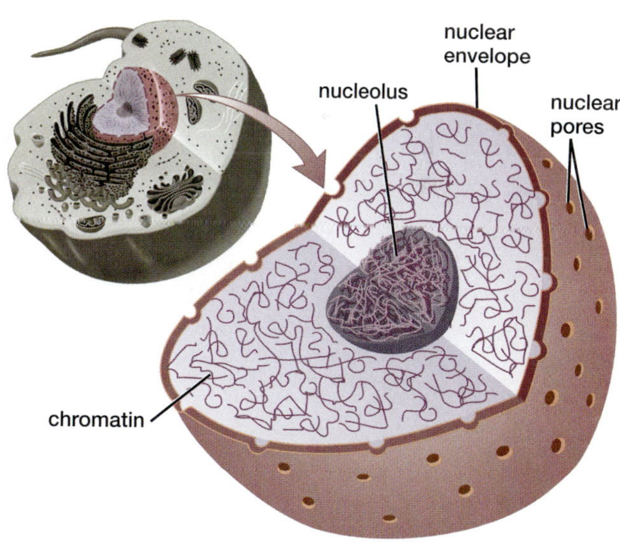

Figure 4-9a Nucleus art

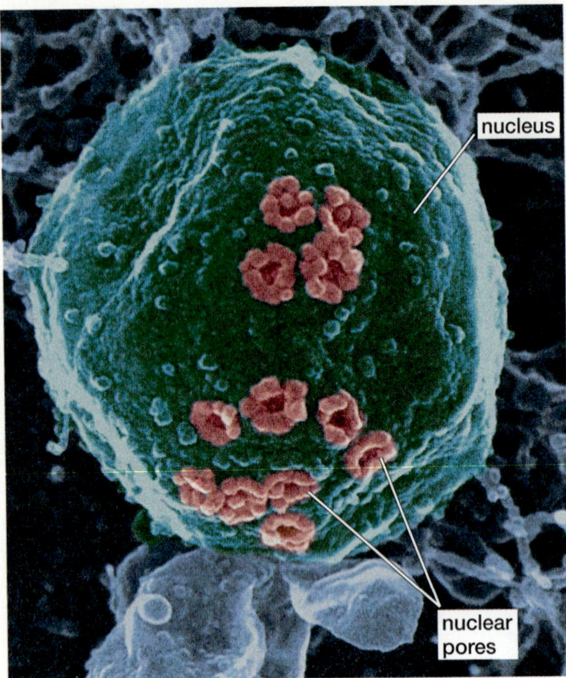

Figure 4-9b Nucleus Electron micrograph

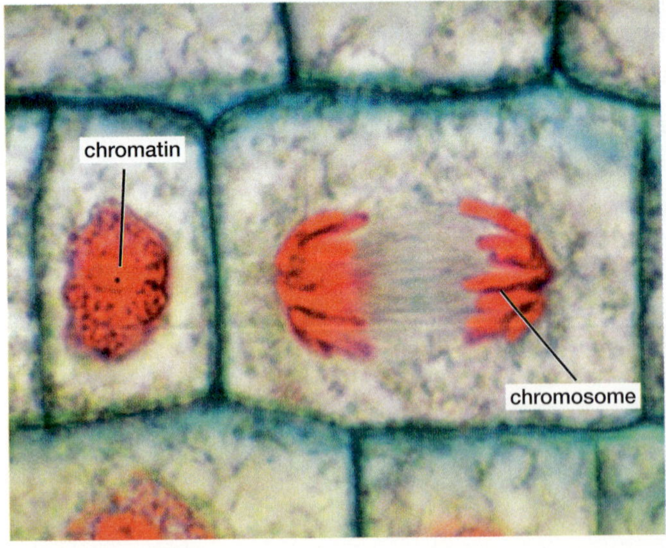

Figure 4-10 Dividing cells

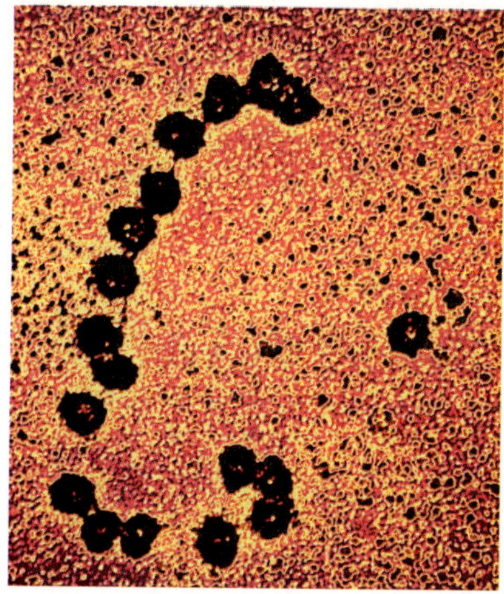

Figure 4-11 Ribosomes

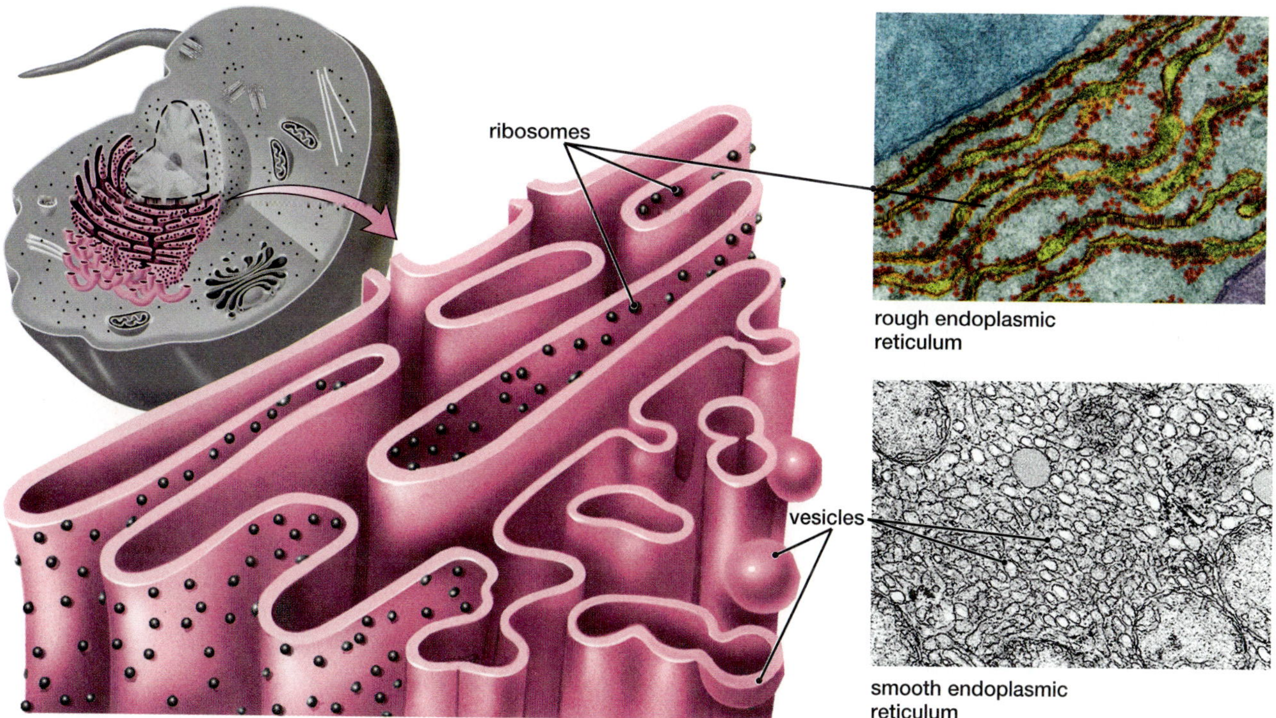

Figure 4-12 Endoplasmic reticulam

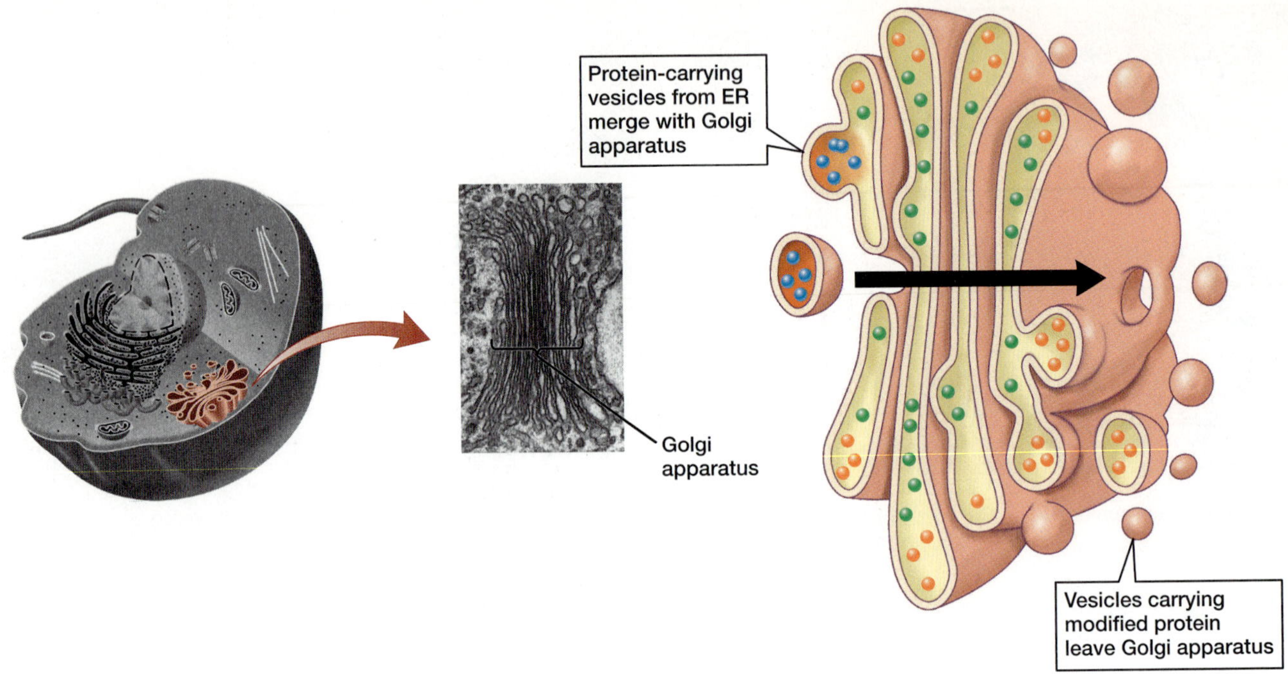

Figure 4-13 Golgi apparatus

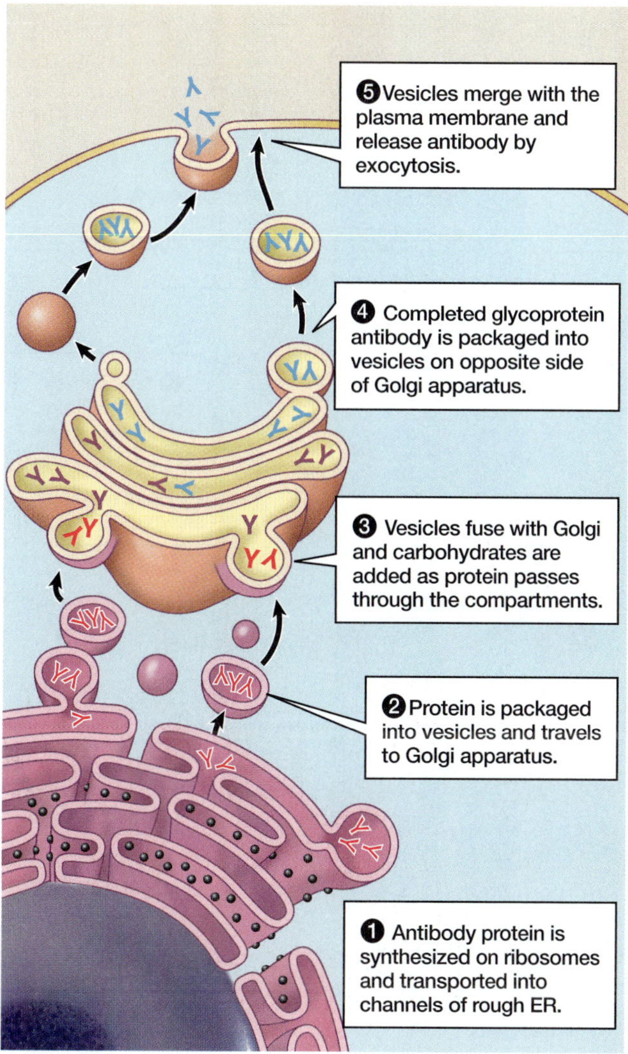

Figure 4-14 Antibody formation

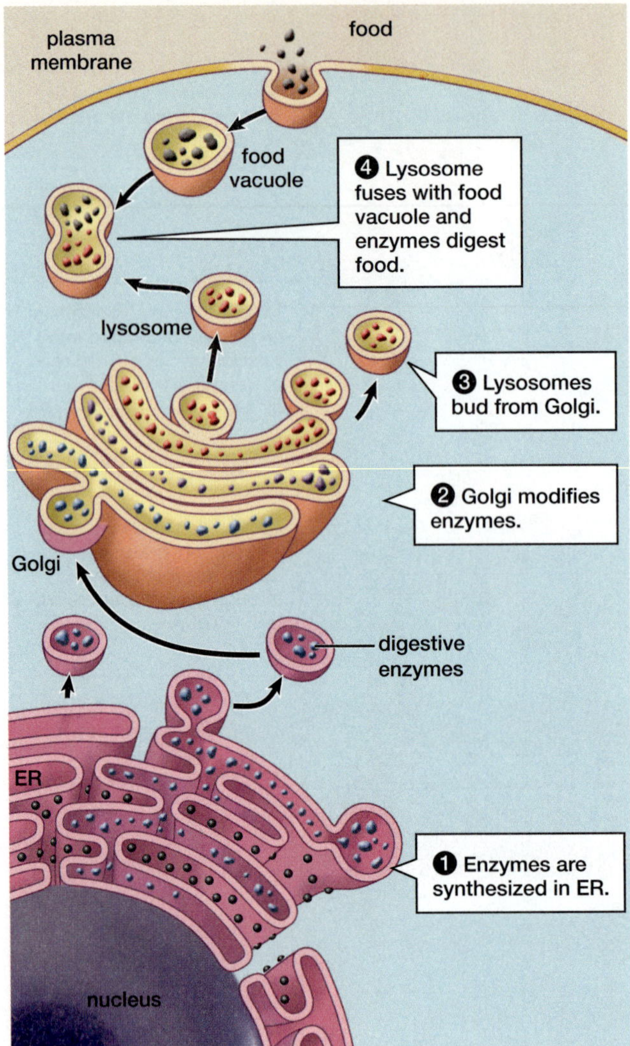

Figure 4-15 Lysosome formation

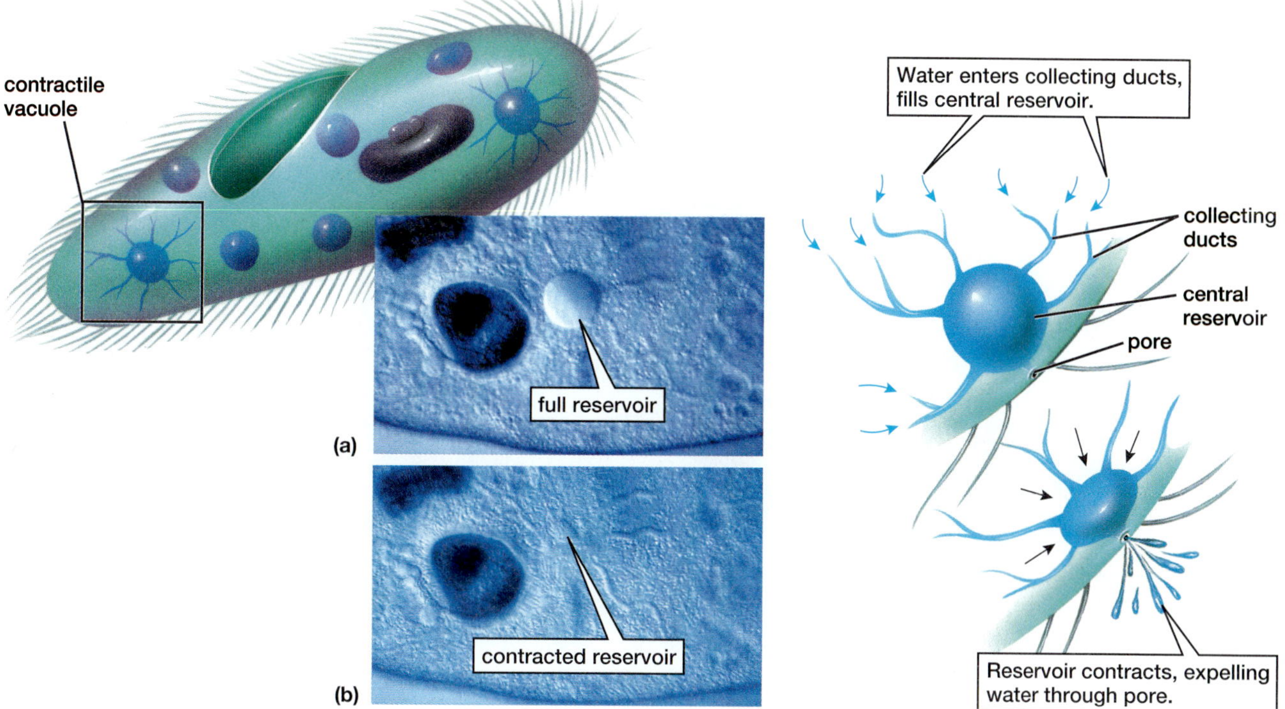

Figure 4-16 Contractile vacuole

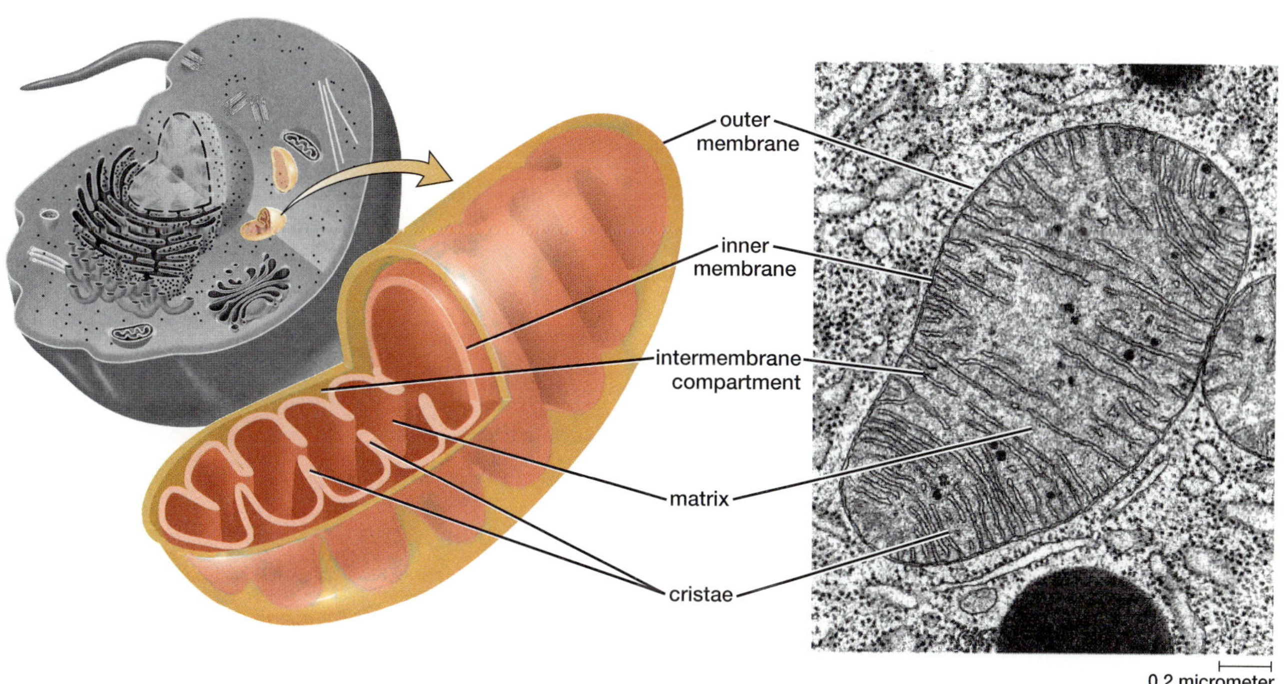

Figure 4-17 Mitochondrion

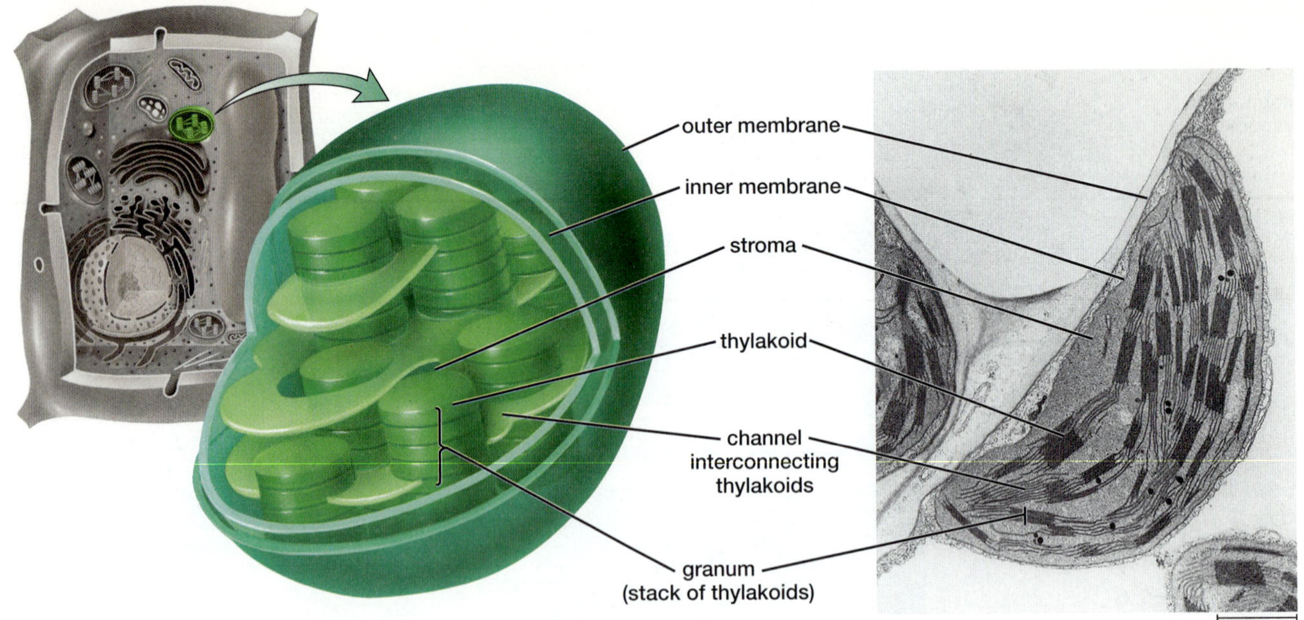

Figure 4-18 Chloroplast

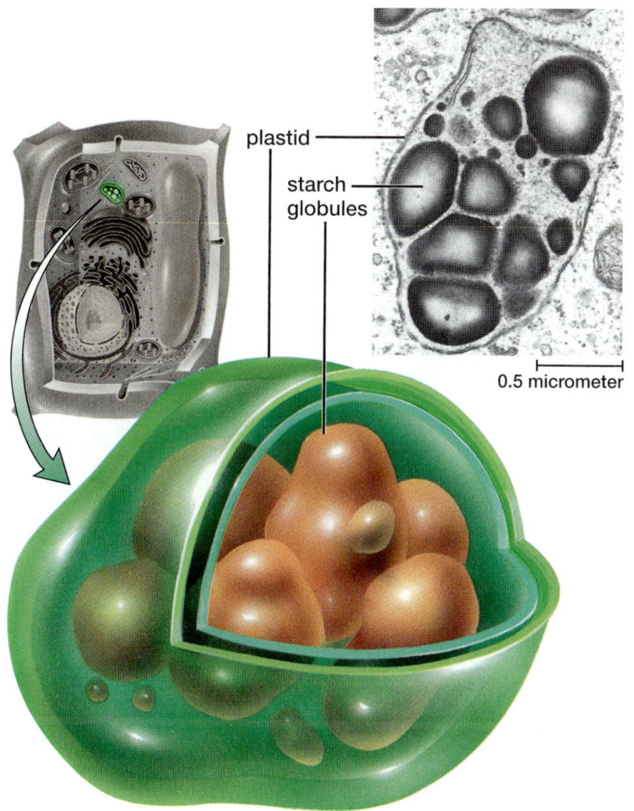

Figure 4-19 Plastid

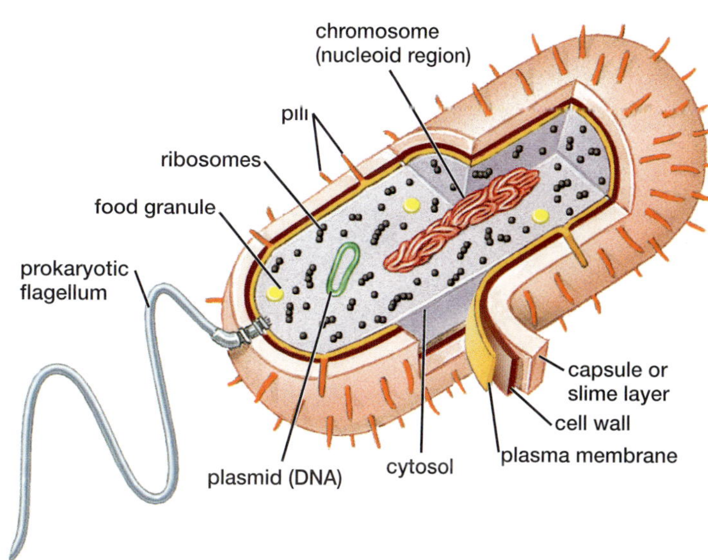

Figure 4-20a Prokaryotic cell

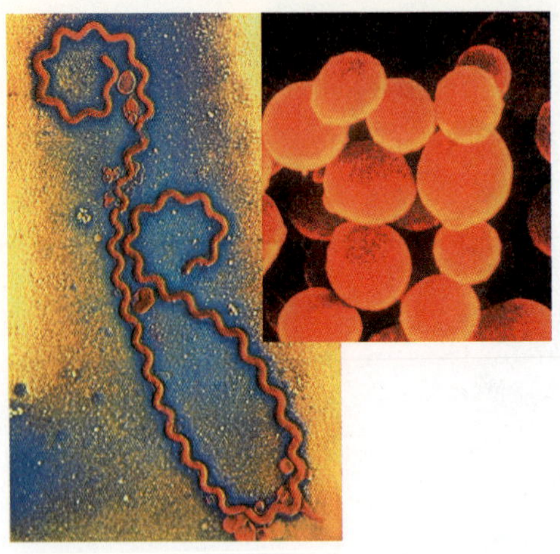

Figure 4-20b Bacterial shapes

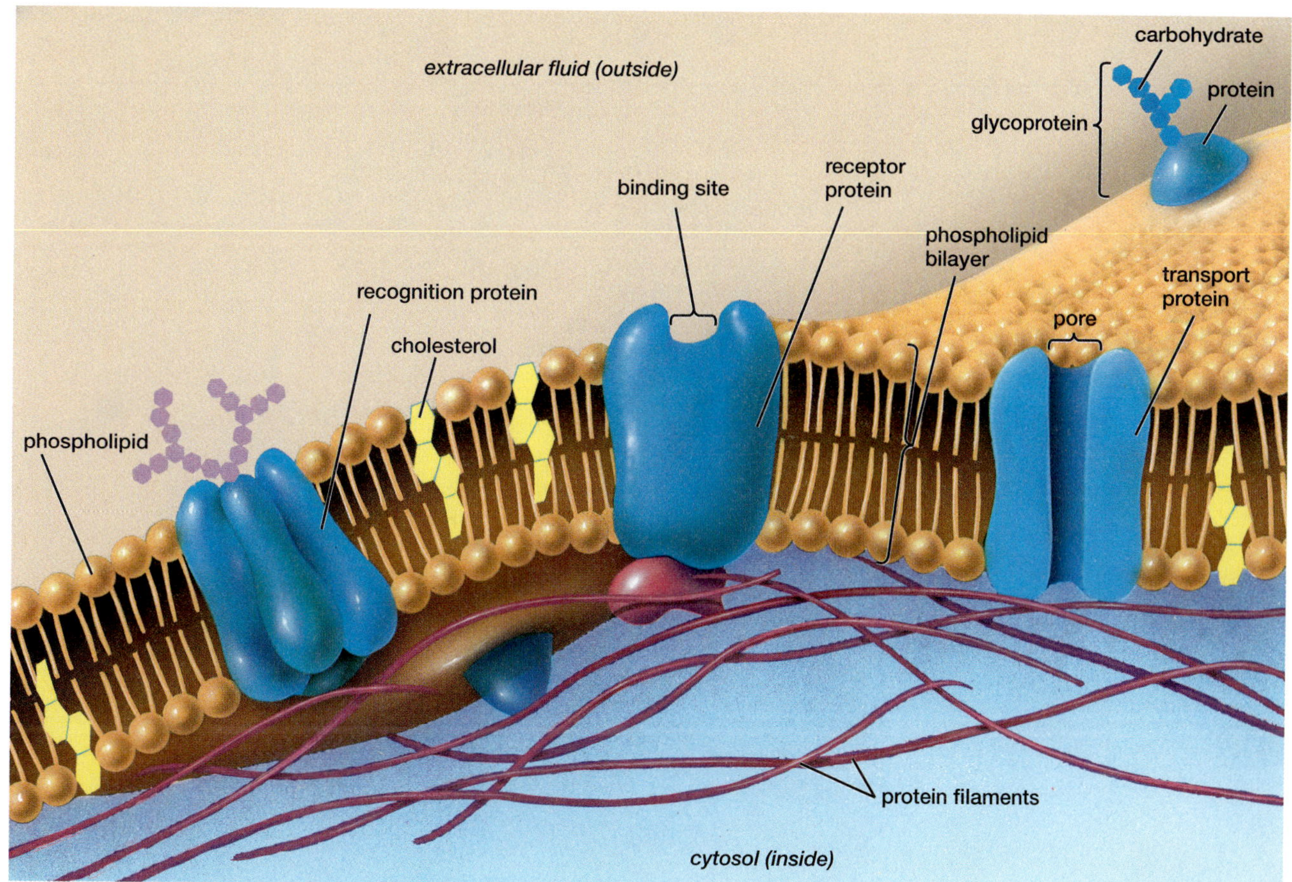

Figure 5-1 Plasma membrane

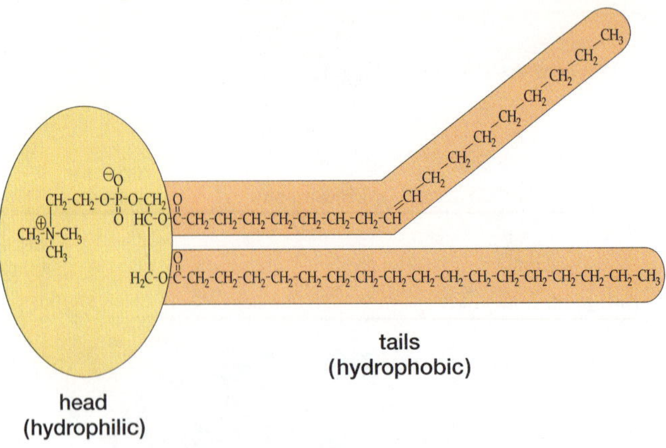

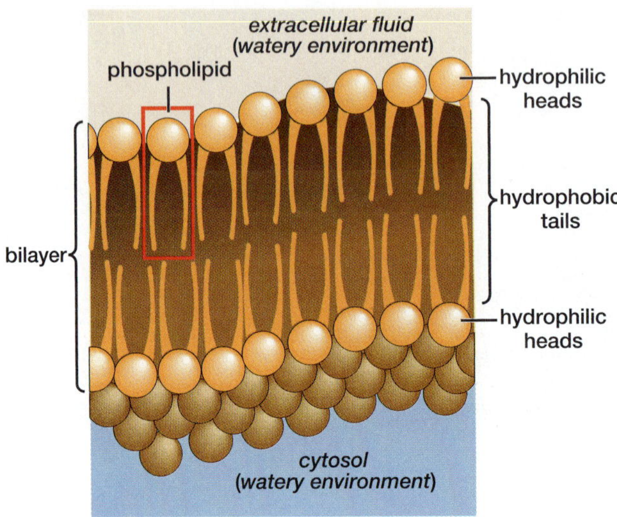

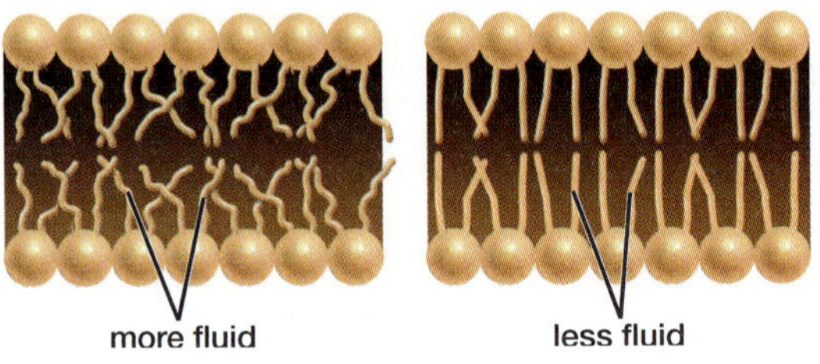

Figures 5-2, 5-3, 5-4 Phospholipid, Membrane bilayer, More and less fluid bilayers

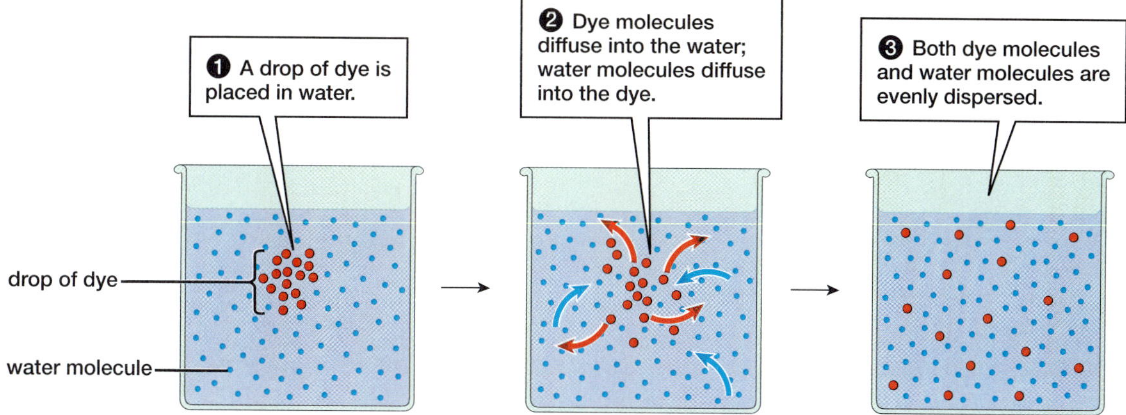

Figure 5-6 Diffusion

(a) Simple diffusion through the phospholipid bilayer

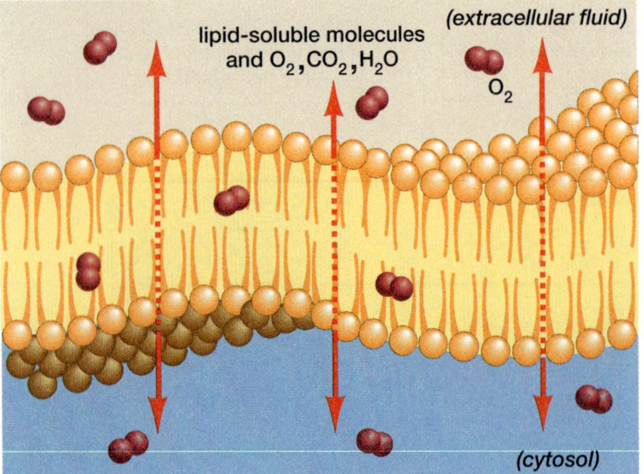

(b) Facilitated diffusion through a channel protein

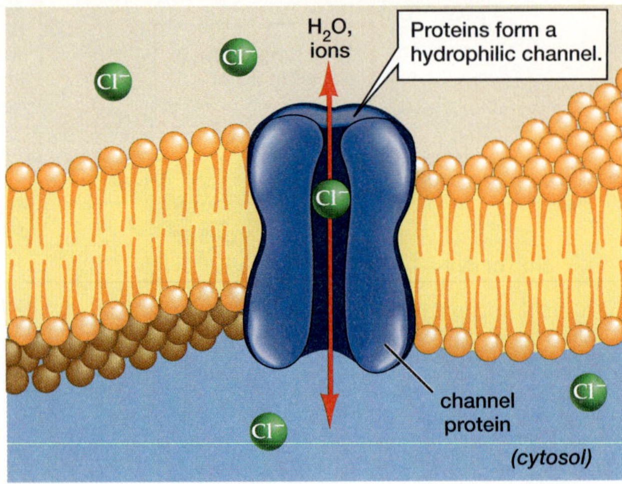

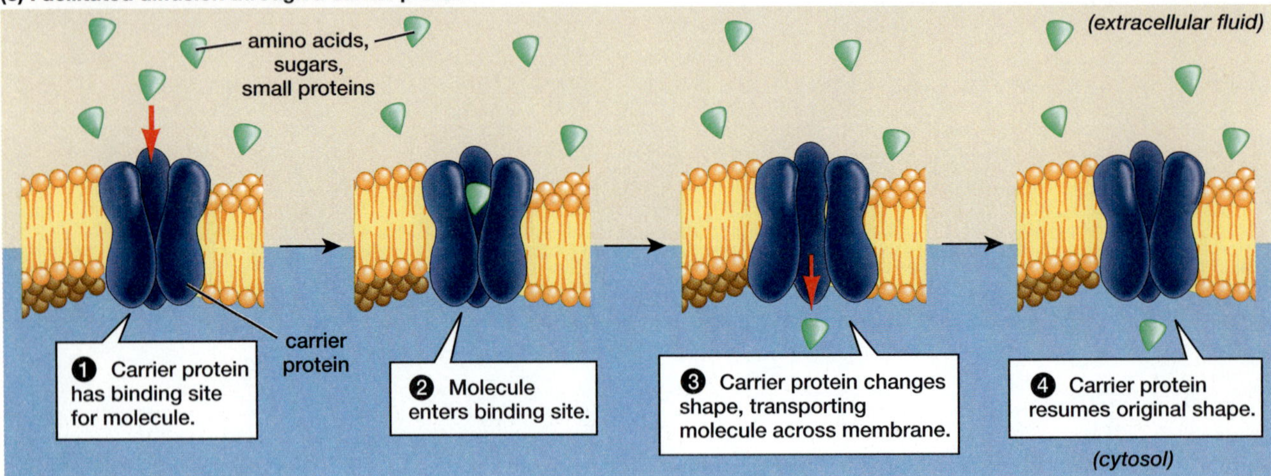

Figure 5-7 Types of diffusion

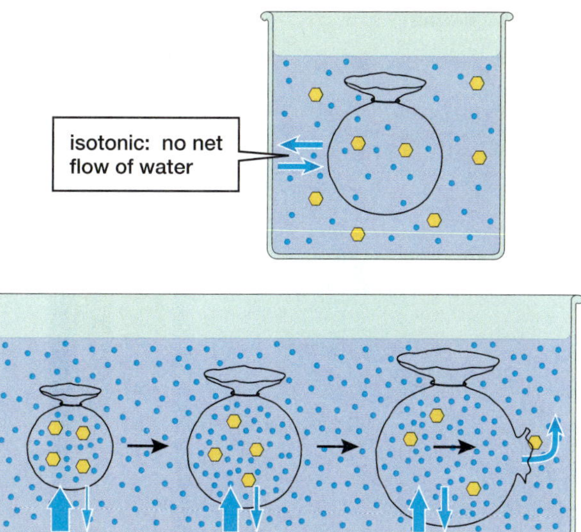

Figures 5-8, 5-9 Isotonic solution, Hypotonic solution

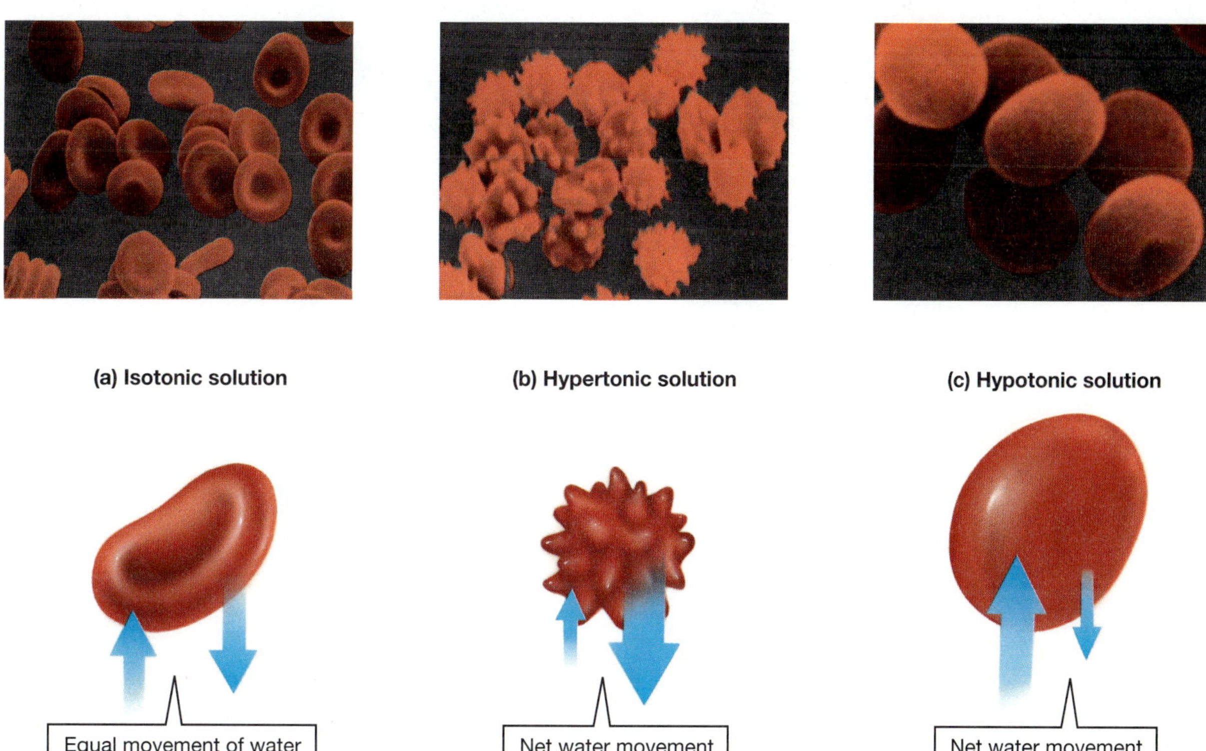

Figure 5-10 Osmosis in blood cells

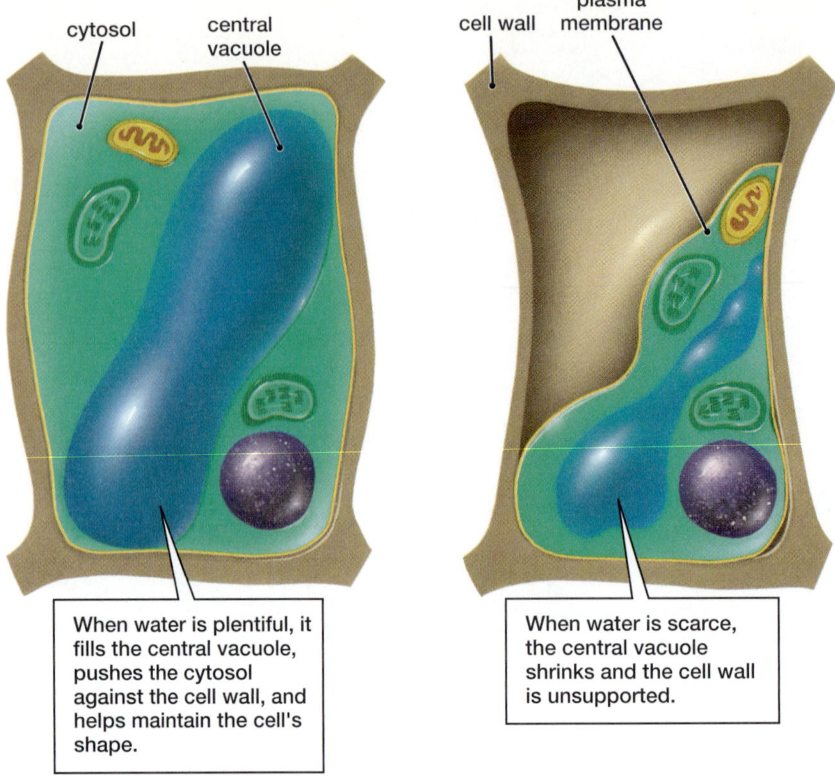

Figure 5-11 Plasmolysis

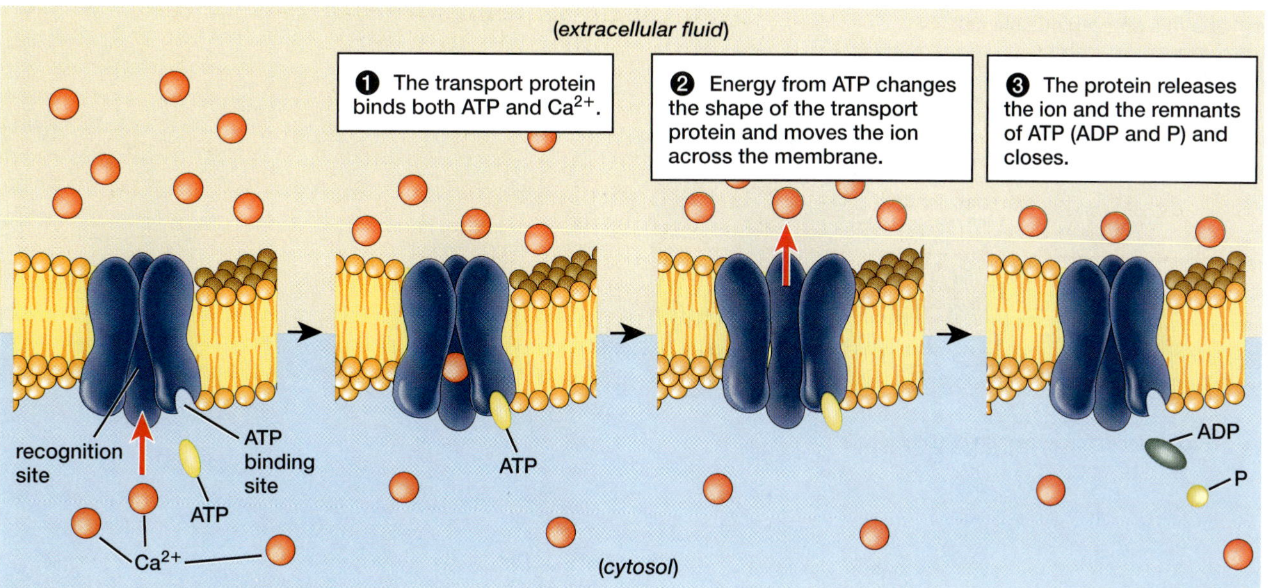

Figure 5-12 Active transport

Pinocytosis

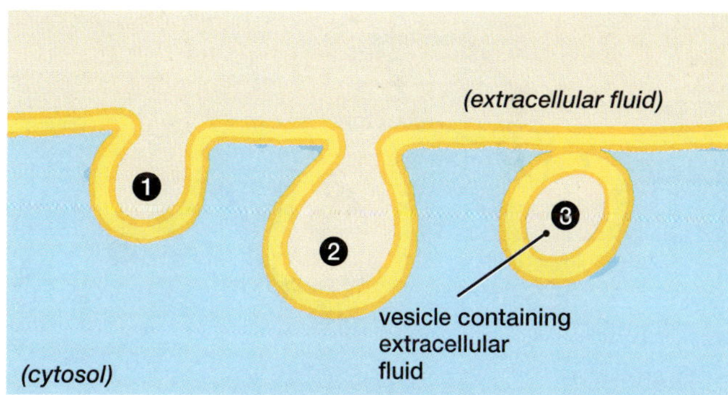

❶ A dimple forms in the plasma membrane, which ❷ deepens and surrounds the extracellular fluid. ❸ The membrane encloses the extracellular fluid, forming a vesicle.

Figure 5-13a Pinocytosis

Pinocytosis in a smooth muscle cell.

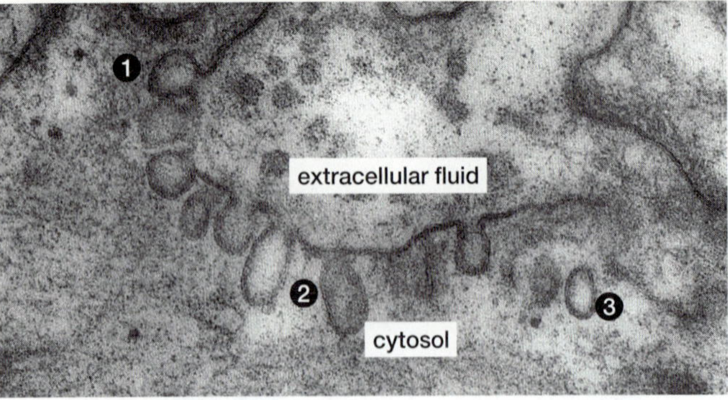

Figure 5-13b Electron micrograph of pinocytosis

Receptor-mediated endocytosis

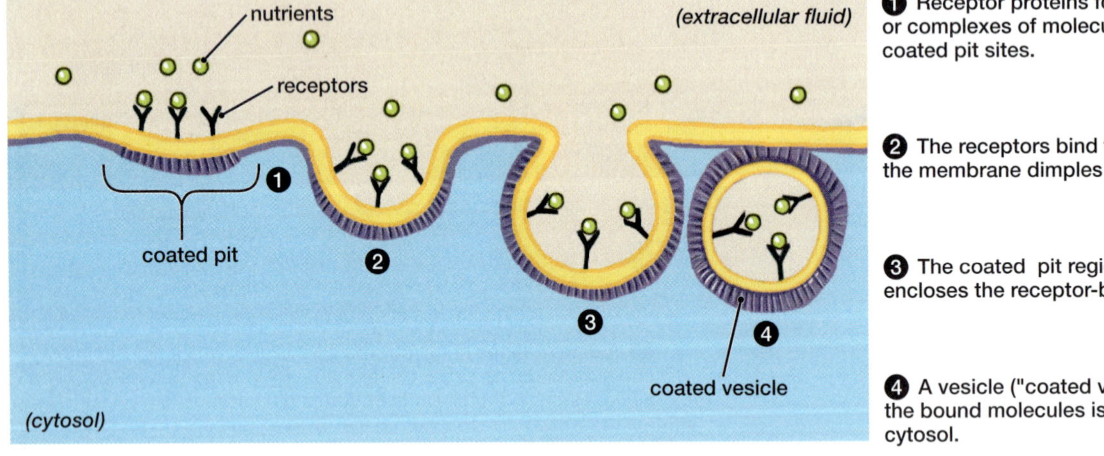

1 Receptor proteins for specific molecules or complexes of molecules are localized at coated pit sites.

2 The receptors bind the molecules and the membrane dimples inward.

3 The coated pit region of the membrane encloses the receptor-bound molecules.

4 A vesicle ("coated vesicle") containing the bound molecules is released into the cytosol.

Figure 5-14 Receptor-mediated endocytosis

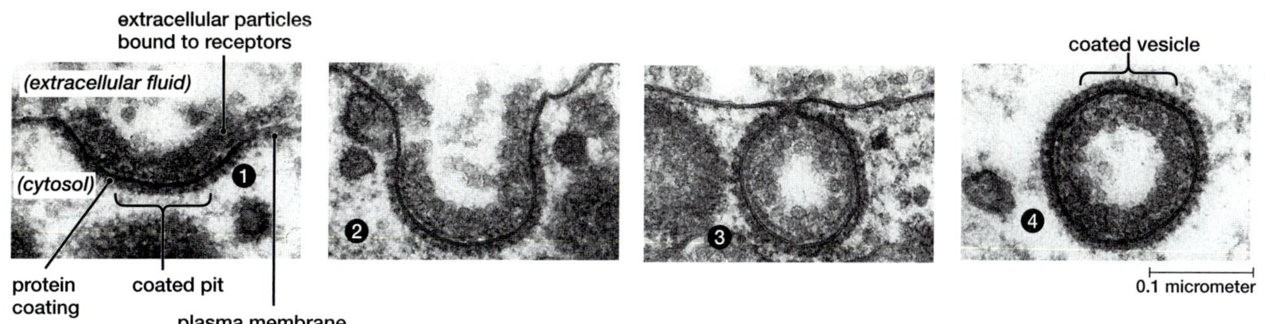

Figure 5-14 Electron micrograph of endocytosis

Phagocytosis

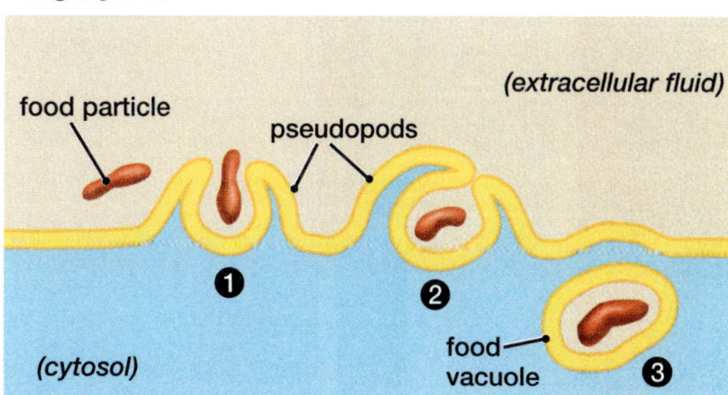

1 The plasma membrane extends pseudopods toward an extracellular particle (for example, food). 2 The ends of the pseudopods fuse, encircling the particle. 3 A vesicle called a food vacuole is formed containing the engulfed particle.

Figure 5-15a Phagocytosis

Amoeba

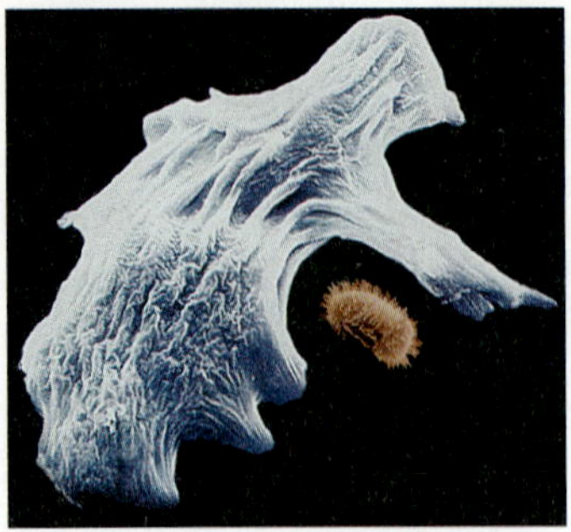

An *Amoeba* (a freshwater protist), engulfs a *Paramecium* using phagocytosis.

Figure 5-15b Ameoba eating

White blood cell

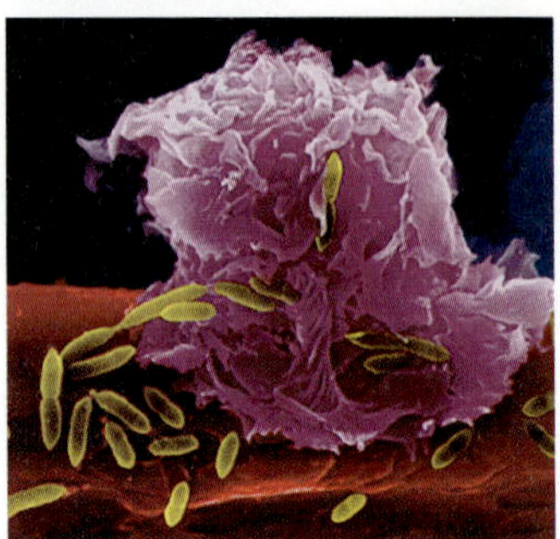

A white blood cell ingests bacteria using phagocytosis.

Figure 5-15c White blood cell eating

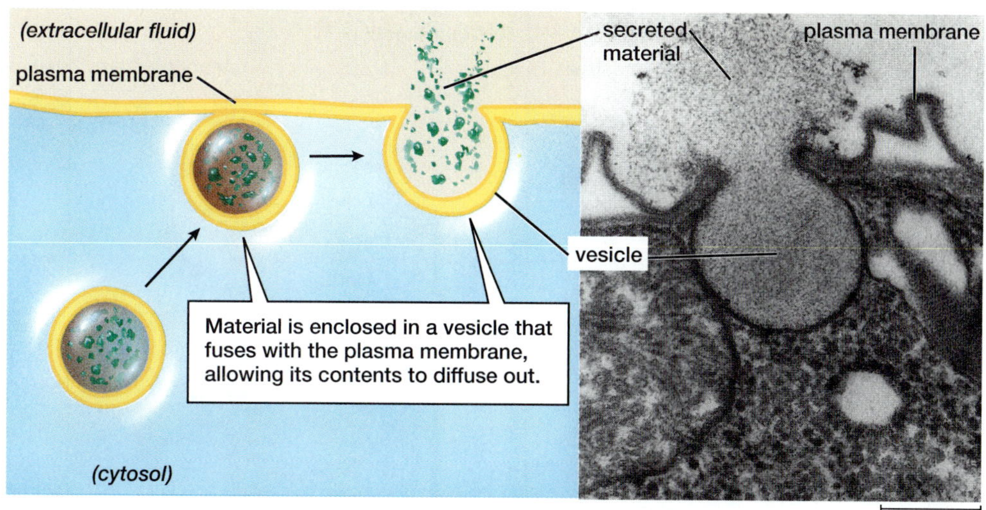

Figure 5-16 Exocytosis

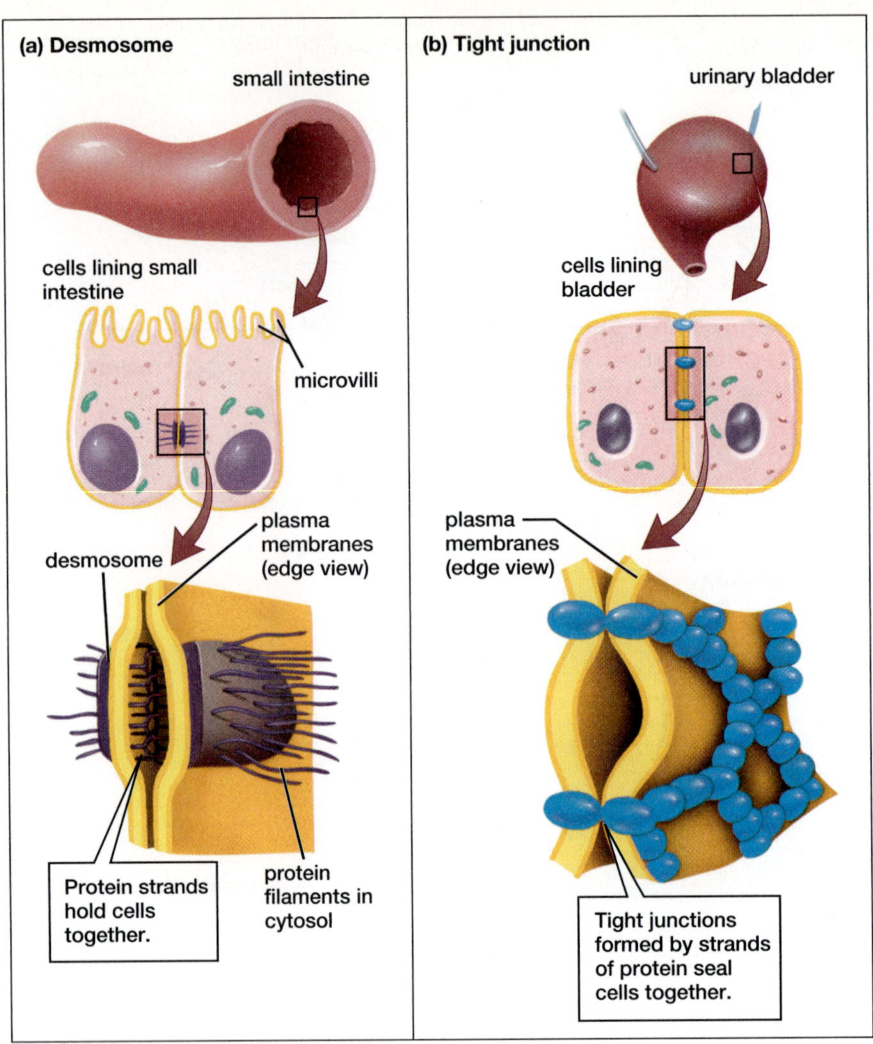

Figure 5-18 Cell attachment structures

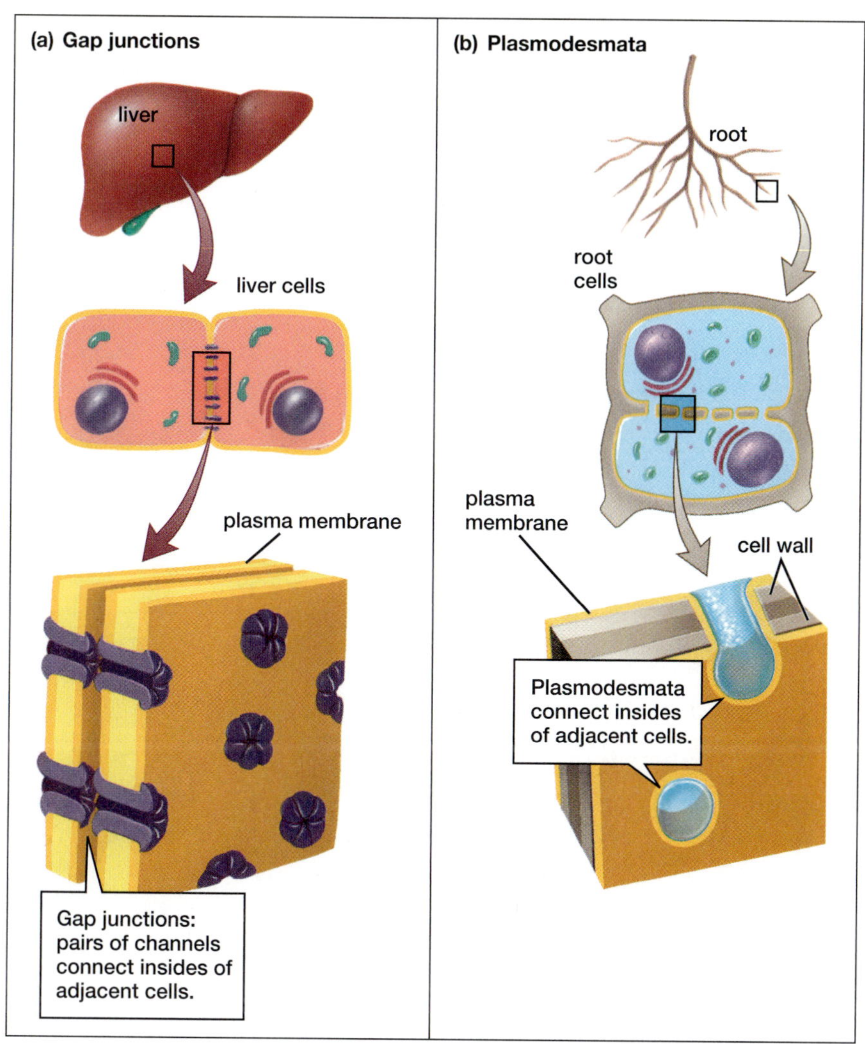

Figure 5-19 Cell communication structures

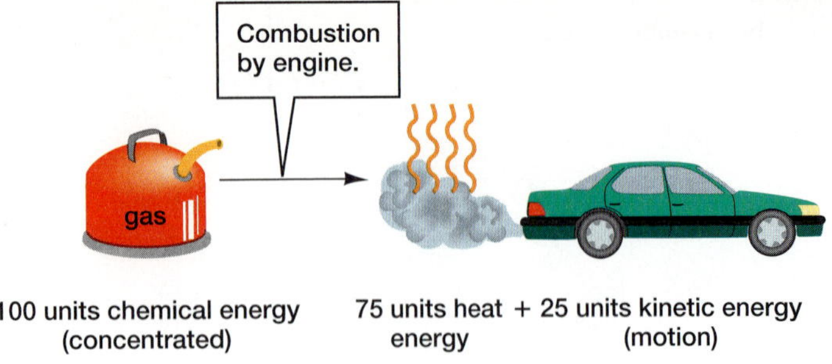

Figure 6-2 Gasoline combustion

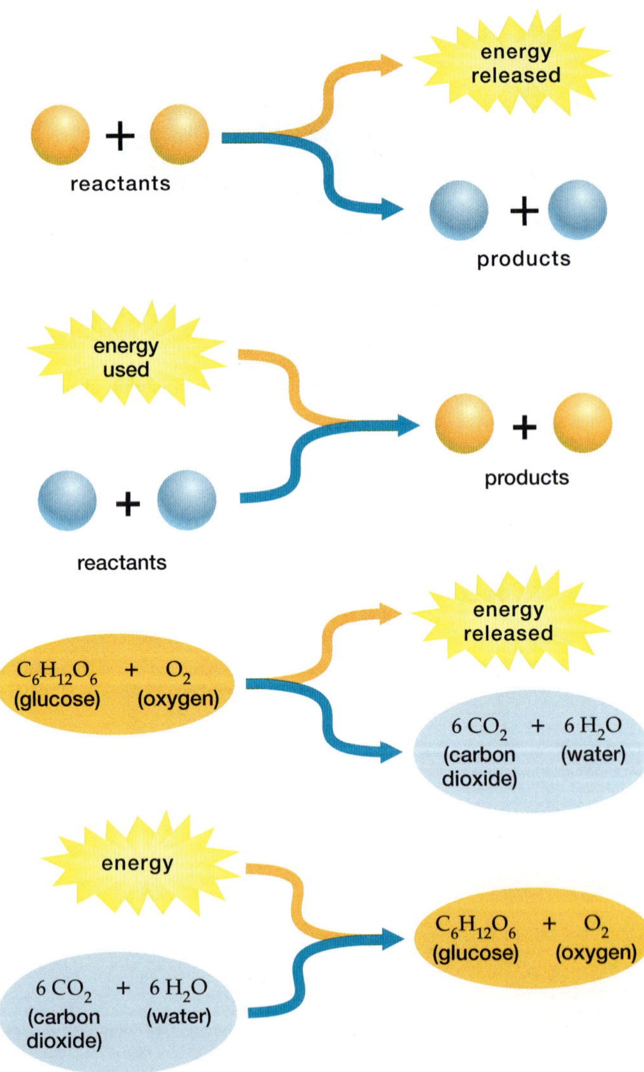

Figures 6-3, 6-4, 6-5, 6-7 Exergonic reaction, Endergonic reaction, Burning glucose, Photosynthesis

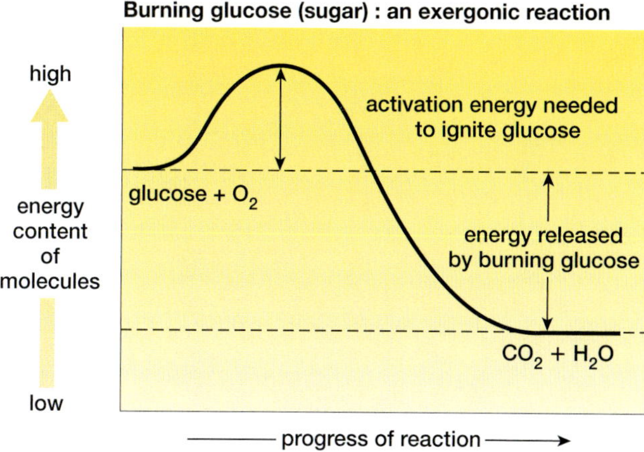

Figure 6-6 Energy relations in Exergonic reaction

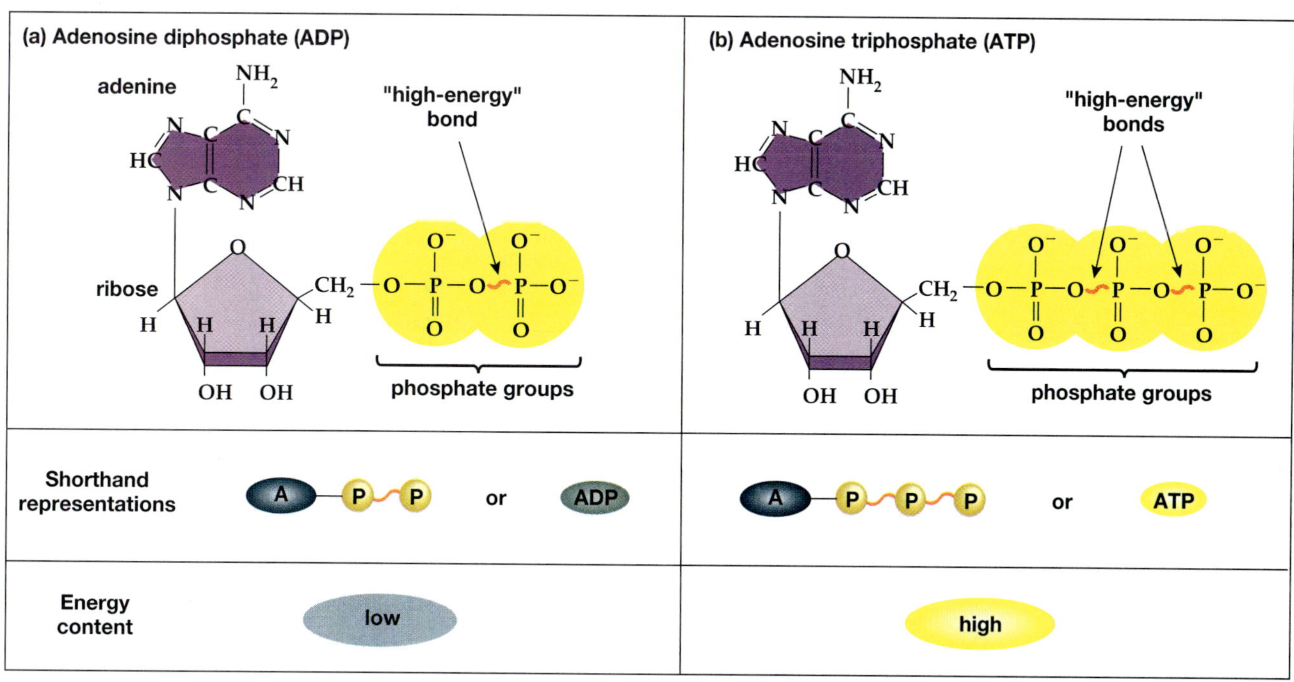

Figure 6-8 ATP and ADP

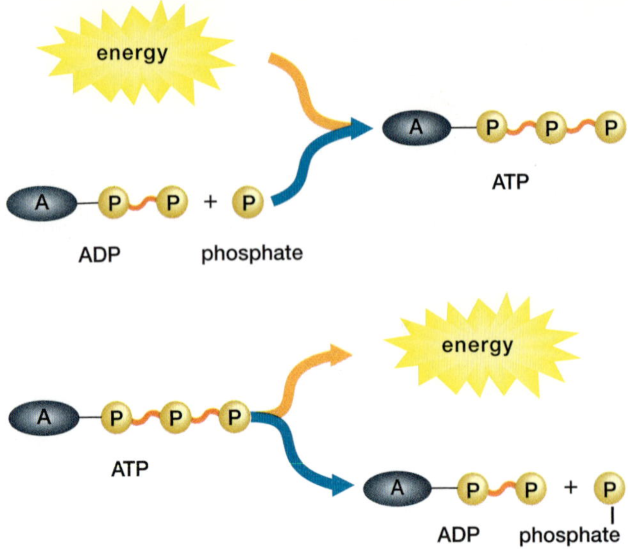

Figures 6-9, 6-10 ATP synthesis and breakdown

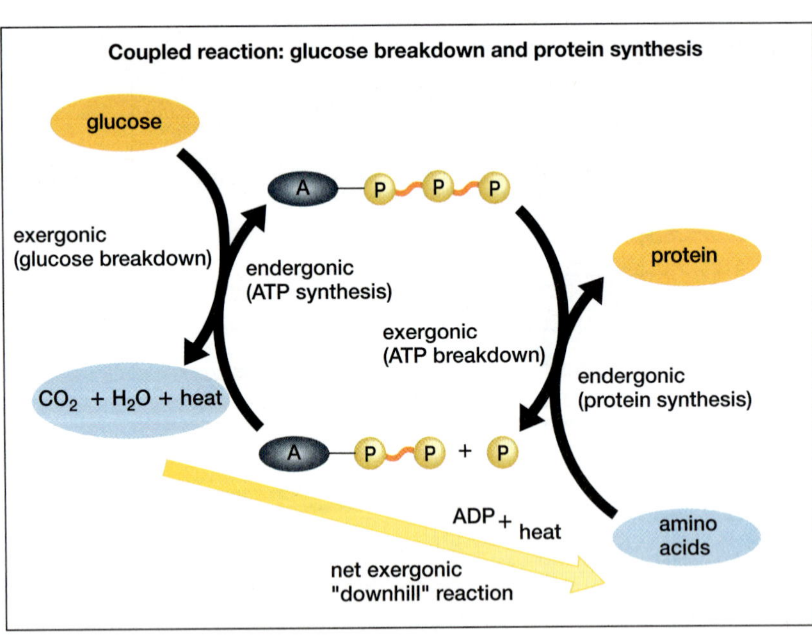

Figure 6-11 Coupled reaction

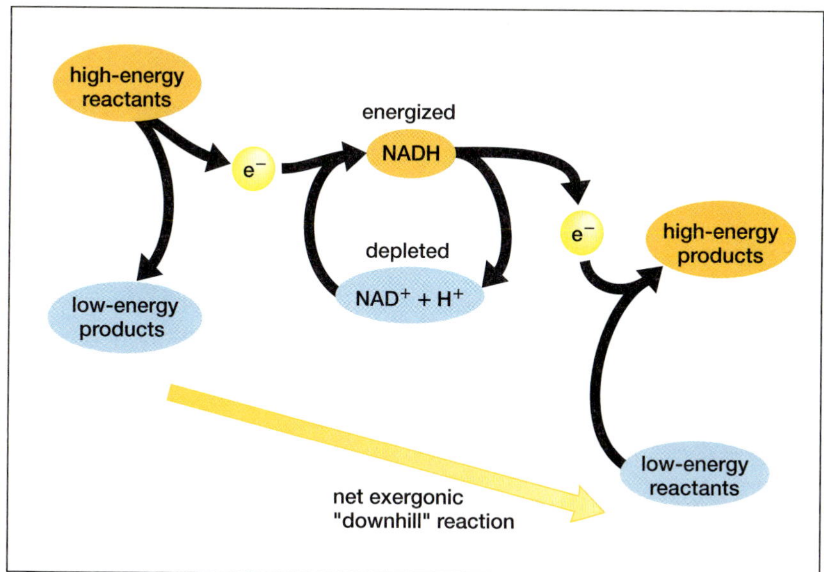

Figure 6-12 Electron carrier

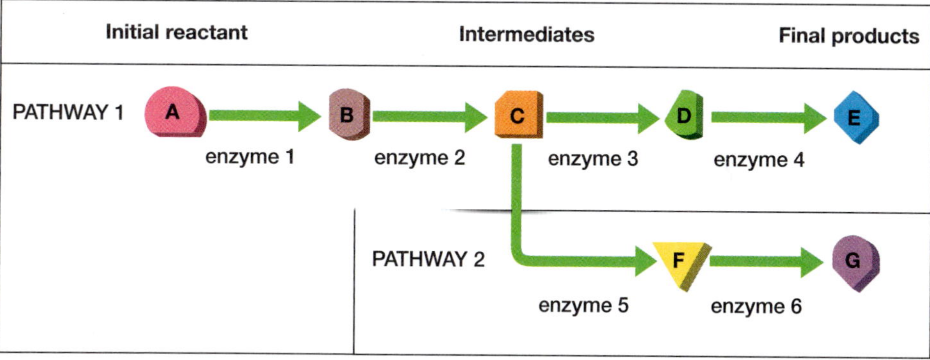

Figure 6-13 Metabolic pathways

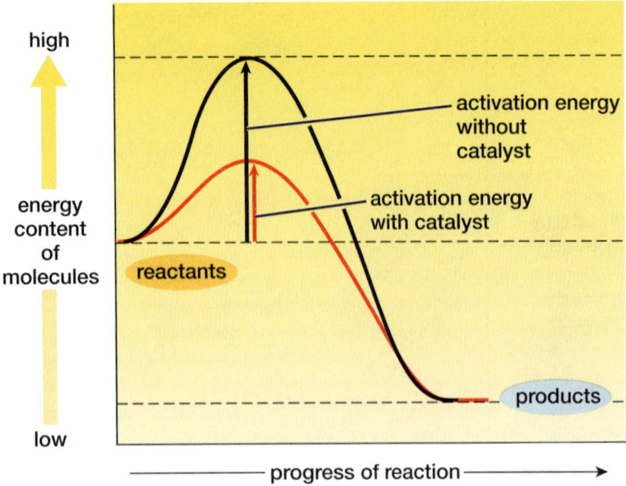

Figure 6-14 Activation energy/catalysts

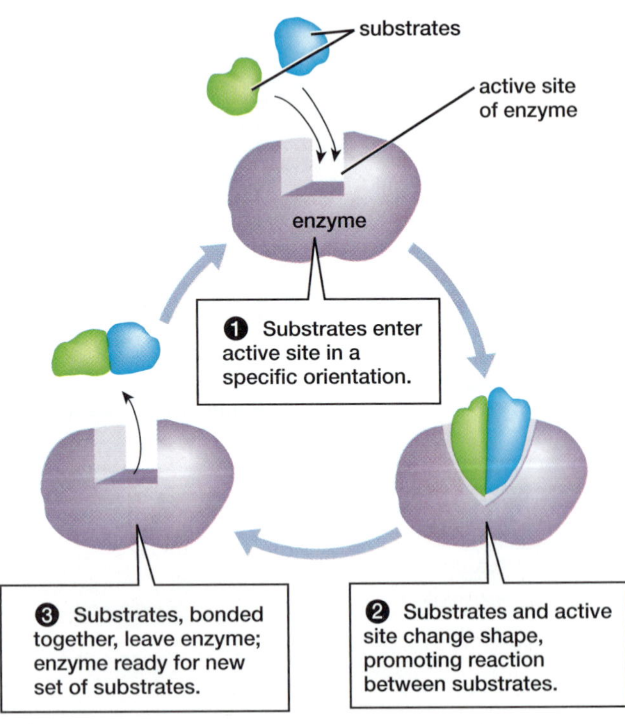

Figure 6-15 Cycle of enzyme-substrate interactions

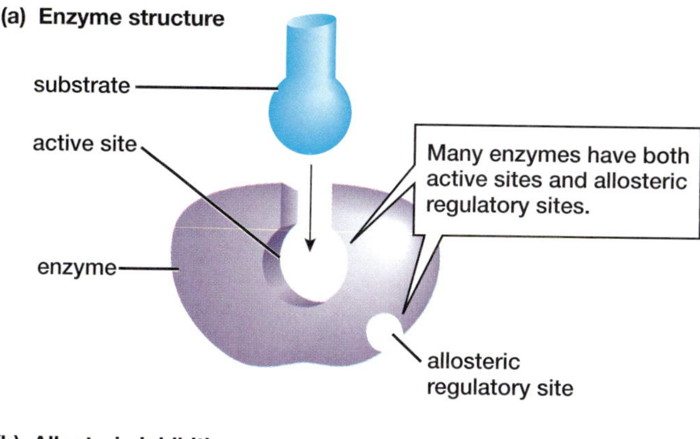

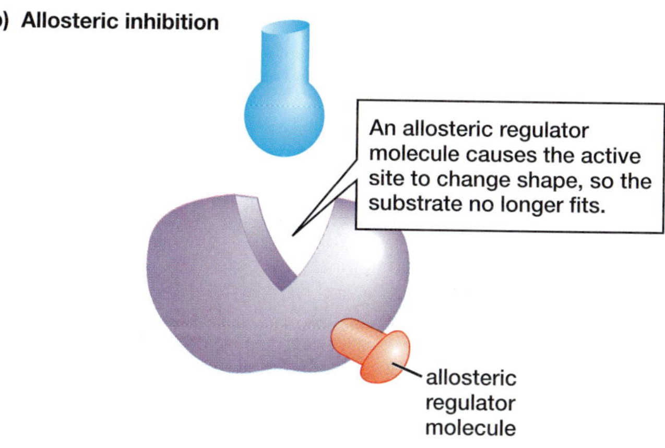

Figure 6-16 Allosteric regulation

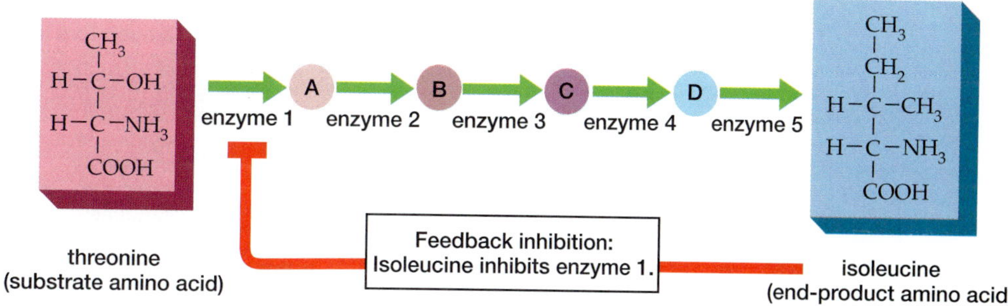

Figure 6-17 Feedback inhibition

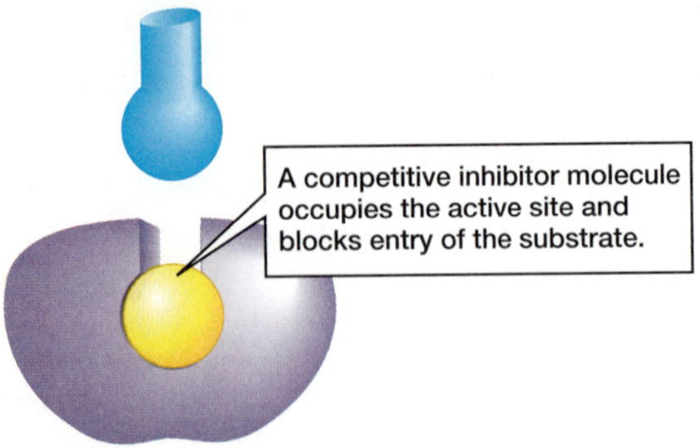

Figure 6-18 Competitive inhibition

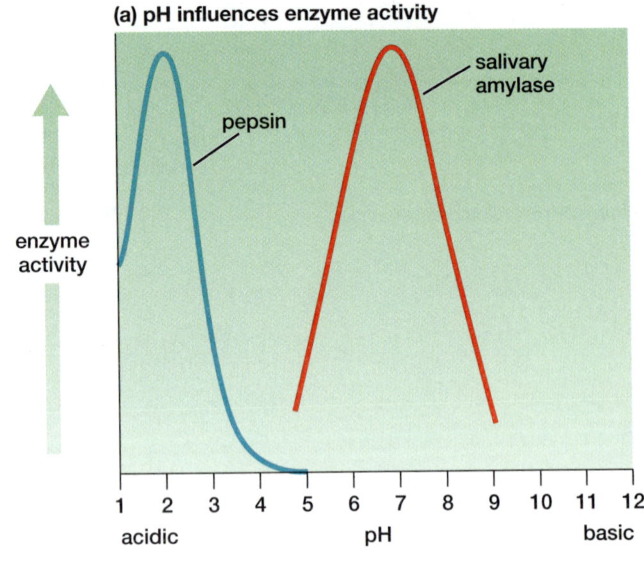

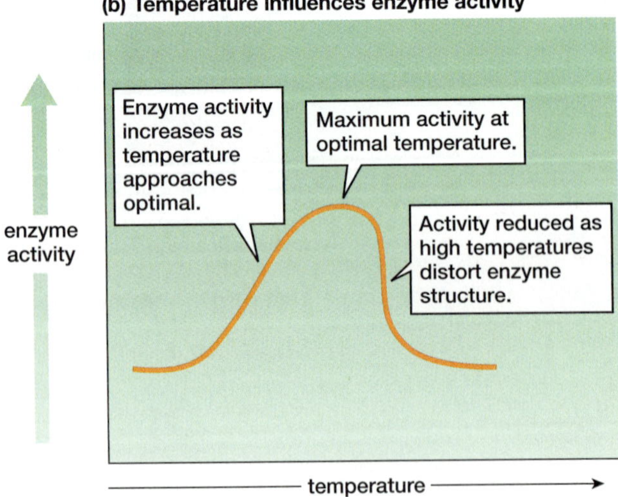

Figure 6-19 Enzyme function graphs

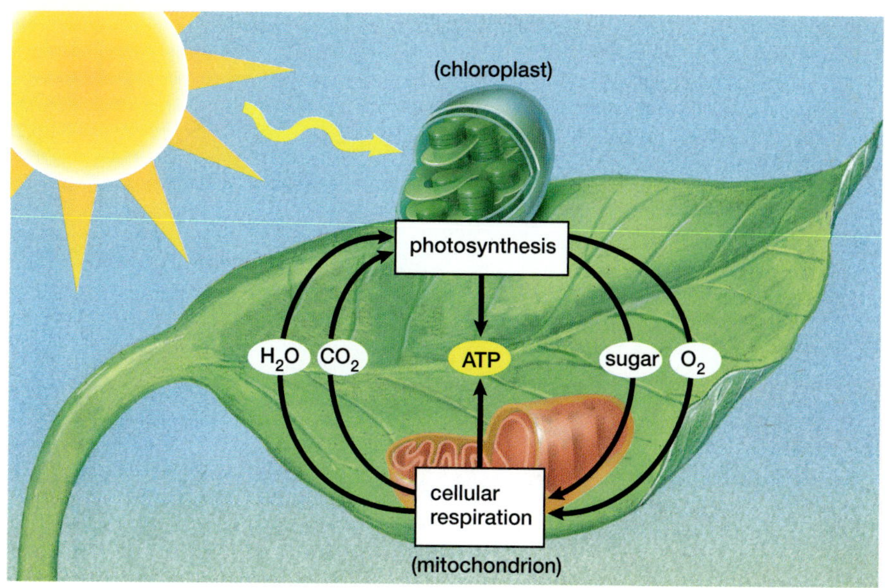

Figure 7-1 Sun and leaf

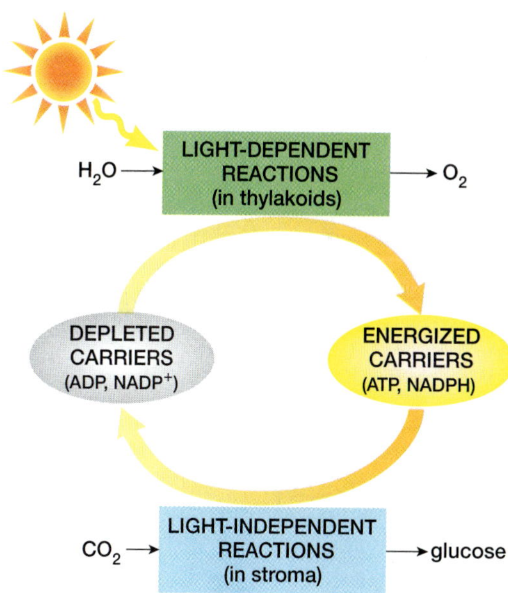

Figure 7-4 Relationship of light and dark reactions

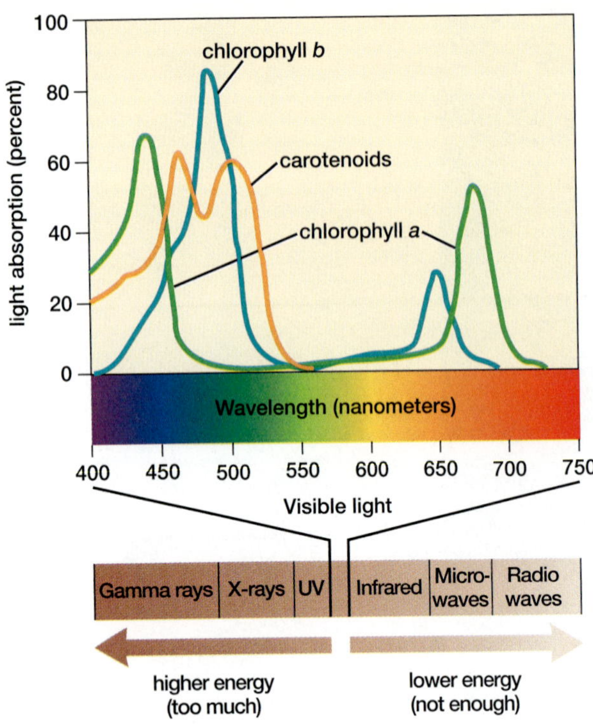

Figure 7-5 Spectrum

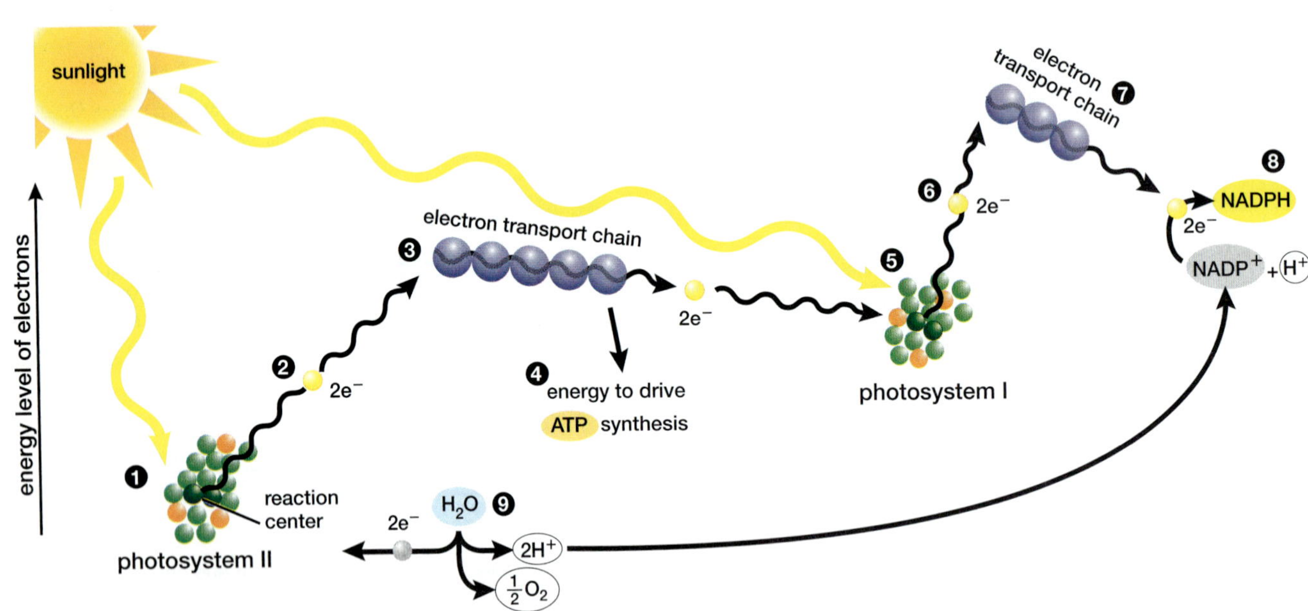

Figure 7-7 Light-dependent reactions-pinball model

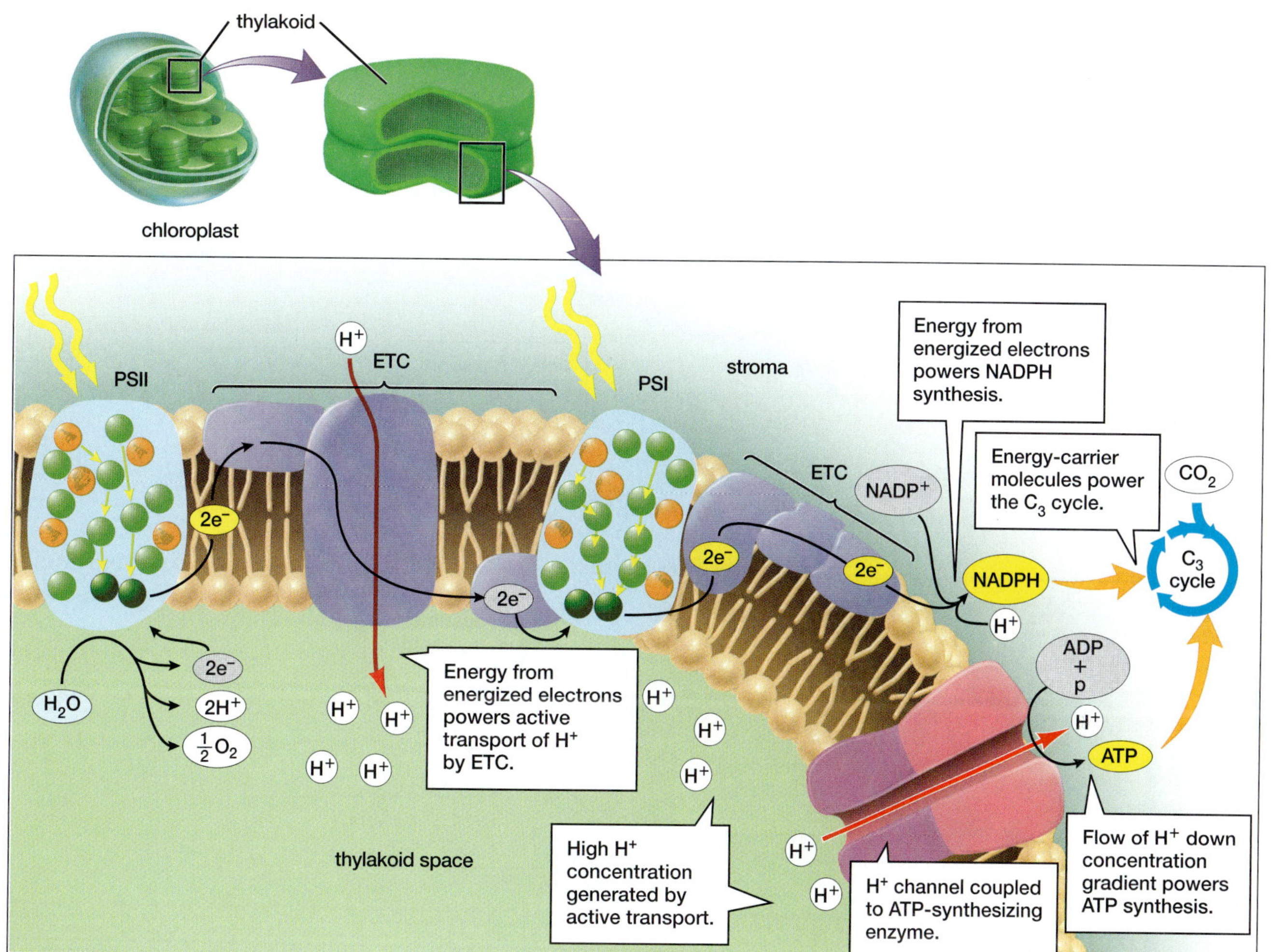

Figure 7-8 Light reactions in thylakoid

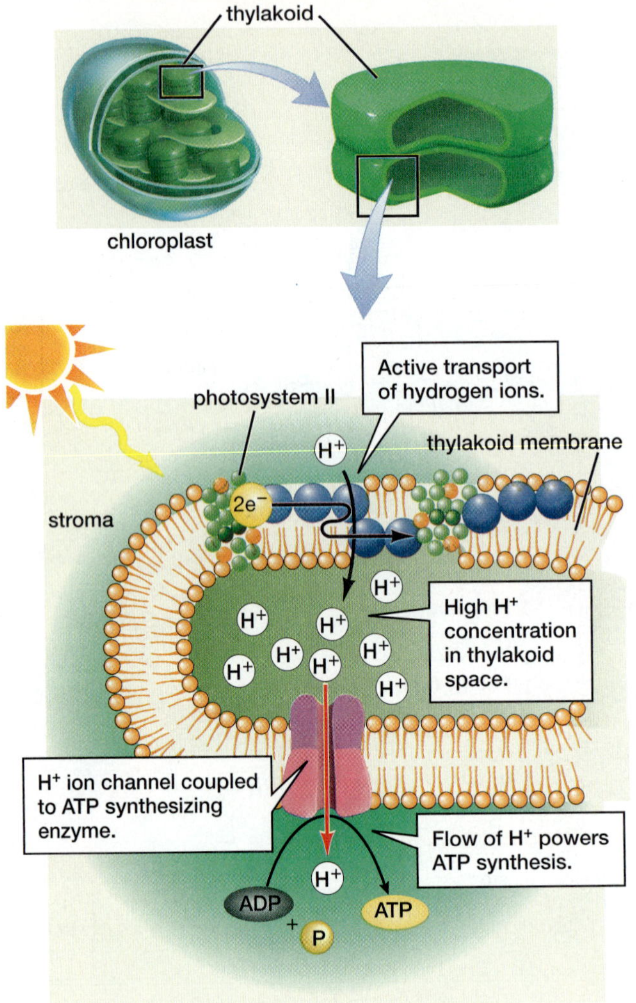

Figure E7-2 Chemiosmosis

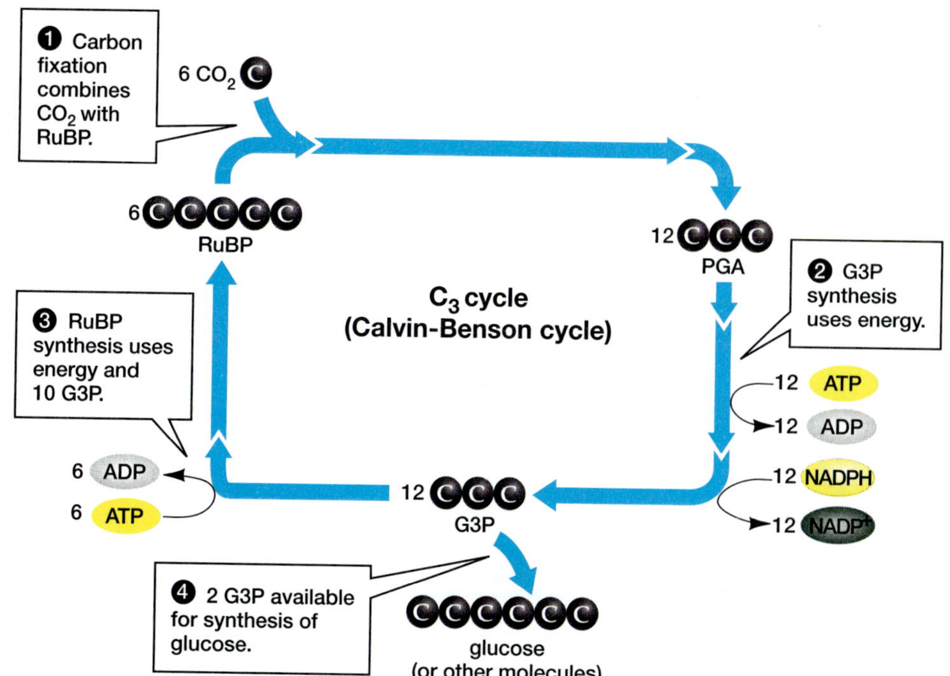

Figure 7-10 C_3 cycle

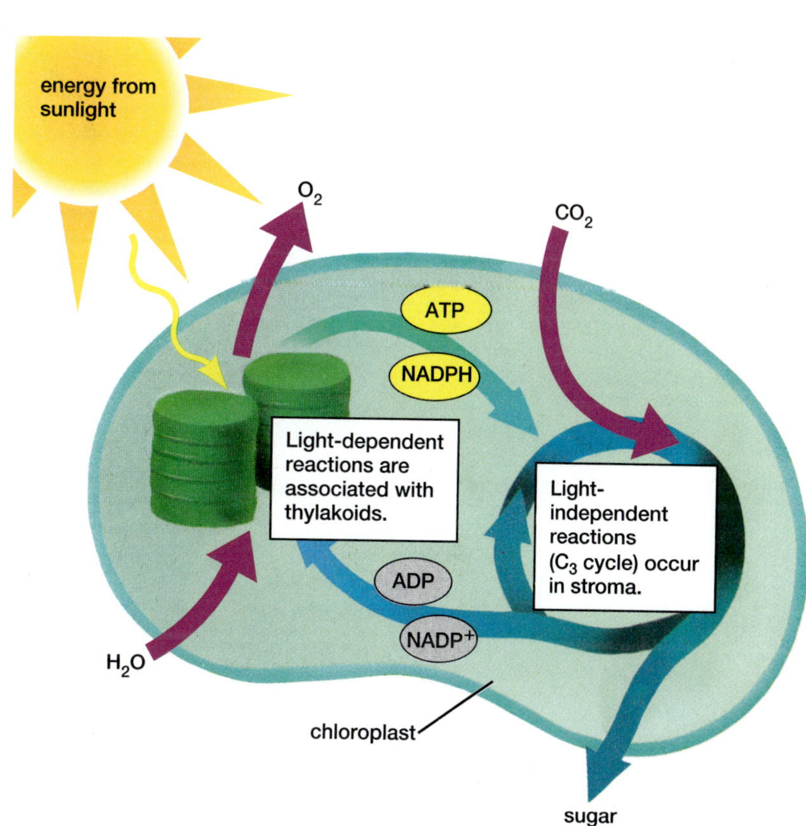

Figure 7.11 Light dependent and light independent reactions

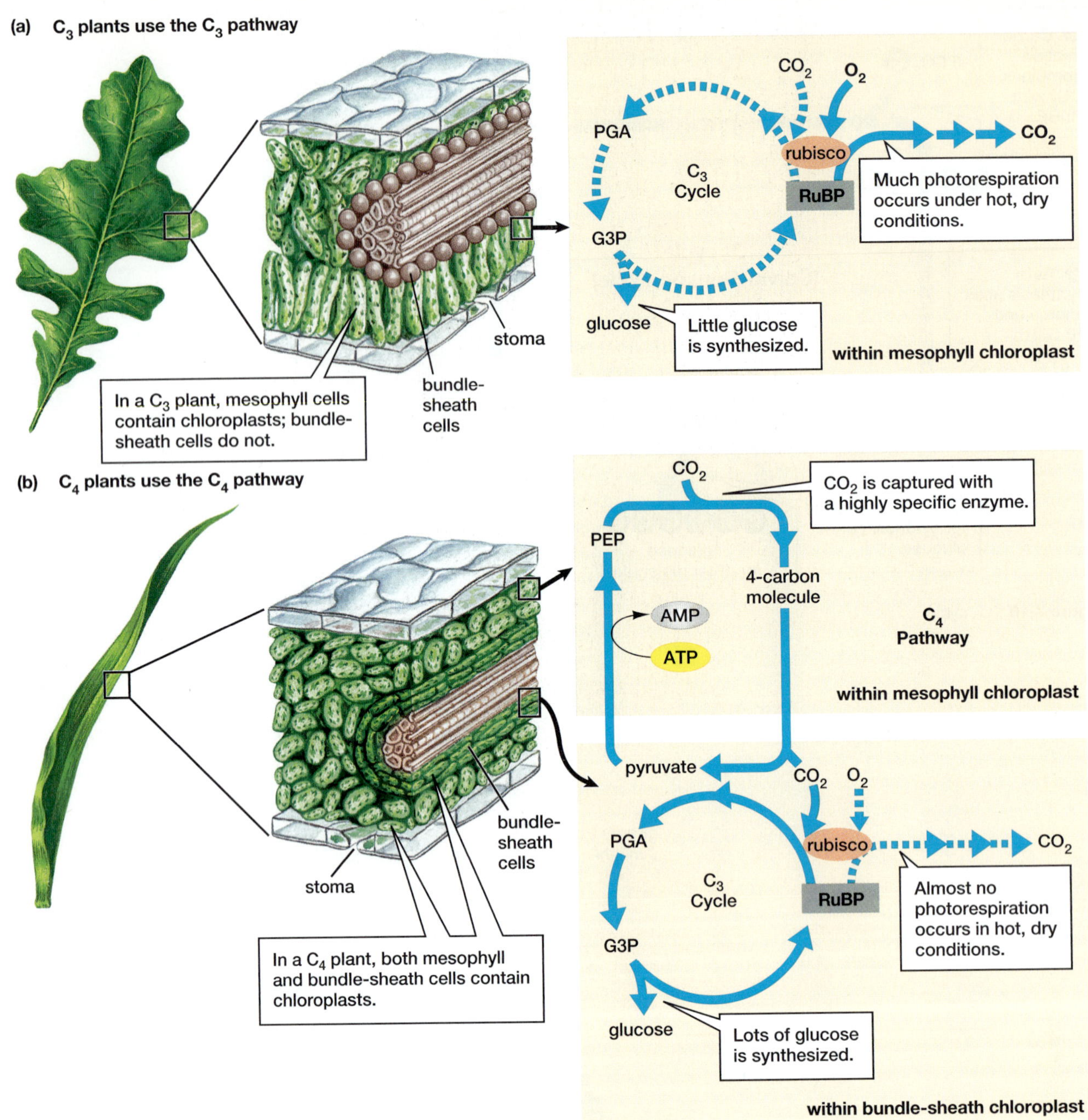

Figure 7-12 C_3 and C_4 compared

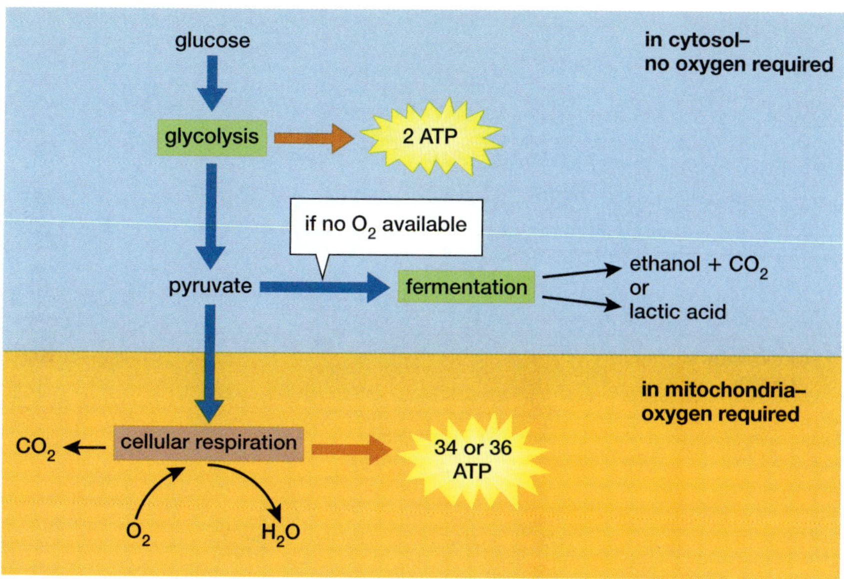

Figure 8-1 Overview of glucose breakdown

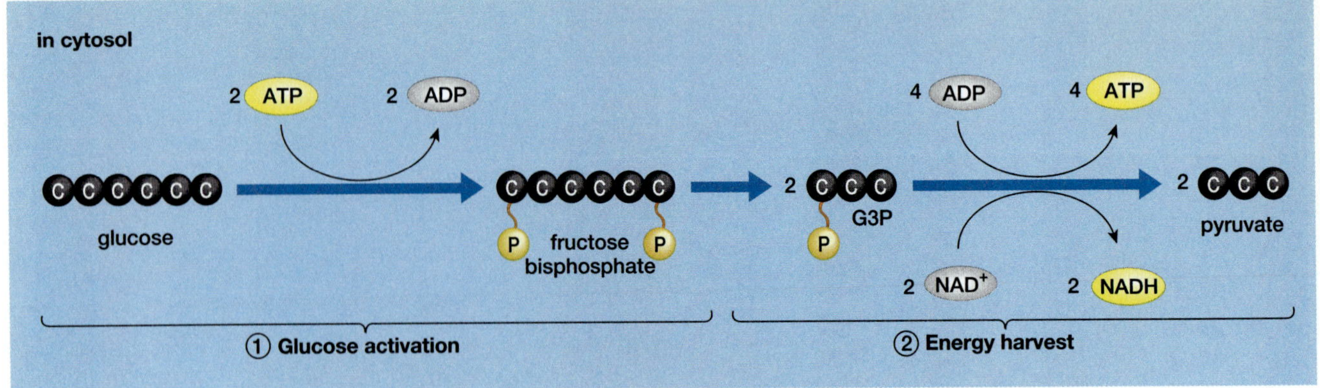

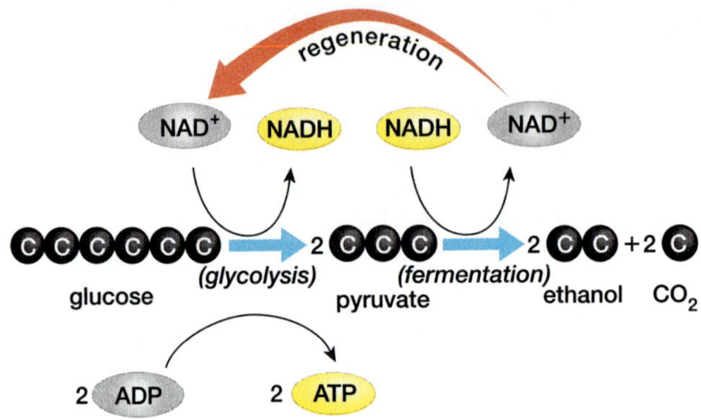

Figures 8-2, 8-5 Stages of glycolysis, Alcoholic fermentation

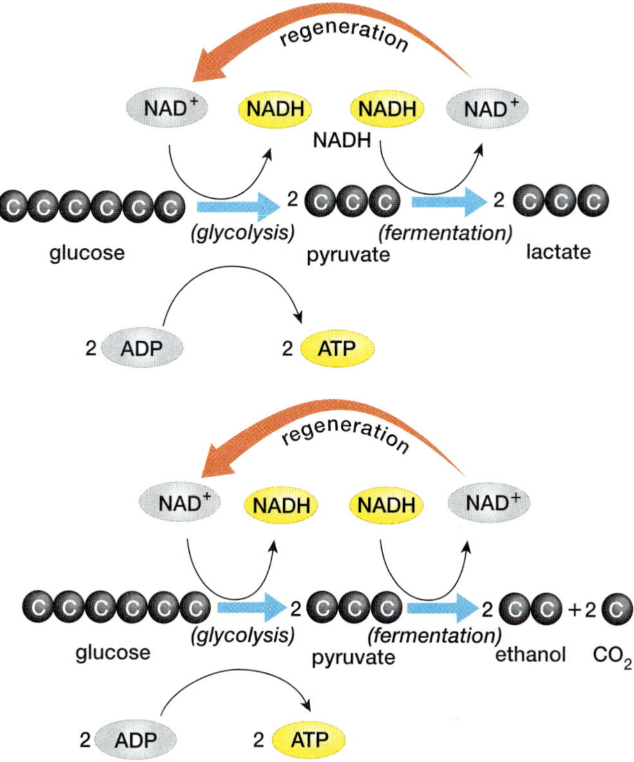

Figures 8-4, 8-5 Lactate fermentation, Alcoholic fermentation

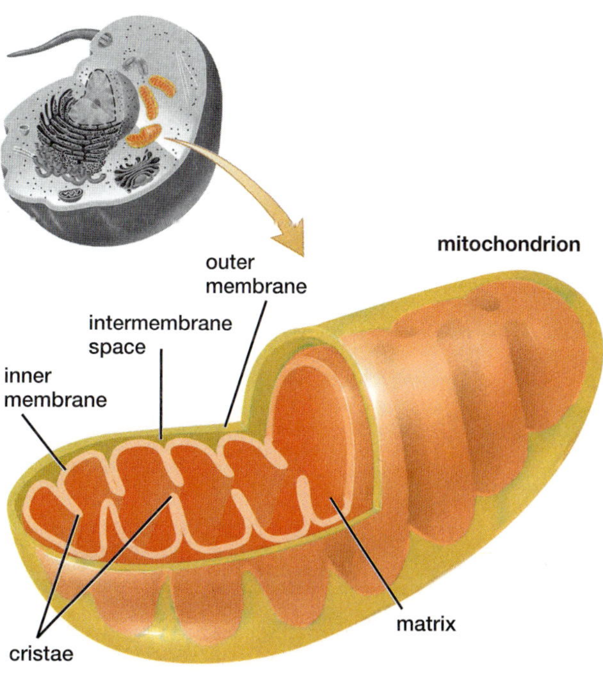

Figure 8-6 Mitochondrion

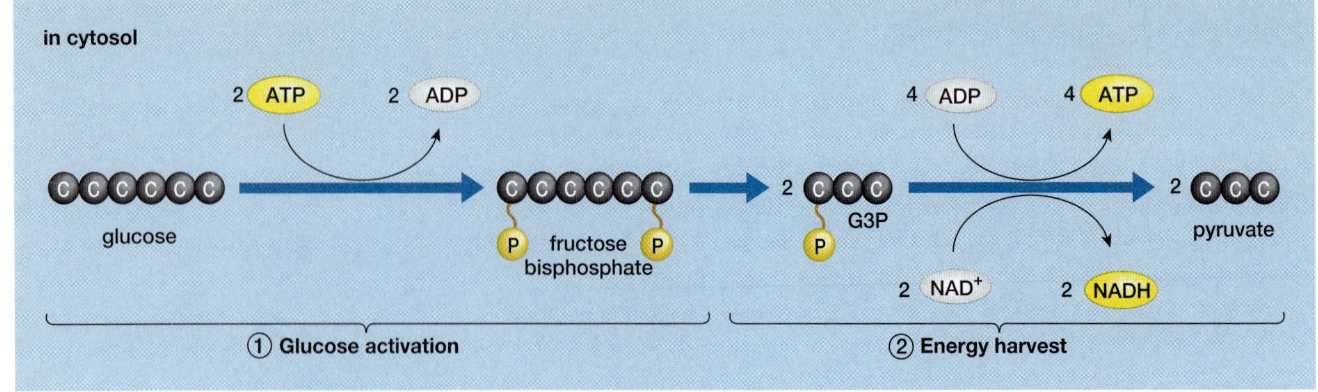

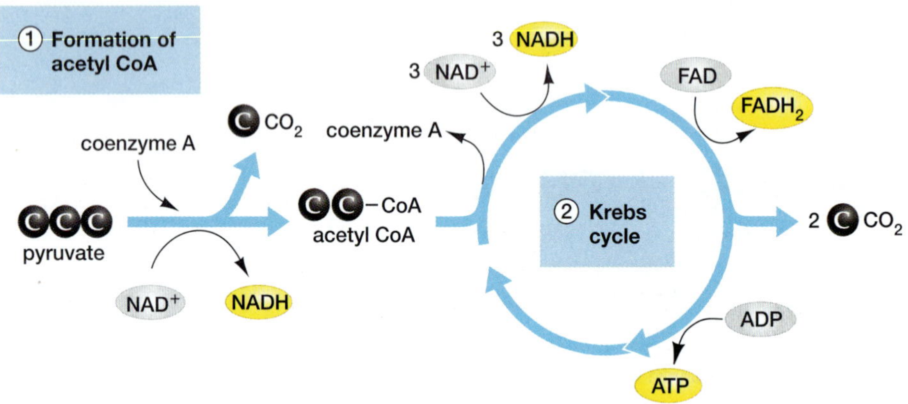

Figures 8-2, 8-7 Stages of glycolysis, Mitochondrial matrix reactions

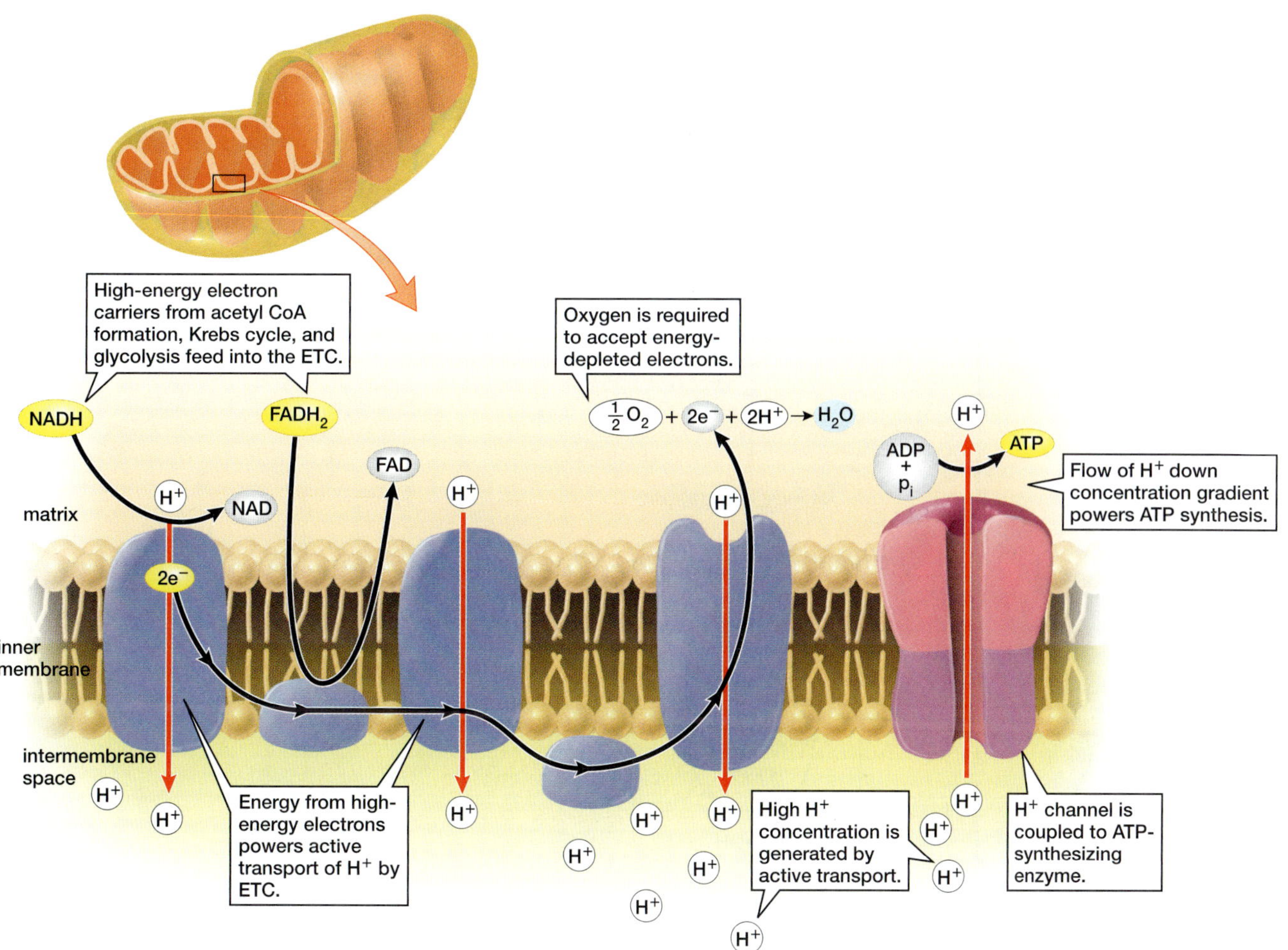

Figure 8-8 Electron transport chain

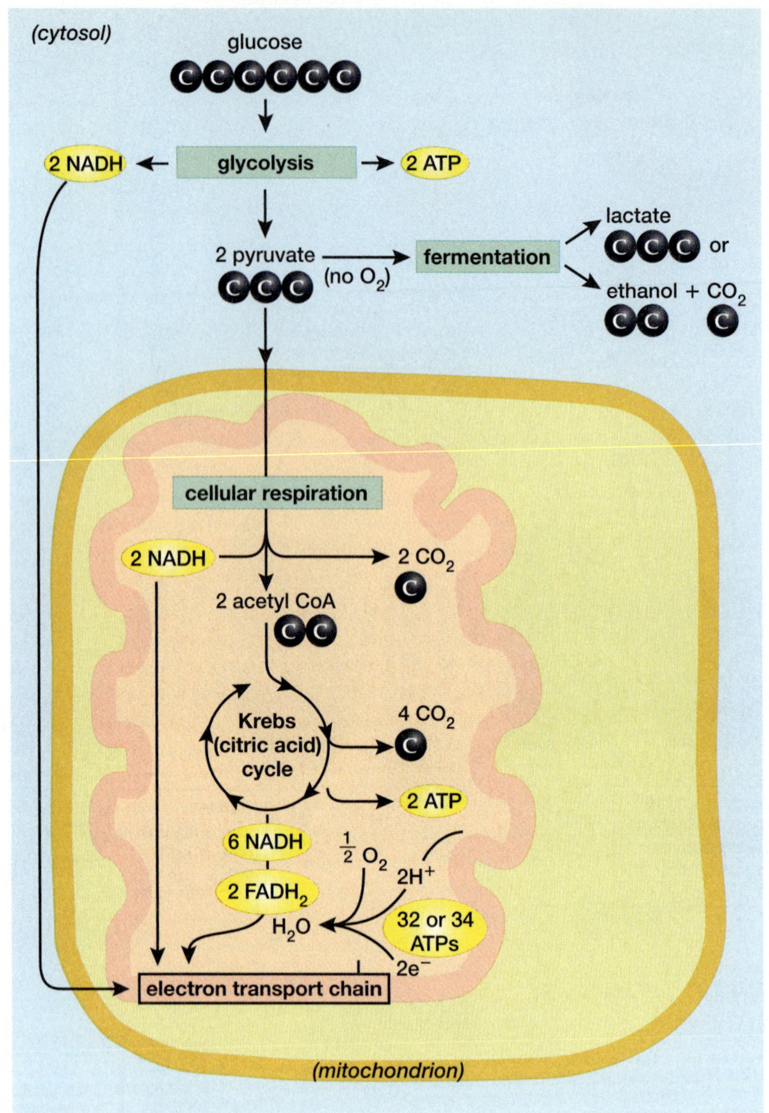

Figure 8-9 Summary glycolysis and cellular respiration

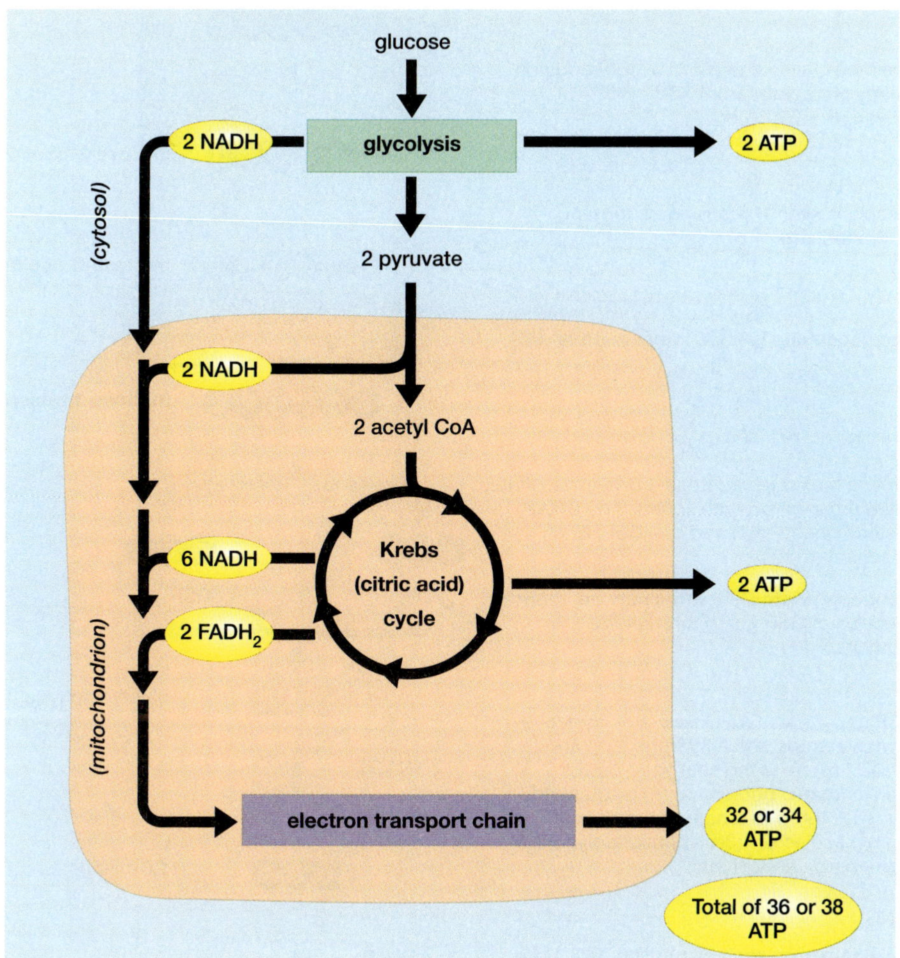

Figure 8-10 Energy harvest

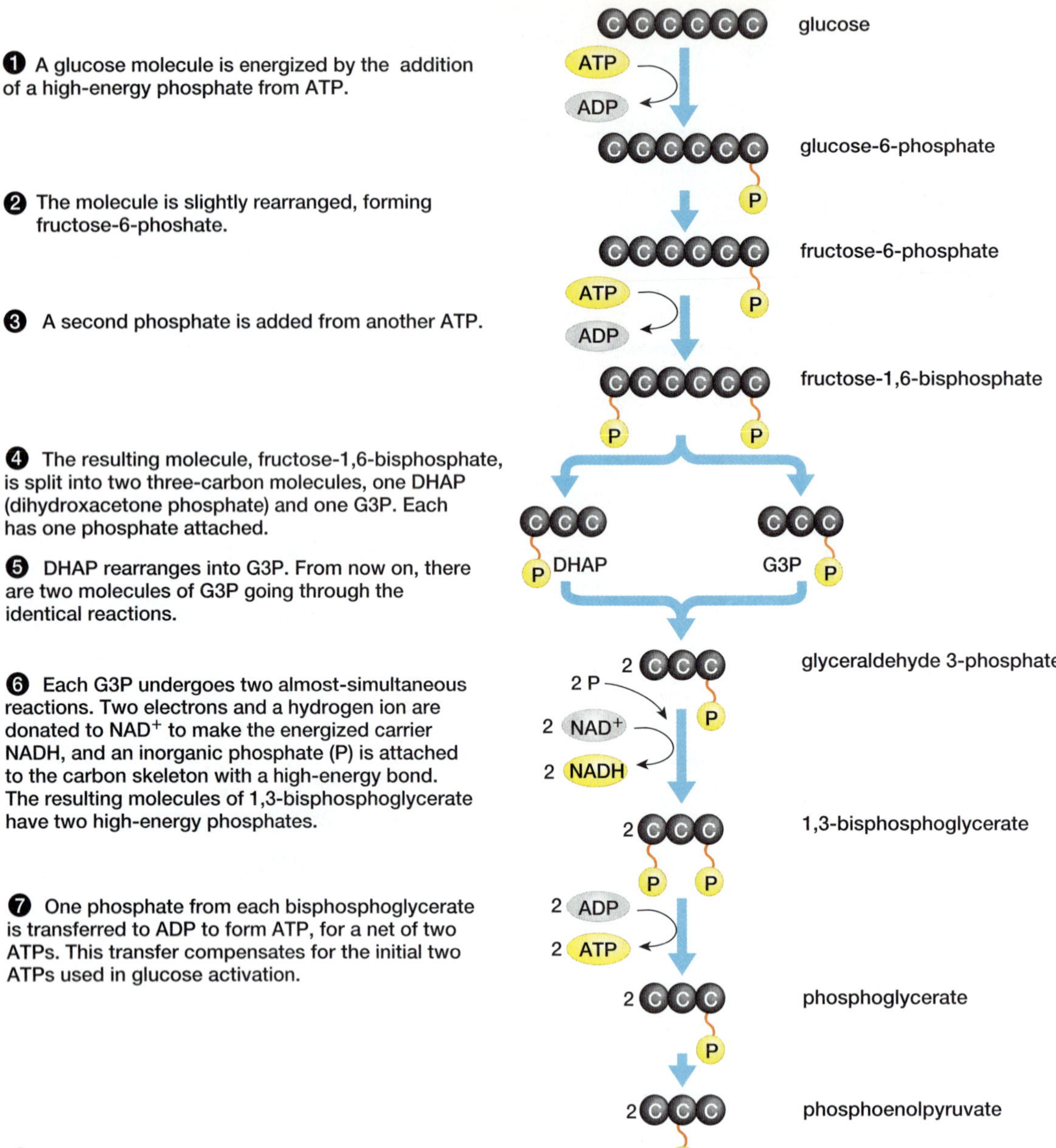

Figure E8-1 Details of glycolysis

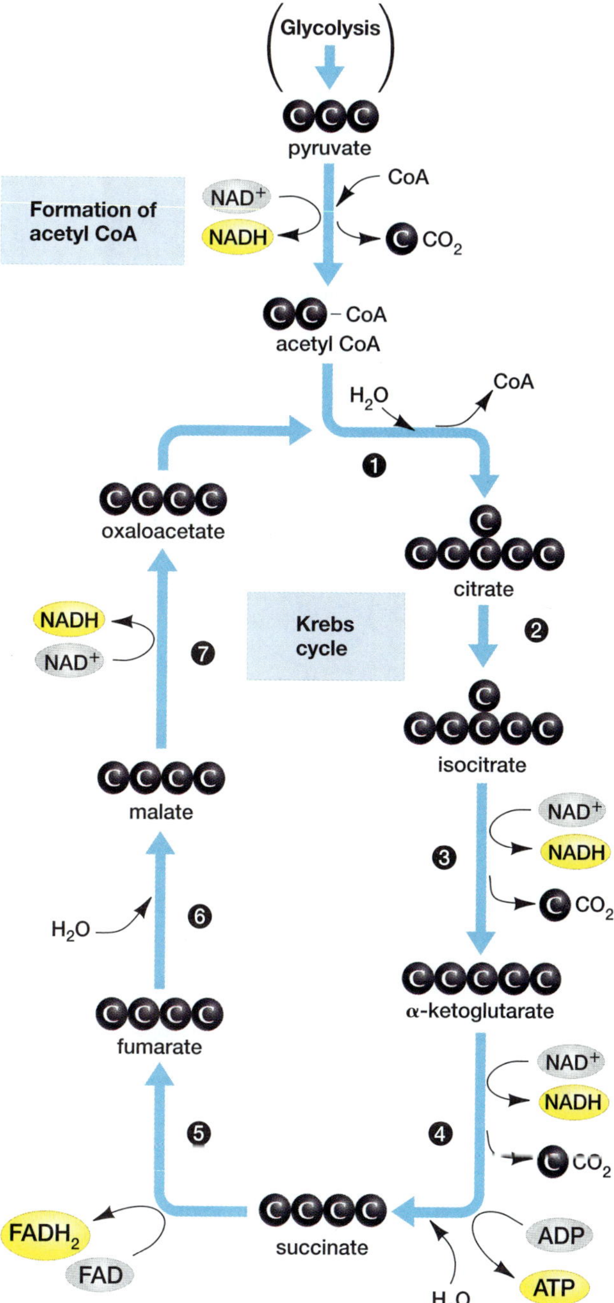

Figure E8-3 Details of cellular respiration

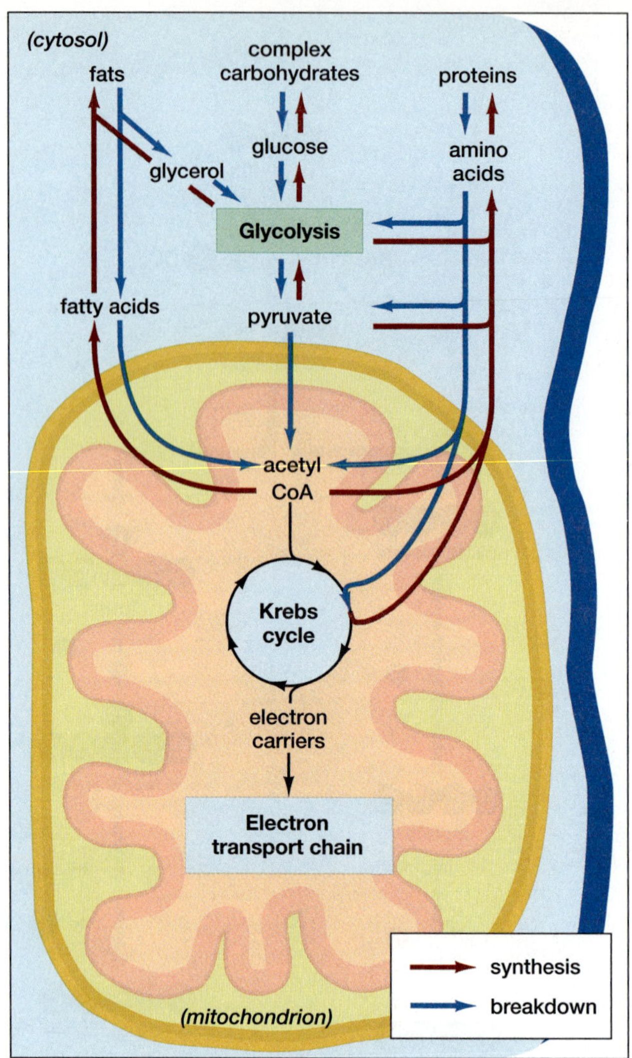

Figure E8-4 Metabolic pathways (Health Watch box)

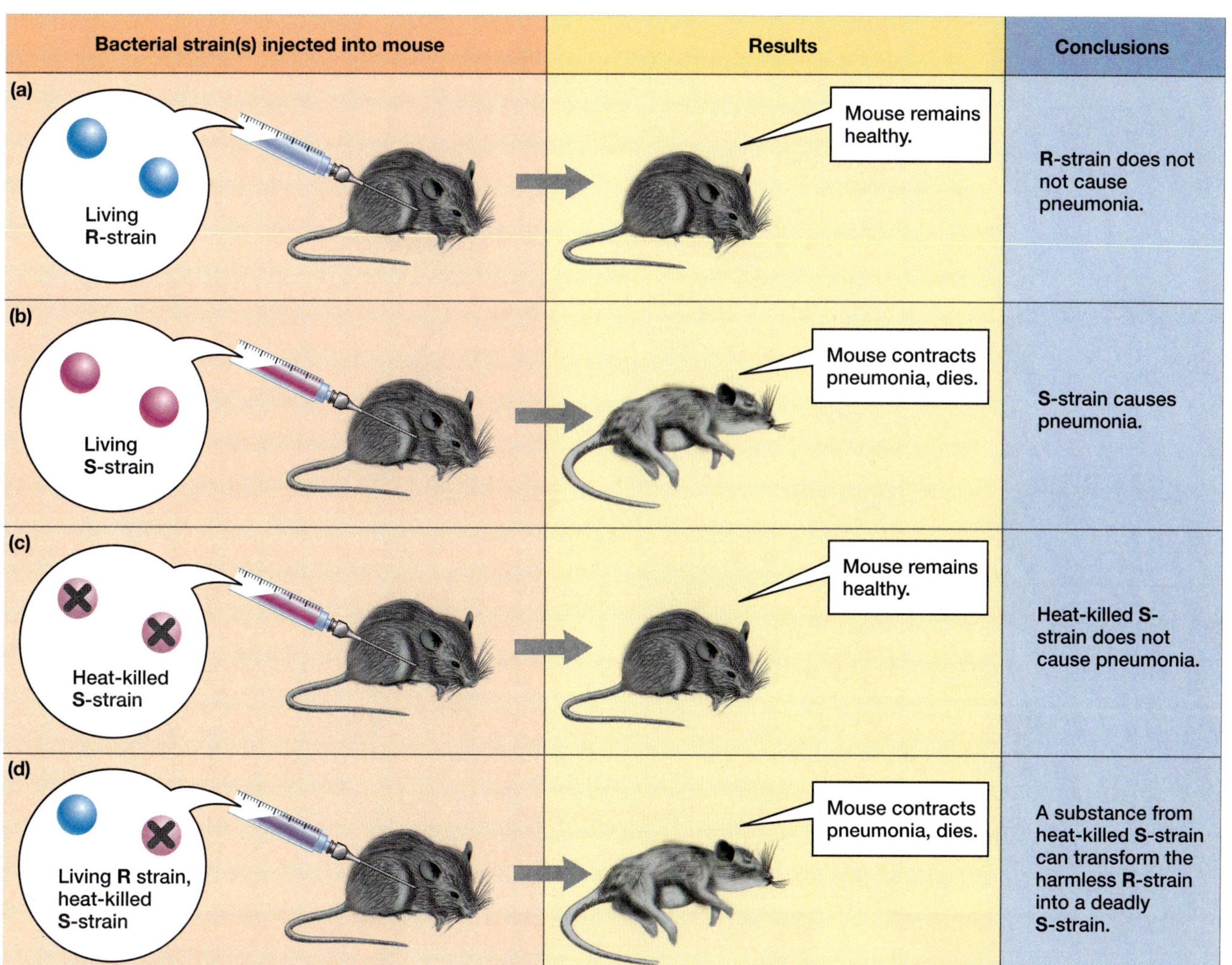

Figure 9-1 Griffith Streptococcus experiment

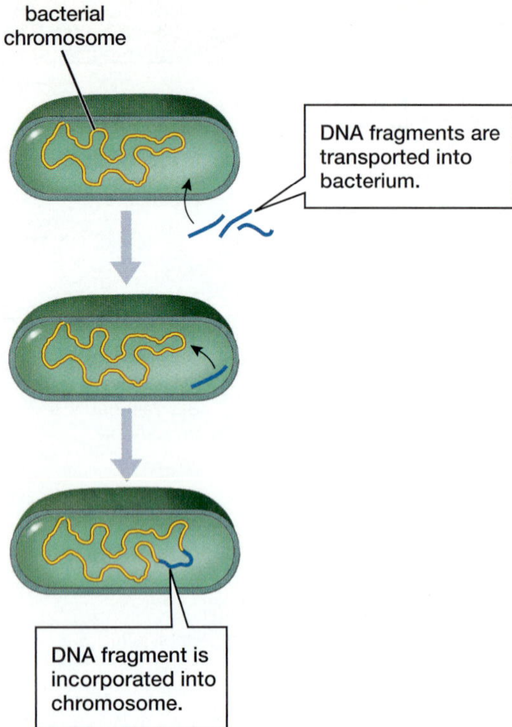

Figure 9-2 Transformation mechanisms

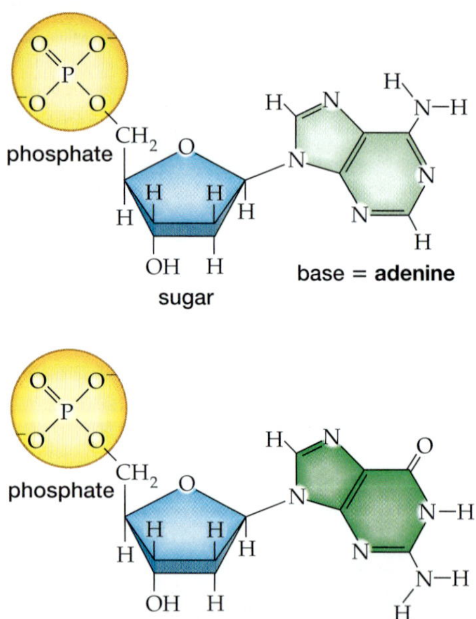

Figure 9-3 part 1 DNA nucleotides (adenine and guanine)

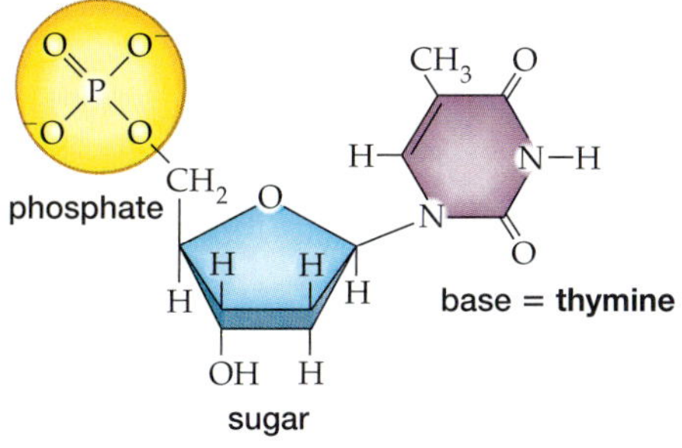

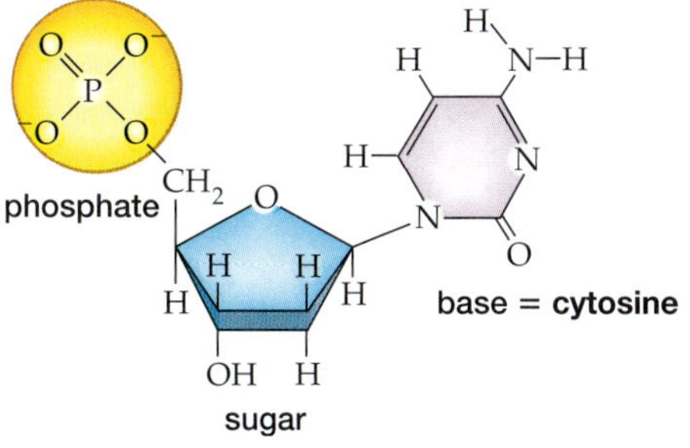

Figure 9-3 part 2 DNA nucleotides (thymine and cytosine)

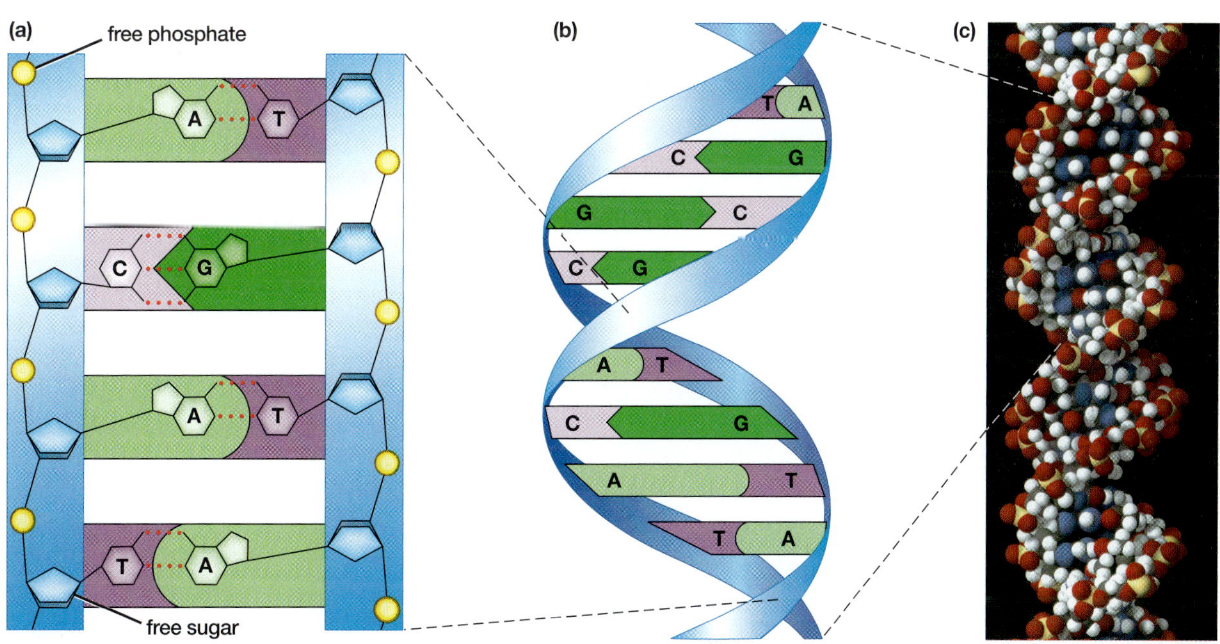

Figure 9-5 DNA structure

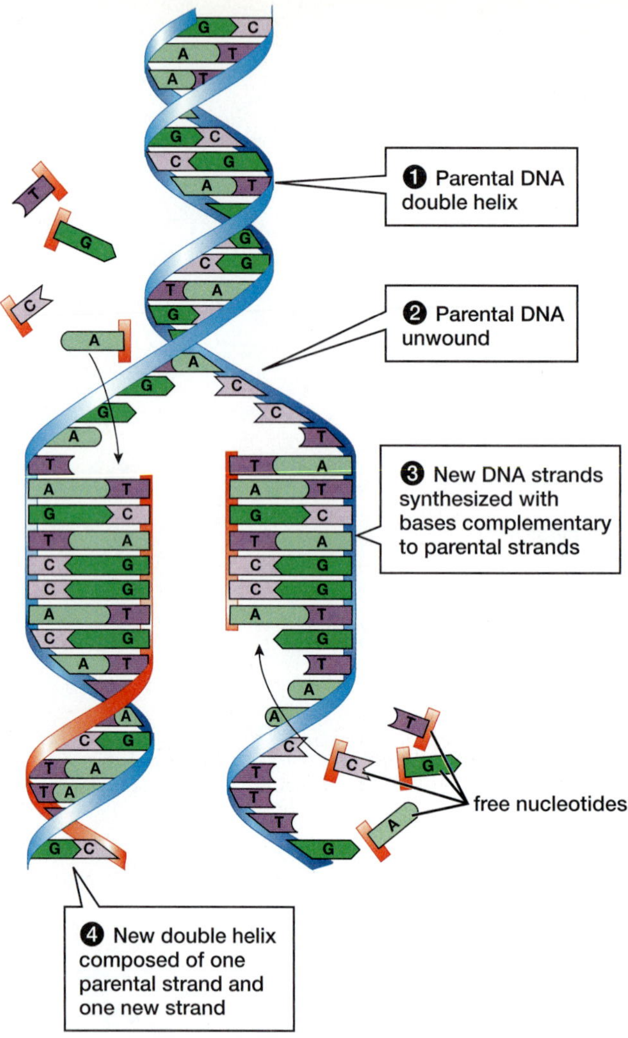

Figure 9-6 DNA replication

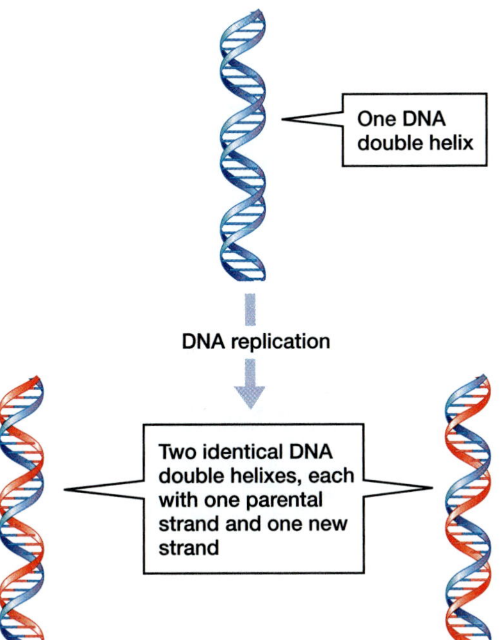

Figure 9-7 Semiconservative DNA replication

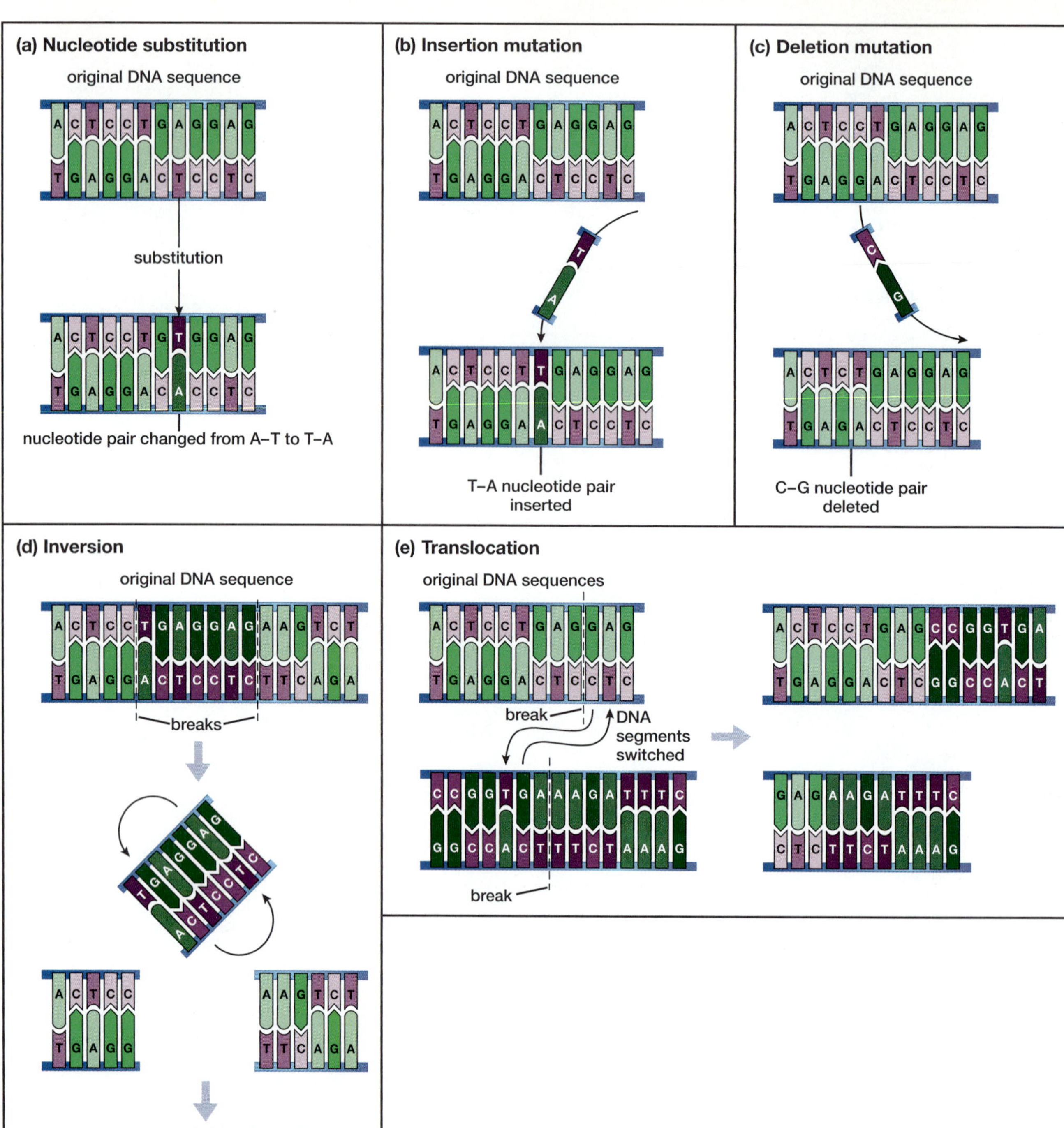

Figure 9-8 Mutations

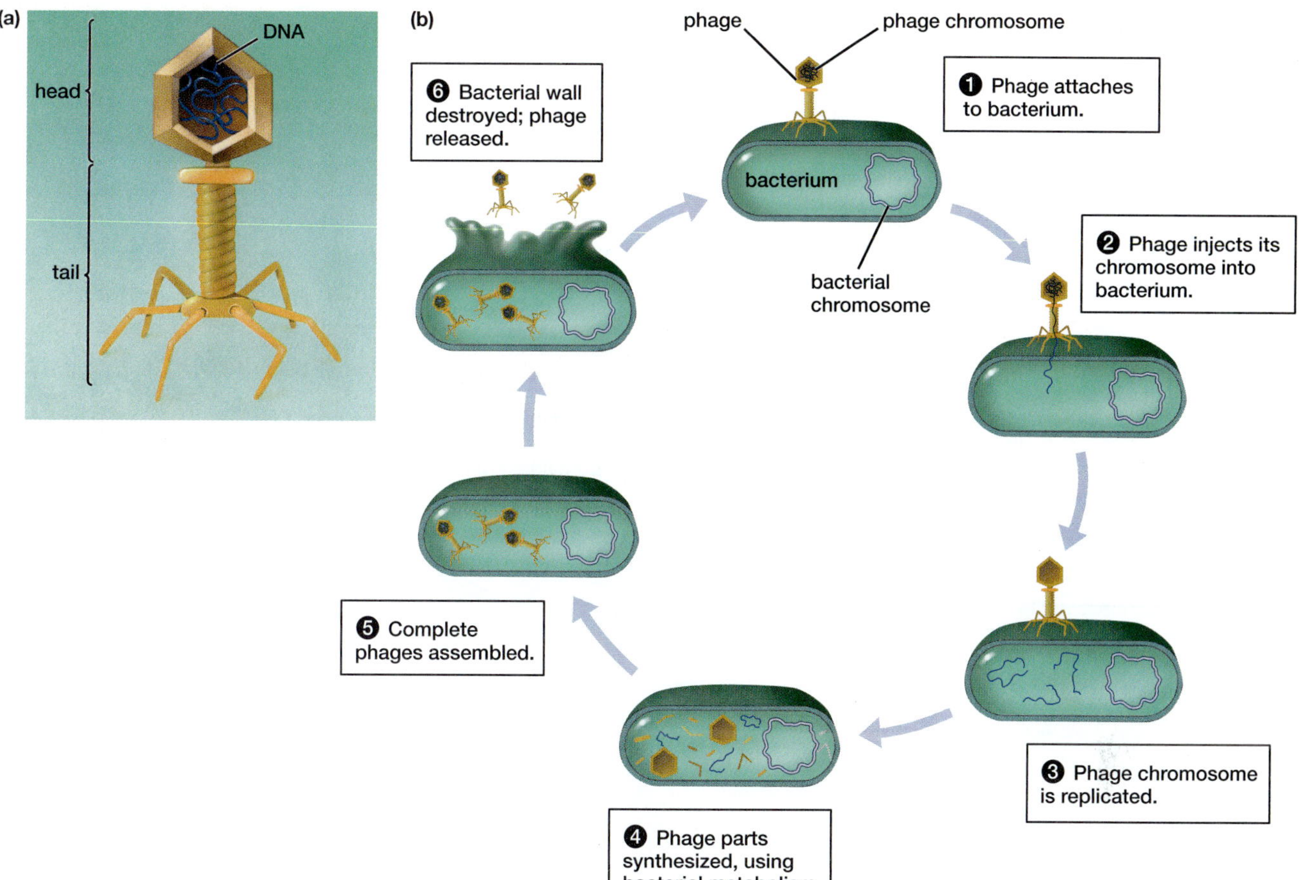

Figure E9-1 Bacteriophages

Observations:
1. Bacteriophage viruses consist of only DNA and protein.
2. Bacteriophage inject their genetic material into bacteria, forcing the bacteria to synthesize more phages.
3. The outer coat of bacteriophages stays outside of the bacteria.
4. DNA contains phosphorus but not sulfur.
   a. DNA can be "labeled" with radioactive phosphorus.
5. Protein contains sulfur but not phosphorus.
   a. Protein can be "labeled" with radioactive sulfur.

Question: Is DNA or protein the genetic material of bacteriophages?

Hypothesis: DNA is the genetic material.

Prediction:
1. If bacteria are infected with bacteriophages containing radioactively labeled DNA, the bacteria will be radioactive.
2. If bacteria are infected with bacteriophages containing radioactively labeled protein, the bacteria will not be radioactive.

Experiment:

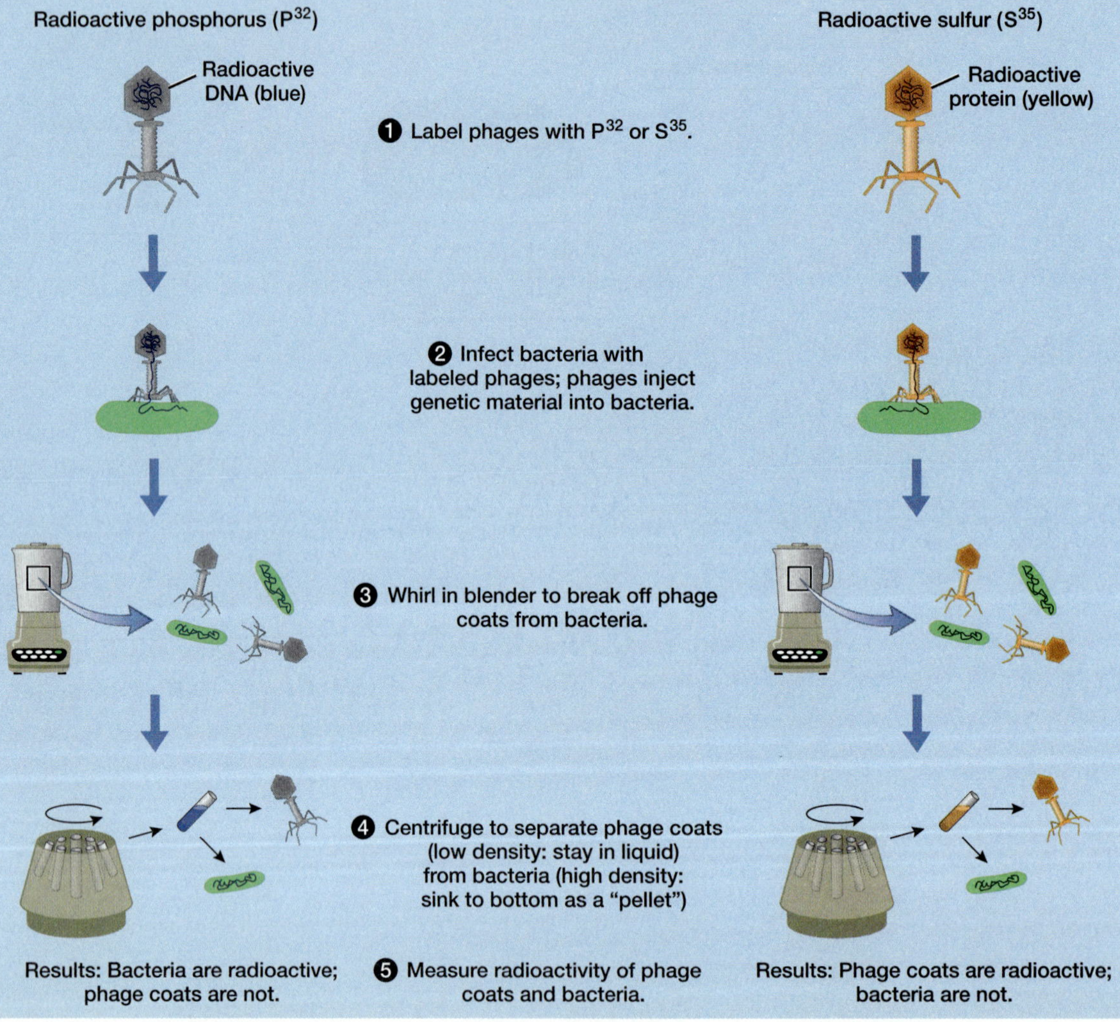

Conclusion: Infected bacteria are labeled with radioactive phosphorus but not with radioactive sulfur, supporting the hypothesis that the genetic material of bacteriophages is DNA, not protein.

Figure E9-2 Hershey-Chase experiment

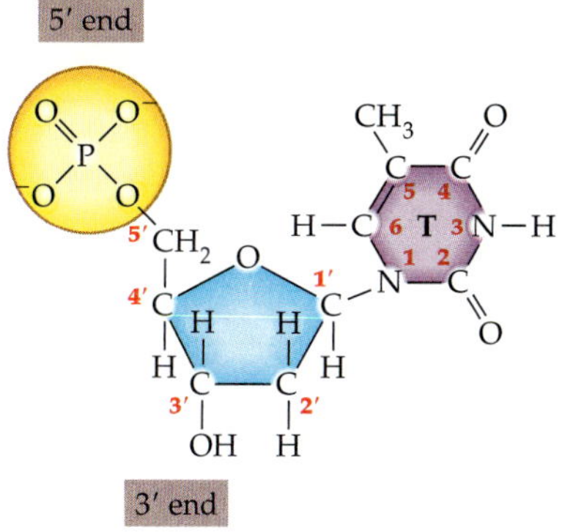

Figure E9-4 Nucleotide with numbering

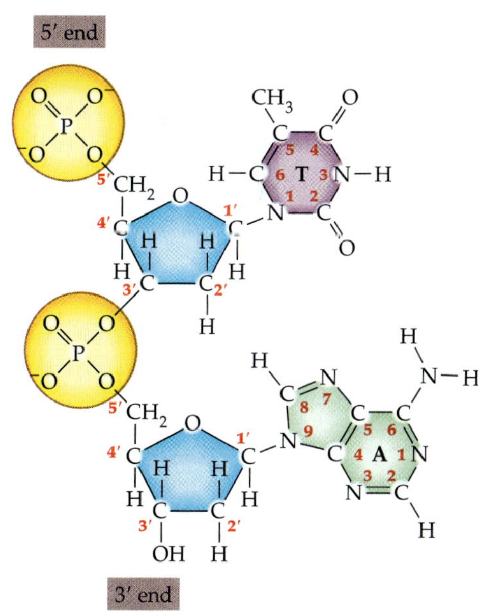

Figure E9-5 Dinucleotide with numbering

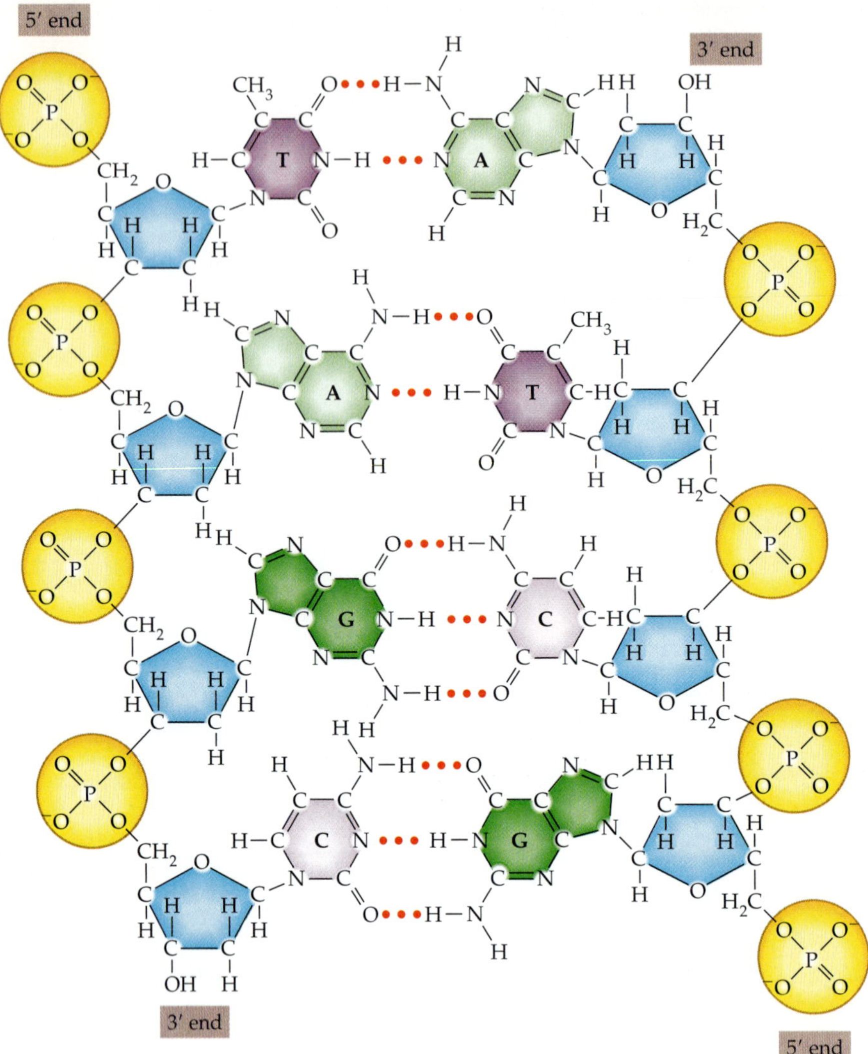

Figure E9-6 Four-nucleotide double strand of DNA, numbered

(a) Growth characteristics of normal and mutant *Neurospora* on simple medium with different supplements show that defects in a single gene lead to defects in a single enzyme.

		Supplements Added to Medium				
		none	ornithine	citrulline	arginine	**Conclusions**
Normal *Neurospora*						Normal *Neurospora* can synthesize arginine, citrulline, and ornithine.
Mutants with single gene defect	A					Mutant A grows only if arginine is added. It cannot synthesize arginine because it has a defect in enzyme 2; gene *A* is needed for synthesis of arginine.
	B					Mutant B grows if either arginine or citrulline are added. It cannot synthesize arginine because it has a defect in enzyme 1. Gene *B* is needed for synthesis of citrulline.

(b) The biochemical pathway for synthesis of the amino acid arginine involves two steps, each catalyzed by a different enzyme.

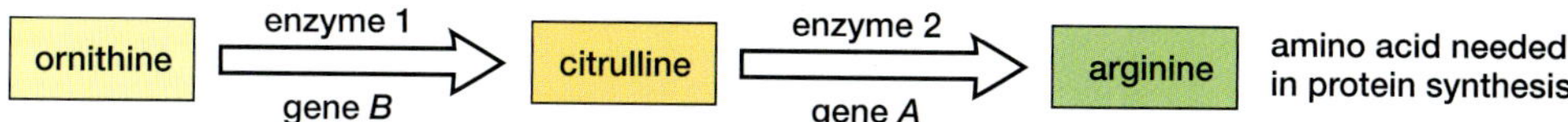

Figure 10-1 Beadle and Tatum *Neurospora* experiment

(a) **Messenger RNA (mRNA)**

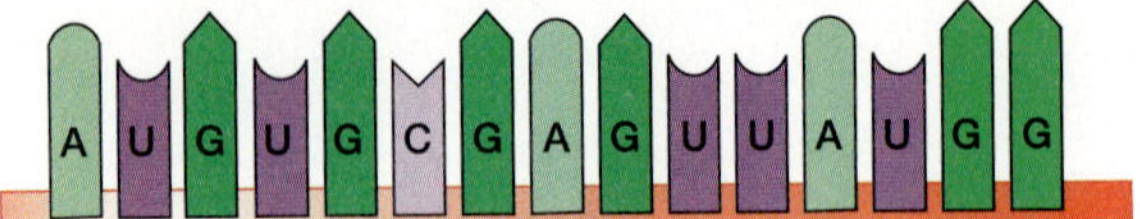

The base sequence of mRNA carries the information for the amino acid sequence of a protein.

(b) **Ribosome: contains ribosomal RNA (rRNA)**

large subunit

catalytic site

1 2

tRNA/amino acid binding sites

small subunit

rRNA combines with proteins to form ribosomes. The small subunit binds mRNA. The large subunit binds tRNA and catalyzes peptide bond formation between amino acids during protein synthesis.

(c) **Transfer RNA (tRNA)**

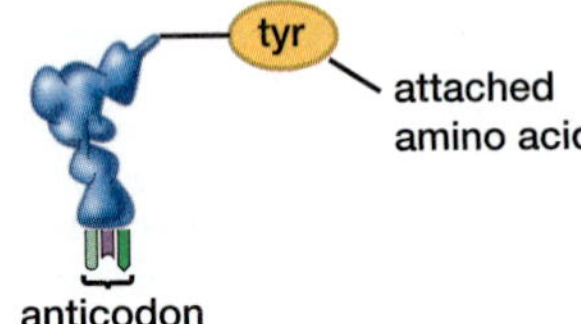

Each tRNA carries a specific amino acid to a ribosome during protein synthesis. The anticodon of tRNA pairs with a codon of mRNA, ensuring that the correct amino acid is incorporated into the protein.

Figure 10-2 Types of RNA

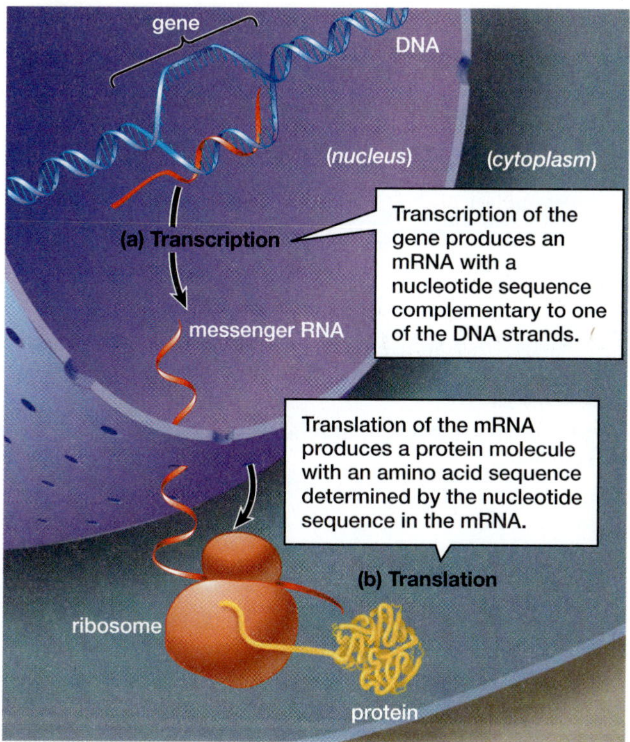

Figure 10-3 Information flow in a cell

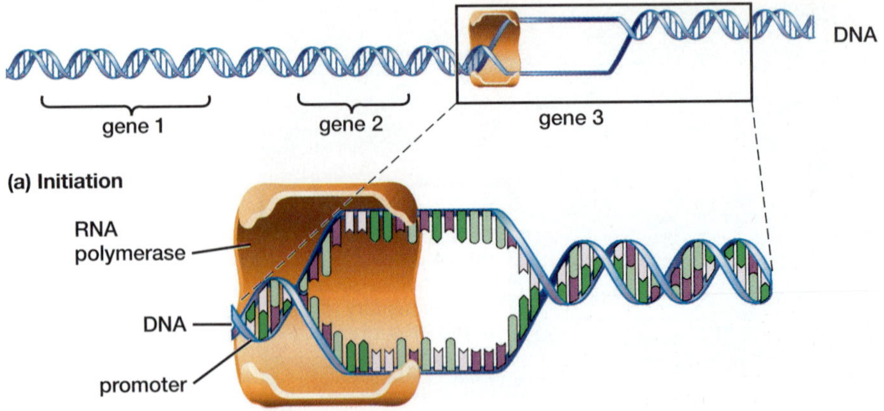

RNA polymerase binds to the promoter region of DNA near the beginning of a gene, separating the double helix near the promoter.

(b) Elongation

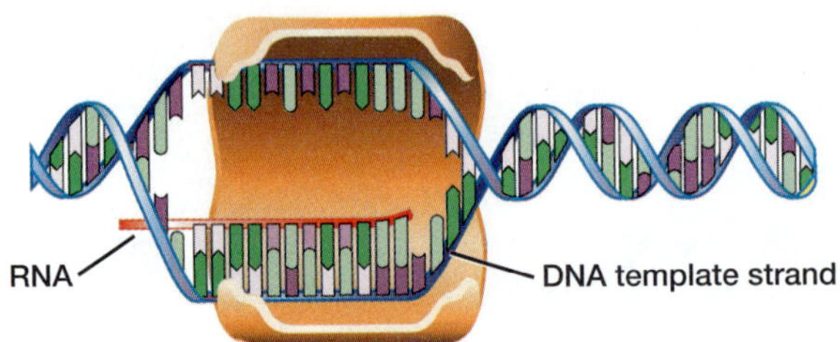

RNA polymerase travels along the DNA template strand (blue), catalyzing the addition of ribose nucleotides into an RNA molecule (pink). The nucleotides in the RNA are complementary to the template strand of the DNA.

(c) Termination

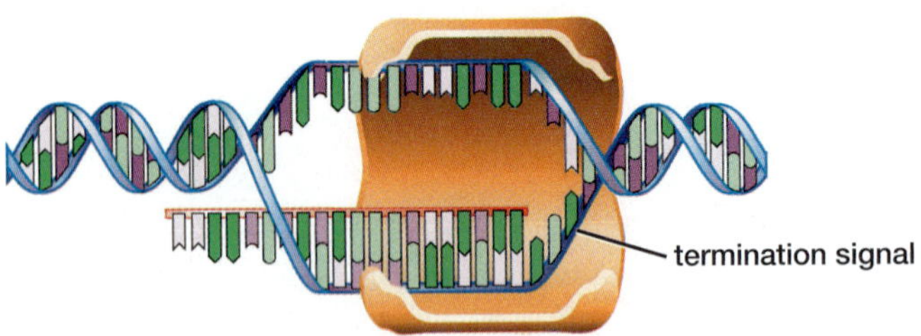

At the end of a gene, RNA polymerase encounters a DNA sequence called a termination signal. RNA polymerase detaches from the DNA and releases the RNA molecule.

(d) Conclusion of transcription

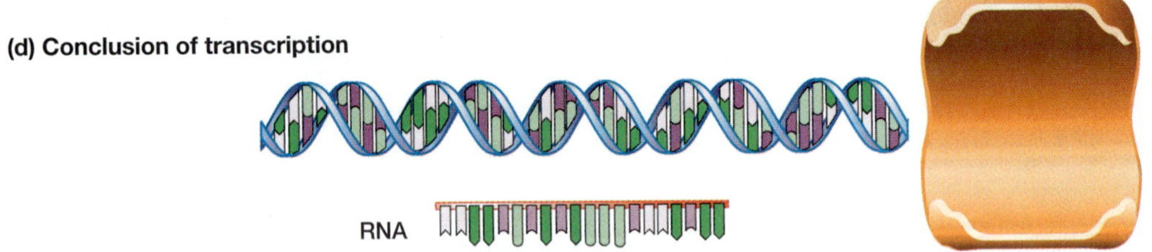

After termination, the DNA completely rewinds into a double helix. The RNA molecule is free to move from the nucleus to the cytoplasm for translation, and RNA polymerase may move to another gene and begin transcription once again.

Figure 10-4 Transcription

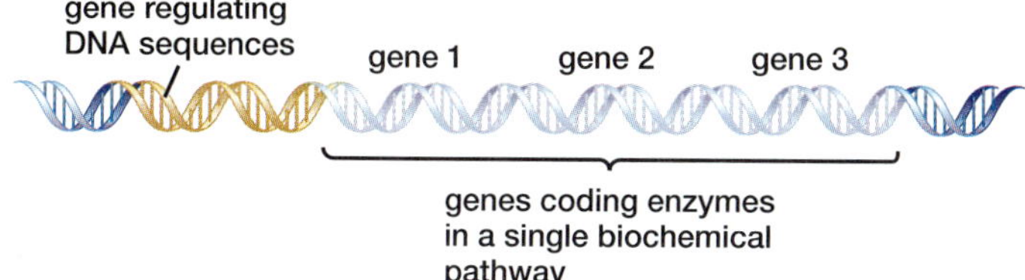

Figure 10-6a Gene structure, transcription/translation in prokaryotes

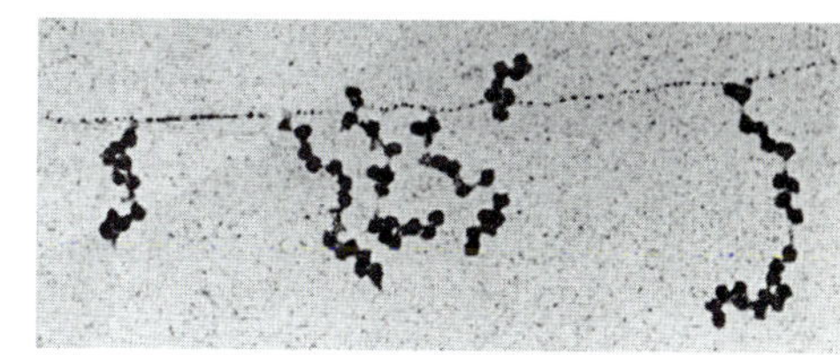

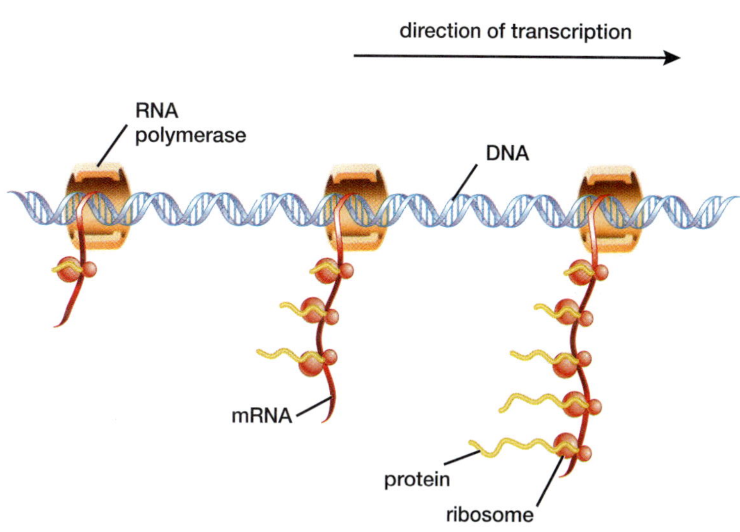

Figure 10-6b Electron micrograph of prokaryotic transcription/translation

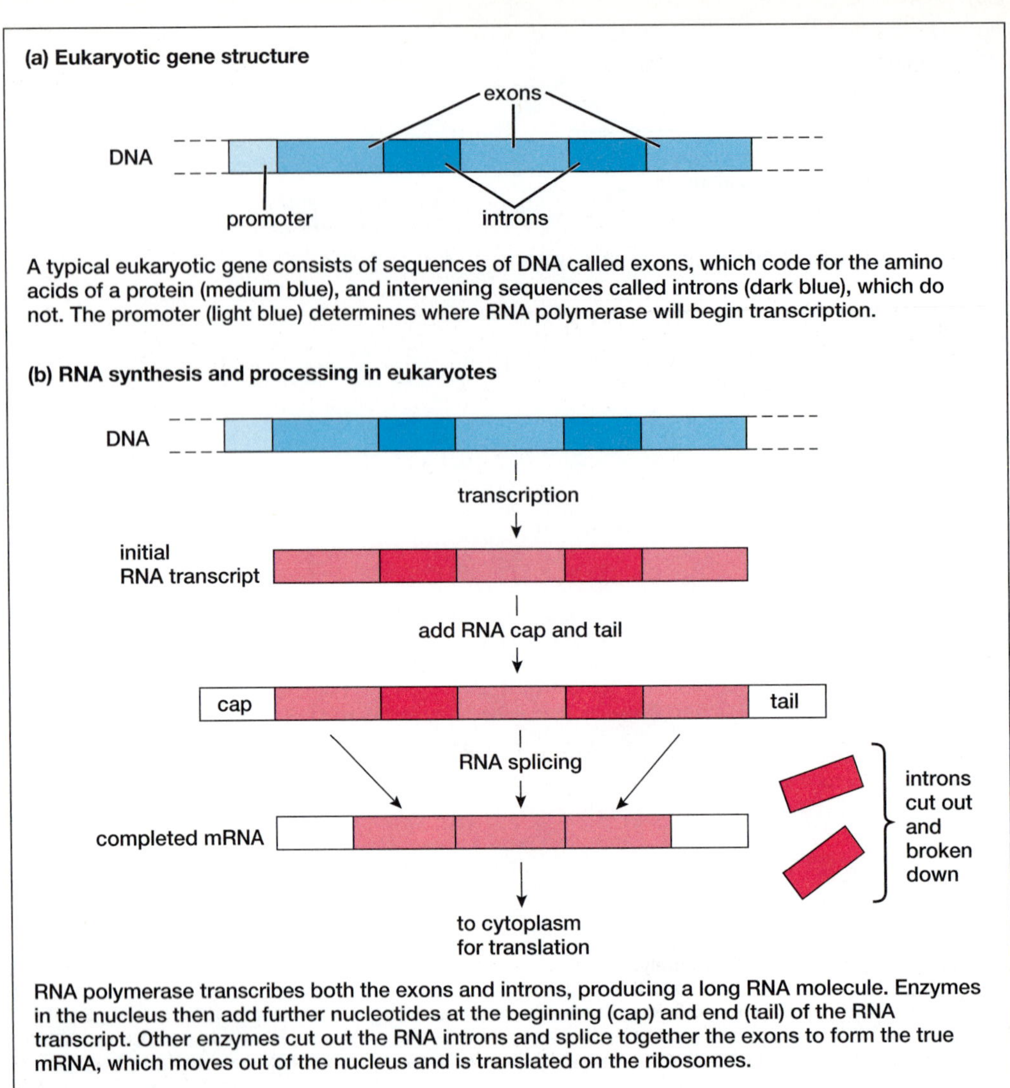

Figure 10-7 Messenger RNA processing in eukaryotes

Initiation:

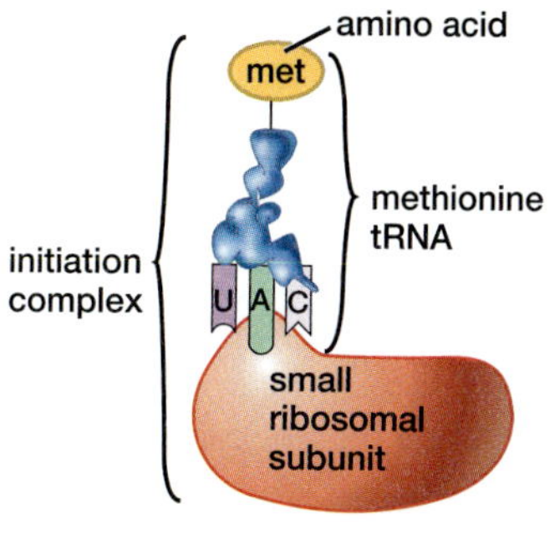

(a) A tRNA with an attached methionine amino acid binds to a small ribosomal subunit, forming an initiation complex.

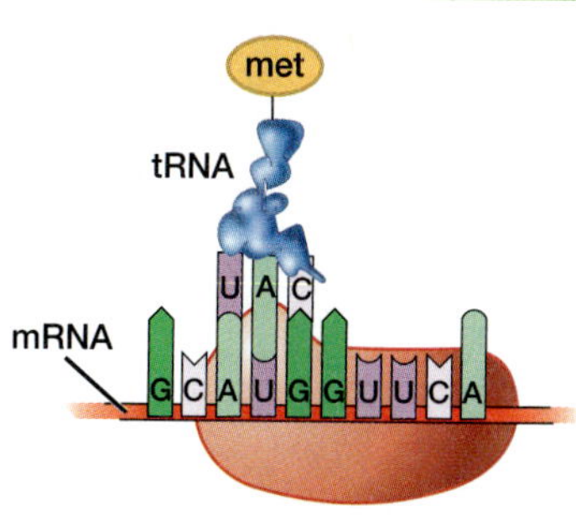

(b) The initiation complex binds to an mRNA molecule. The methionine (met) tRNA anticodon (UAC) base-pairs with the start codon (AUG) of the mRNA.

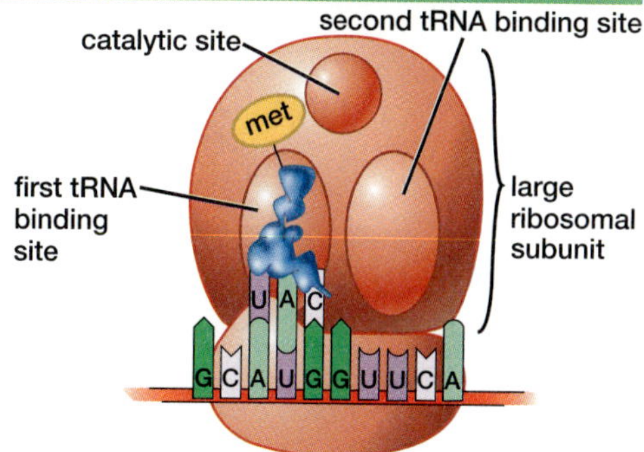

(c) The large ribosomal subunit binds to the small subunit. The methionine tRNA binds to the first tRNA site on the large subunit.

Elongation:

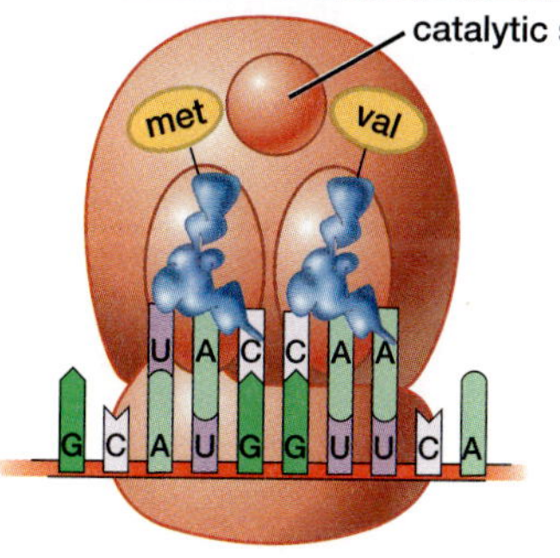

(d) The second codon of mRNA (GUU) base-pairs with the anticodon (CAA) of a second tRNA carrying the amino acid valine (val). This tRNA binds to the second tRNA site on the large subunit.

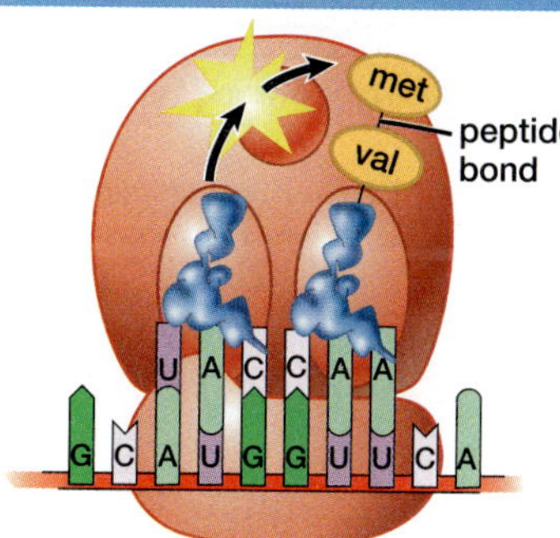

(e) The catalytic site on the large subunit catalyzes the formation of a peptide bond linking the amino acids methionine and valine. The two amino acids are now attached to the tRNA in the second binding position.

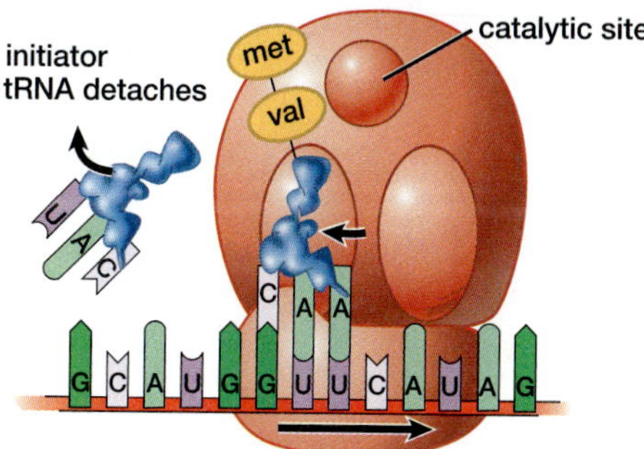

(f) The "empty" tRNA is released and the ribosome moves down the mRNA, one codon to the right. The tRNA that is attached to the two amino acids is now in the first tRNA binding site and the second tRNA binding site is empty.

Termination:

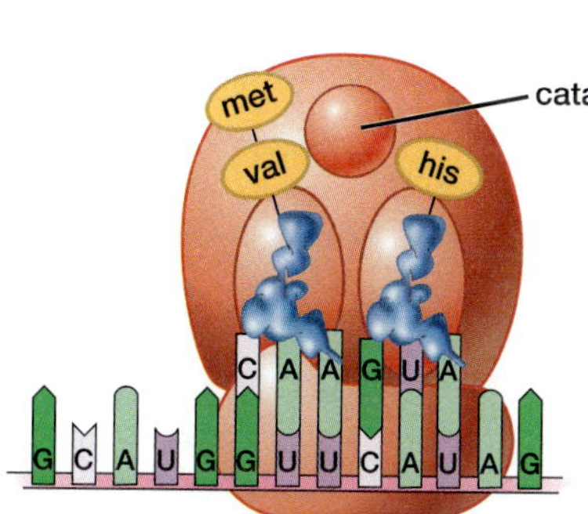

(g) The third codon of mRNA (CAU) base-pairs with the anticodon (GUA) of a tRNA carrying the amino acid histidine (his). This tRNA enters the second tRNA binding site on the large subunit.

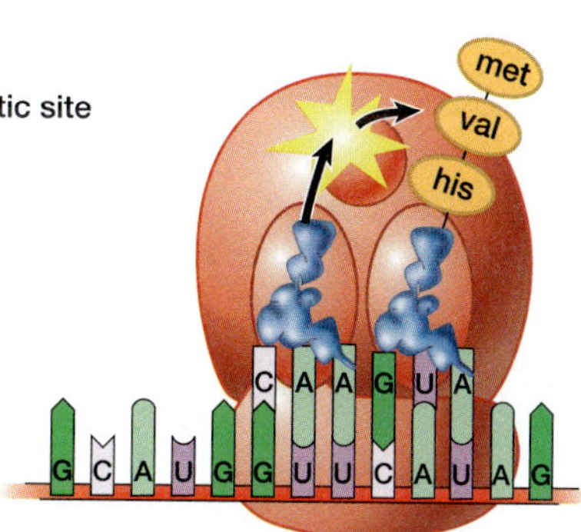

(h) The catalytic site forms a new peptide bond between valine and histidine. A three-amino-acid chain is now attached to the tRNA in the second binding site. The tRNA in the first site leaves, and the ribosome moves one codon over on the mRNA.

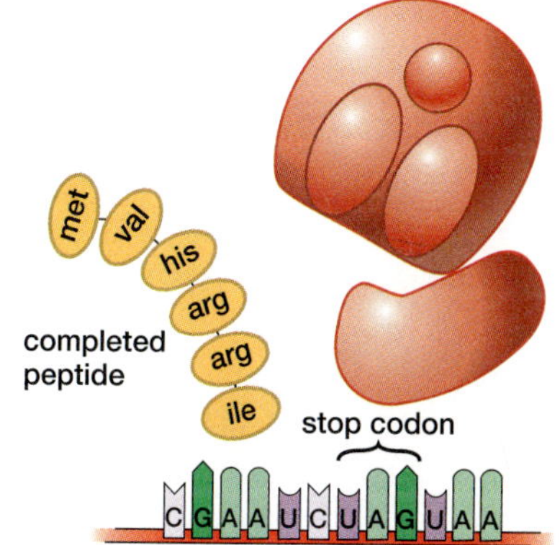

(i) This process repeats until a stop codon is reached; the mRNA and the completed peptide are released from the ribosome, and the subunits separate.

Figure 10-8 Translation

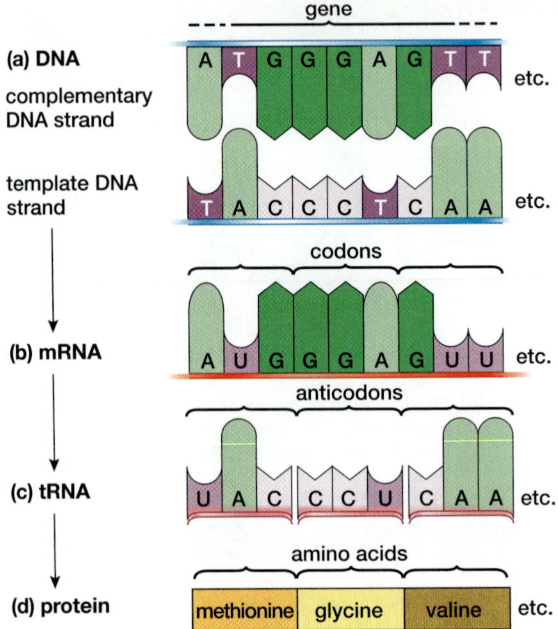

Figure 10-9 Summary of DNA—RNA--protein

(a) Structure of the lactose operon

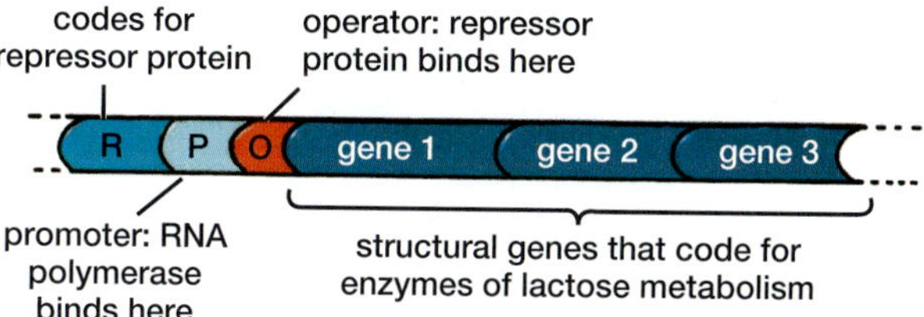

The lactose operon consists of a regulatory gene, a promoter, an operator, and three structural genes that code for enzymes involved in lactose metabolism. The regulatory gene codes for a protein, called a repressor, which can bind to the operator site under certain circumstances.

(b) Lactose absent

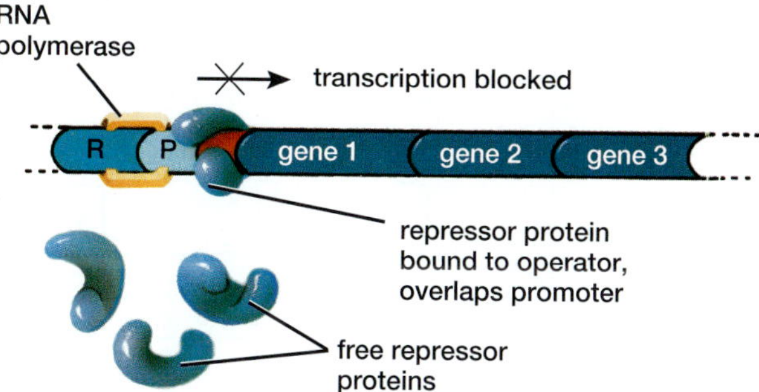

When lactose is not present, repressor proteins bind to the operator of the lactose operon. When RNA polymerase binds to the promoter, the repressor protein blocks access to the structural genes, which therefore cannot be transcribed.

(c) Lactose present

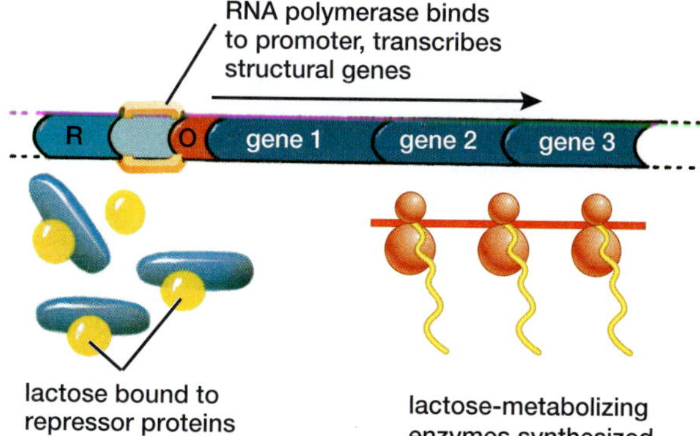

When lactose is present, it binds to the repressor protein. The lactose-repressor complex cannot bind to the operator, so RNA polymerase has free access to the promoter. The RNA polymerase transcribes the three structural genes coding for the lactose-metabolizing enzymes.

Figure 10-10 Lactose operon regulation

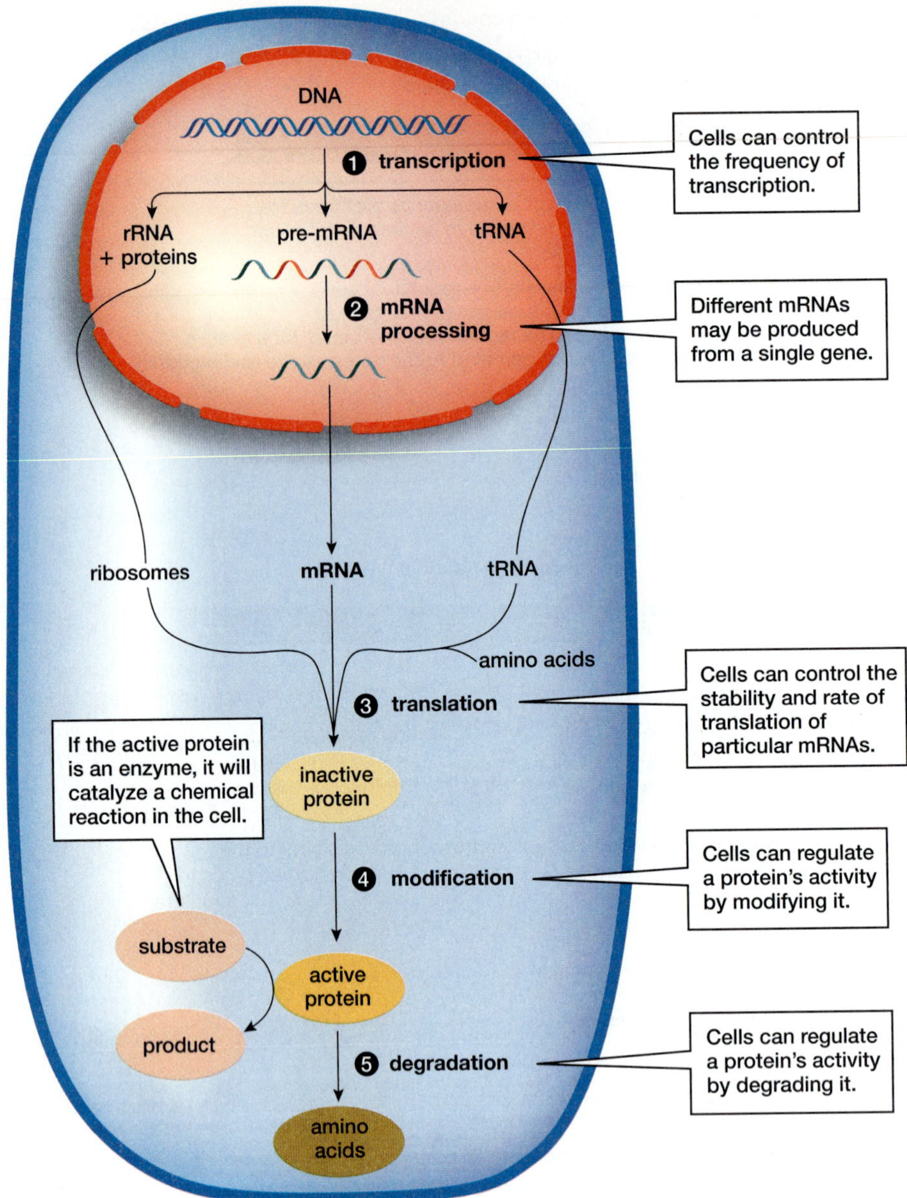

Figure 10-11 Gene regulation in a cell

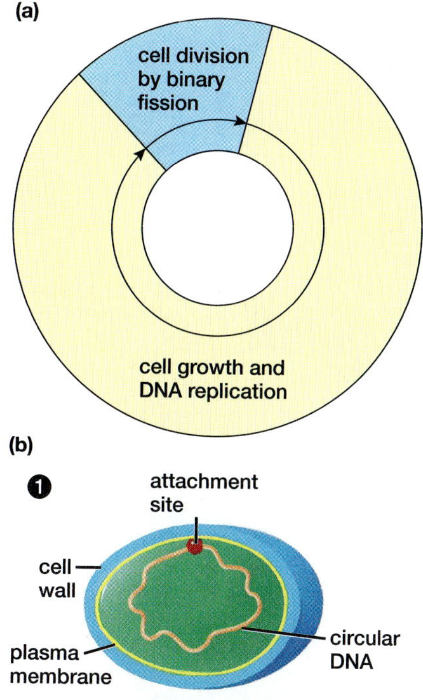

The circular DNA double helix is attached to the plasma membrane at one point.

❷

The DNA replicates and the two DNA double helices attach to the plasma membrane at nearby points.

❸

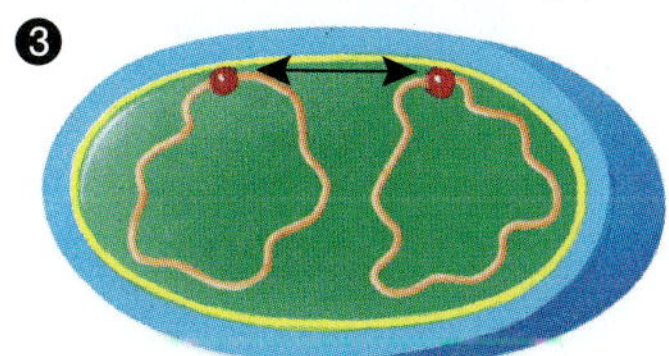

New plasma membrane is added between the attachment points, pushing them further apart.

❹

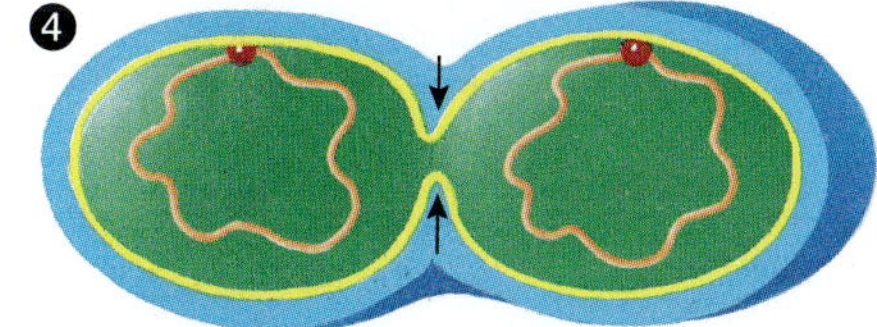

The plasma membrane grows inward at the middle of the cell.

❺

The parent cell divides into two daughter cells.

Figure 11-2 Prokaryotic cell cycle

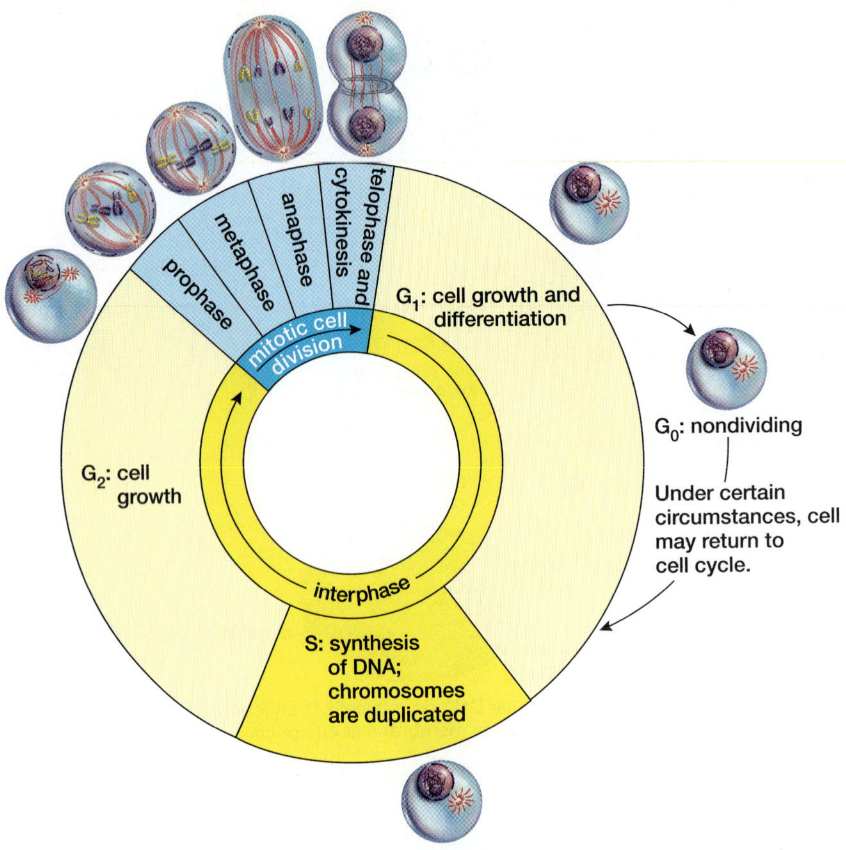

Figure 11-3 Eukaryotic cell cycle

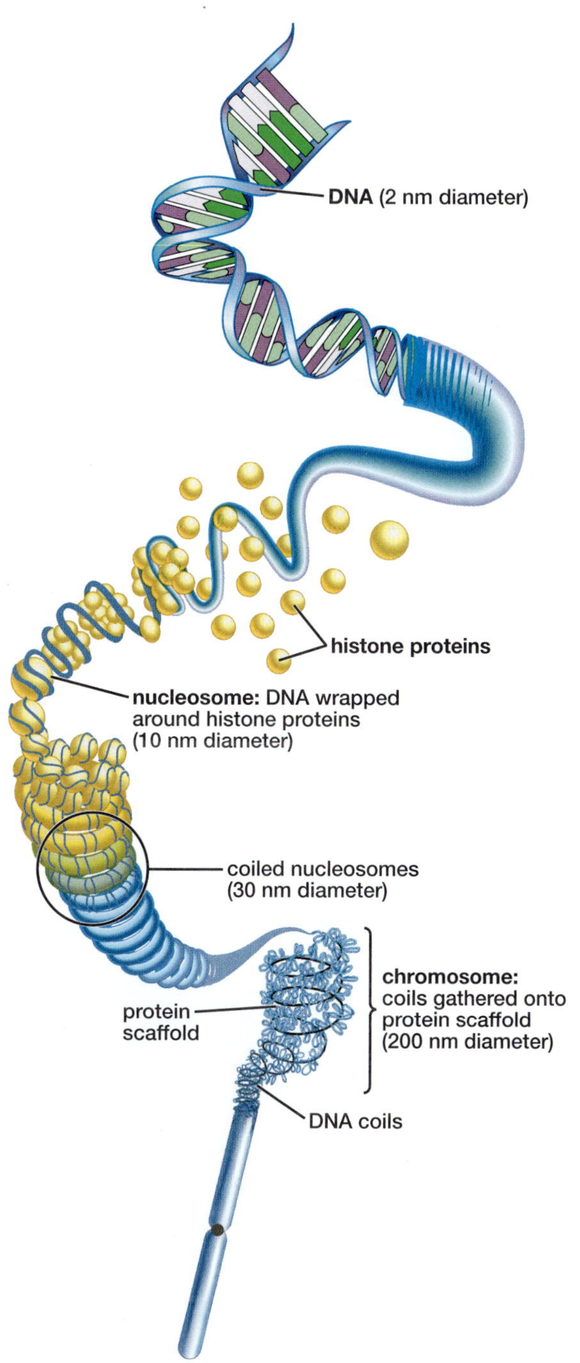

Figure 11-4 Chromosome structure

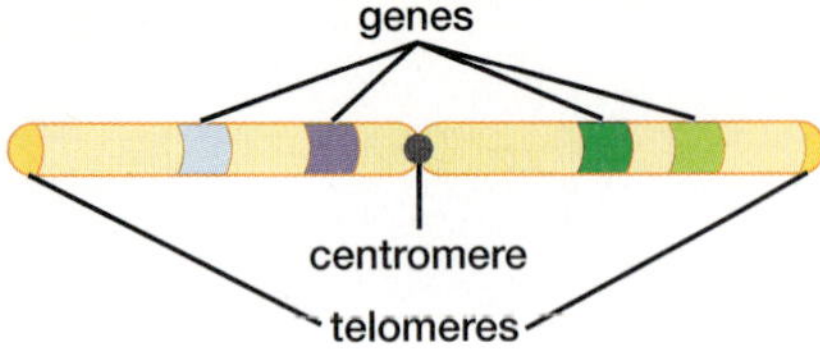

Figure 11-5 Schematic eukaryotic chromosome

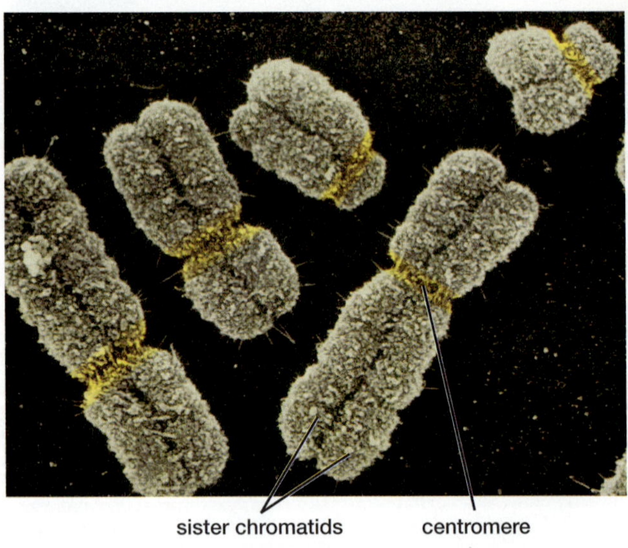

Figure 11-6 Human chromosomes

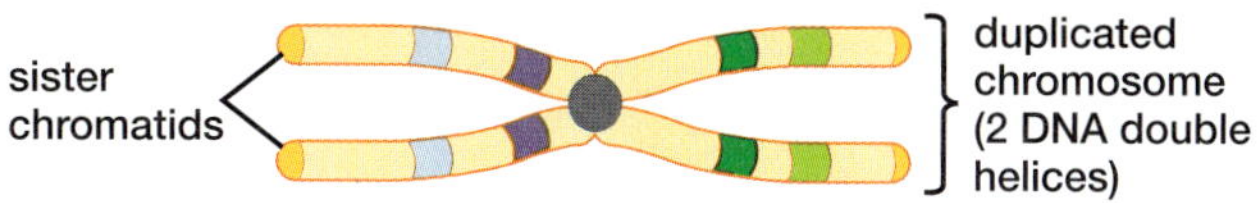

Figure 11-7 Replicated chromosome

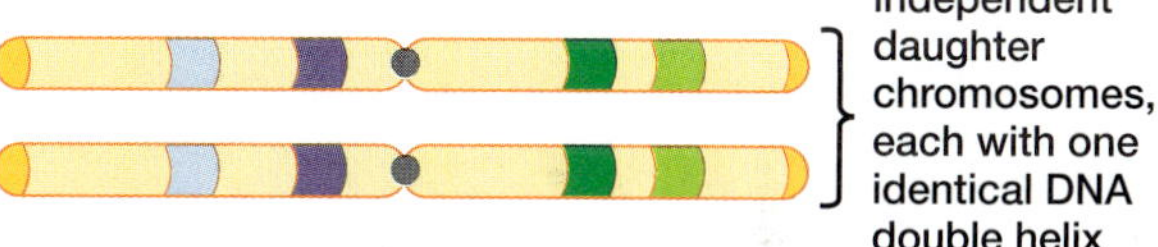

Figure 11-8 Separated chromosomes

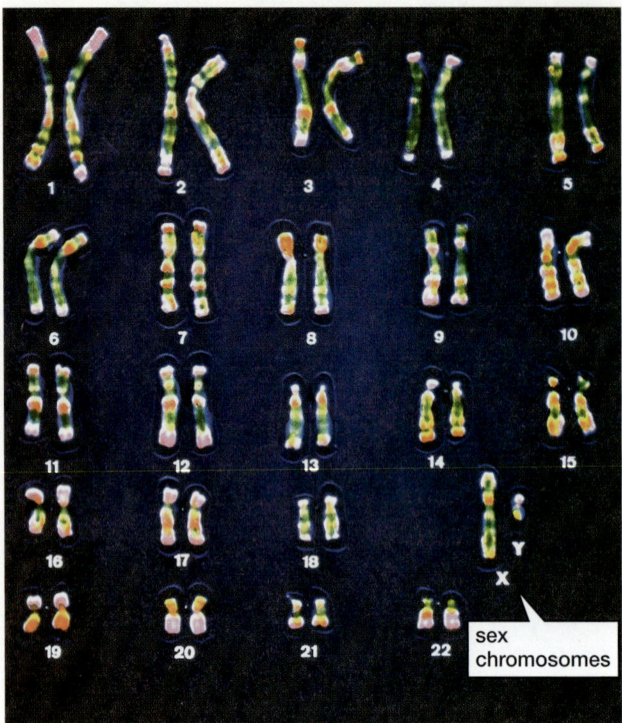

Figure 11-9 Human male karyotype

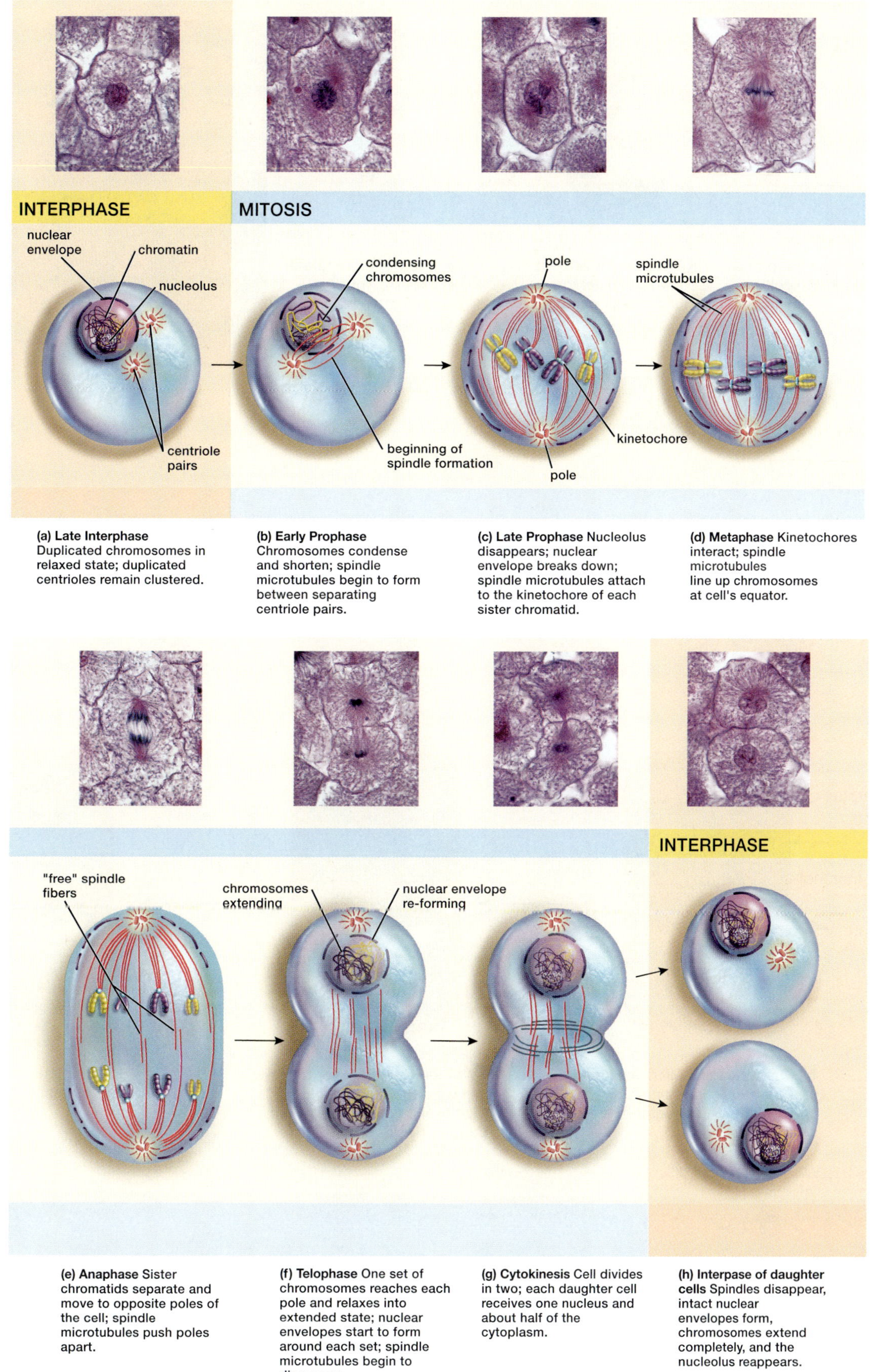

(a) Late Interphase Duplicated chromosomes in relaxed state; duplicated centrioles remain clustered.

(b) Early Prophase Chromosomes condense and shorten; spindle microtubules begin to form between separating centriole pairs.

(c) Late Prophase Nucleolus disappears; nuclear envelope breaks down; spindle microtubules attach to the kinetochore of each sister chromatid.

(d) Metaphase Kinetochores interact; spindle microtubules line up chromosomes at cell's equator.

(e) Anaphase Sister chromatids separate and move to opposite poles of the cell; spindle microtubules push poles apart.

(f) Telophase One set of chromosomes reaches each pole and relaxes into extended state; nuclear envelopes start to form around each set; spindle microtubules begin to disappear.

(g) Cytokinesis Cell divides in two; each daughter cell receives one nucleus and about half of the cytoplasm.

(h) Interpase of daughter cells Spindles disappear, intact nuclear envelopes form, chromosomes extend completely, and the nucleolus reappears.

Figure 11-10 Mitotic cell division, light micrographs and accompanying drawings

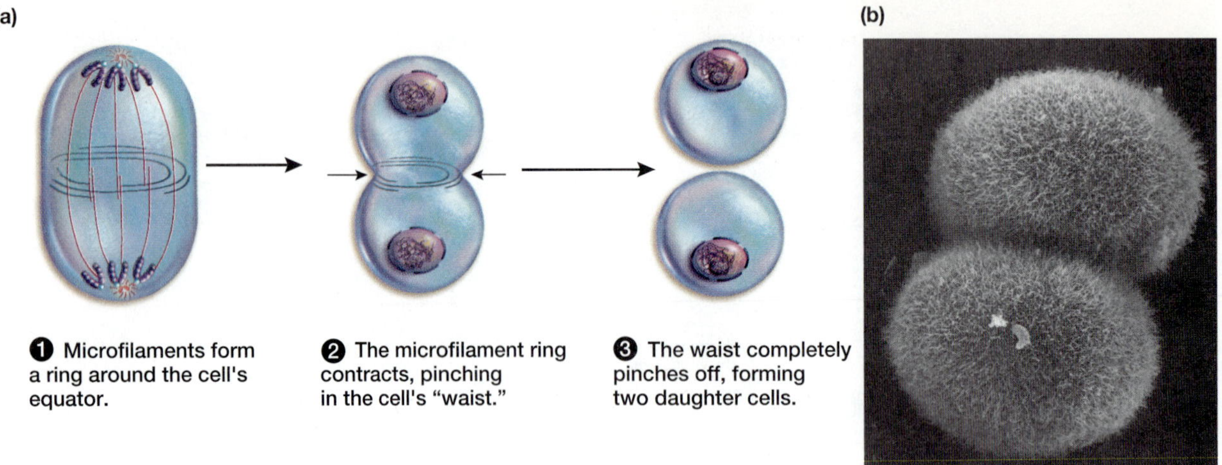

Figure 11-11 Cytokinesis in animal cell

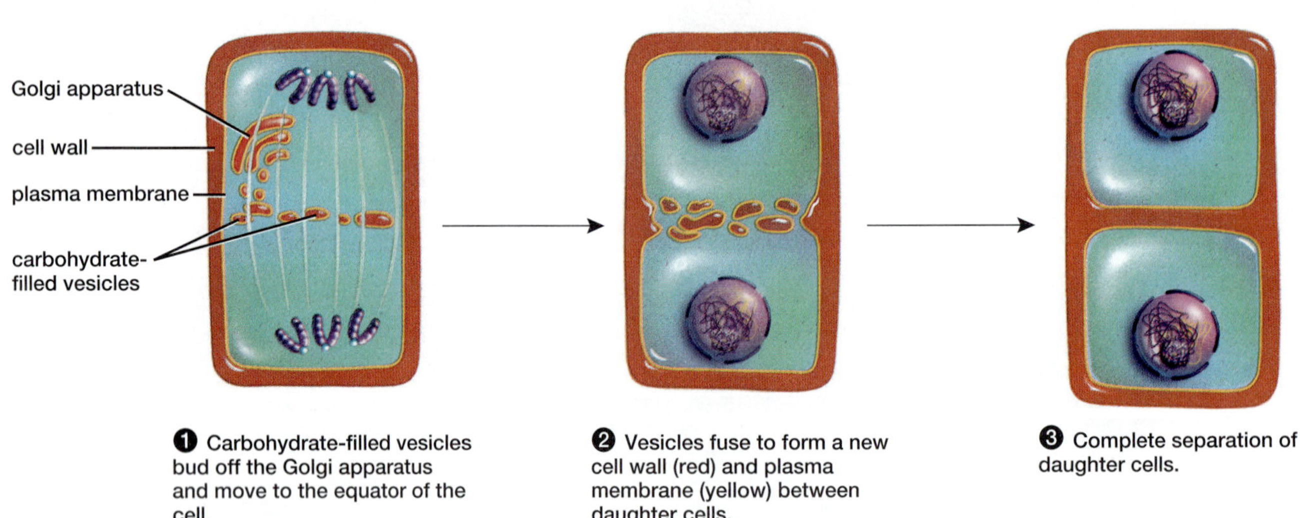

Figure 11-12 Cytokinesis in plant cell

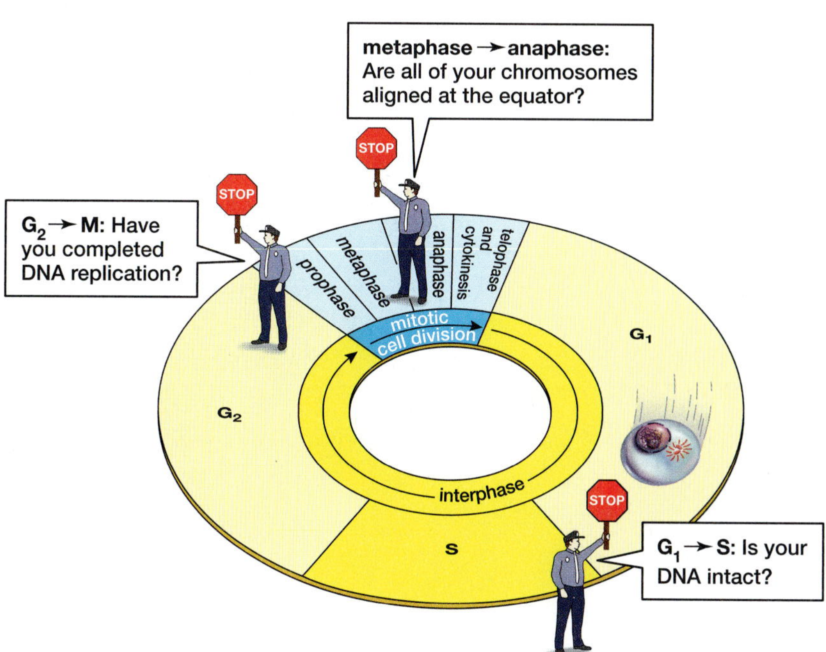

Figure 11-13 Cell cycle control analogy

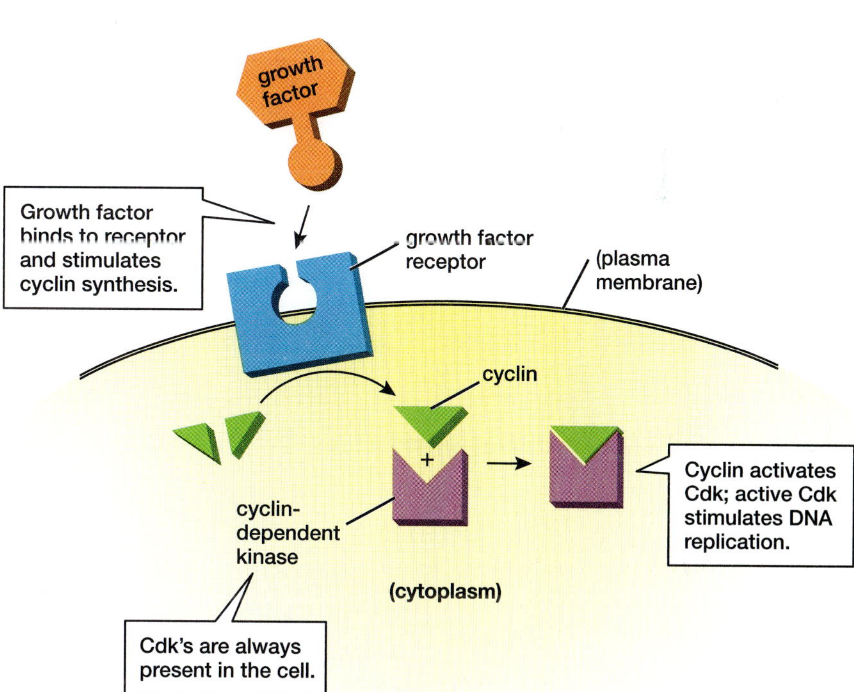

Figure 11-14 Cyclin control of checkpoint

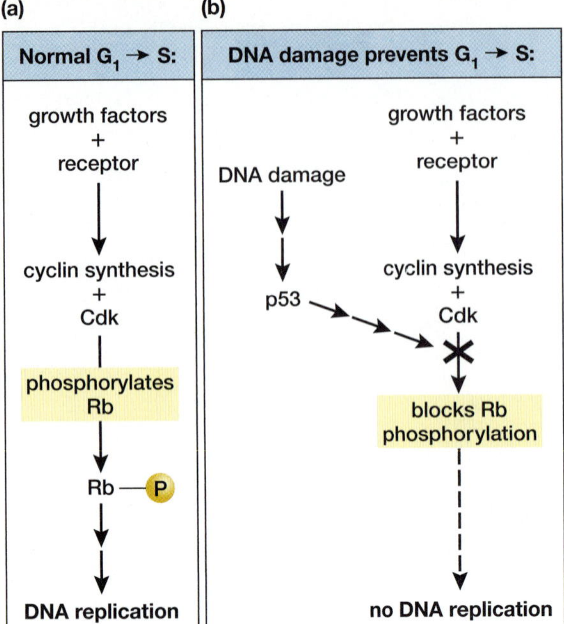

Figure 11-15 Rb and p53 control checkpoints

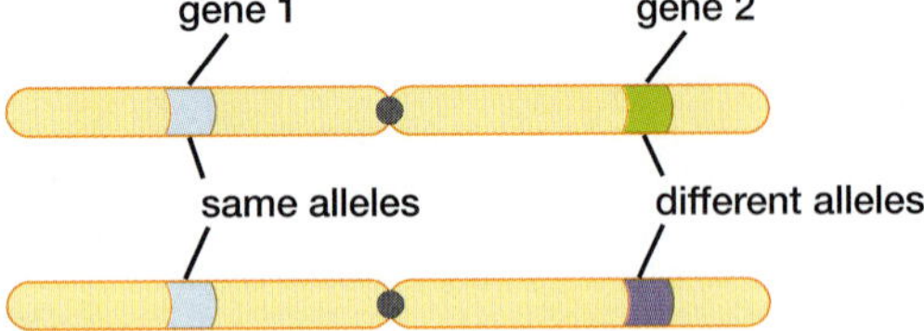

Figure 11-16 Chromosomes with different alleles of same genes

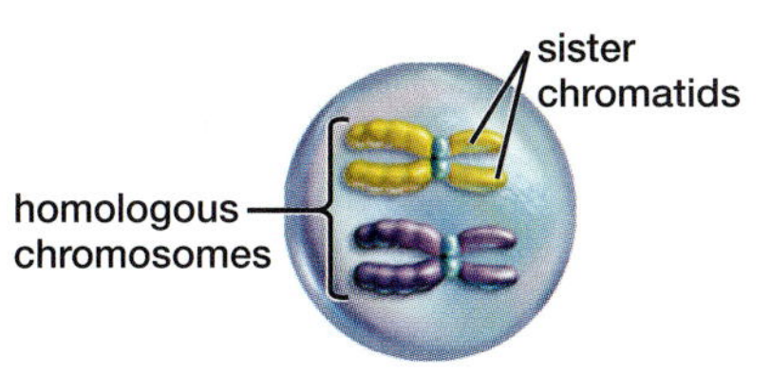

Figure 11-17 Cell with replicated chromosomes

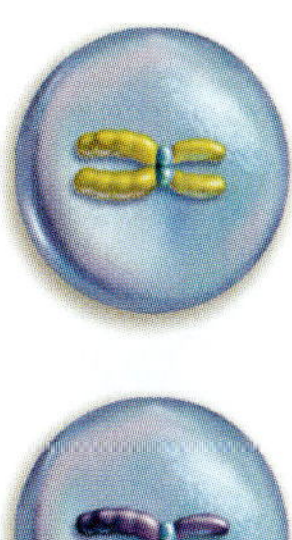

Figure 11-18 Pair of cells with replicated chromosomes

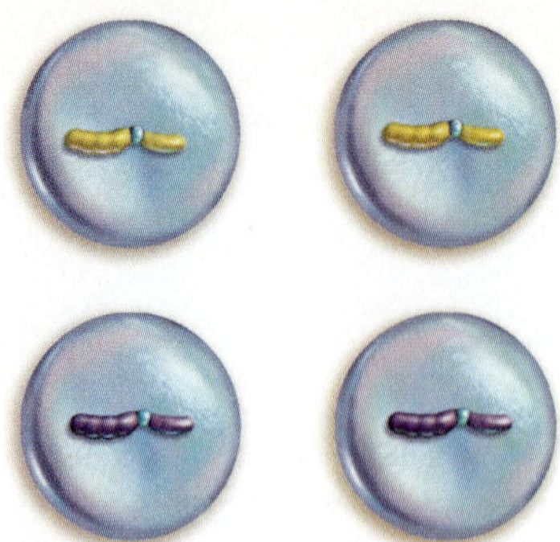

Figure 11-19 Four cells, with unreplicated chromosomes

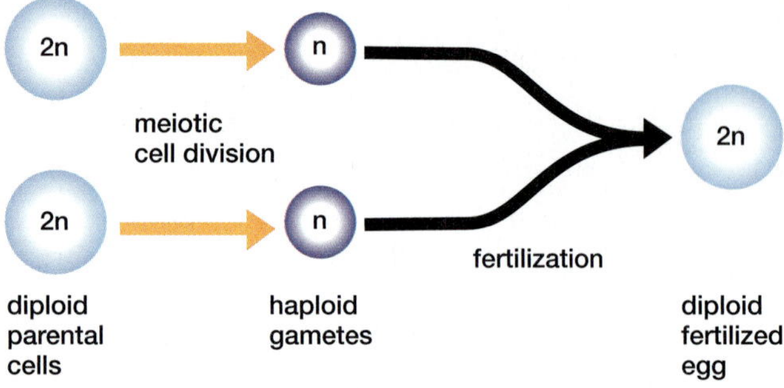

Figure 11-20 Schematic of meiosis and fertilization

MEIOSIS I

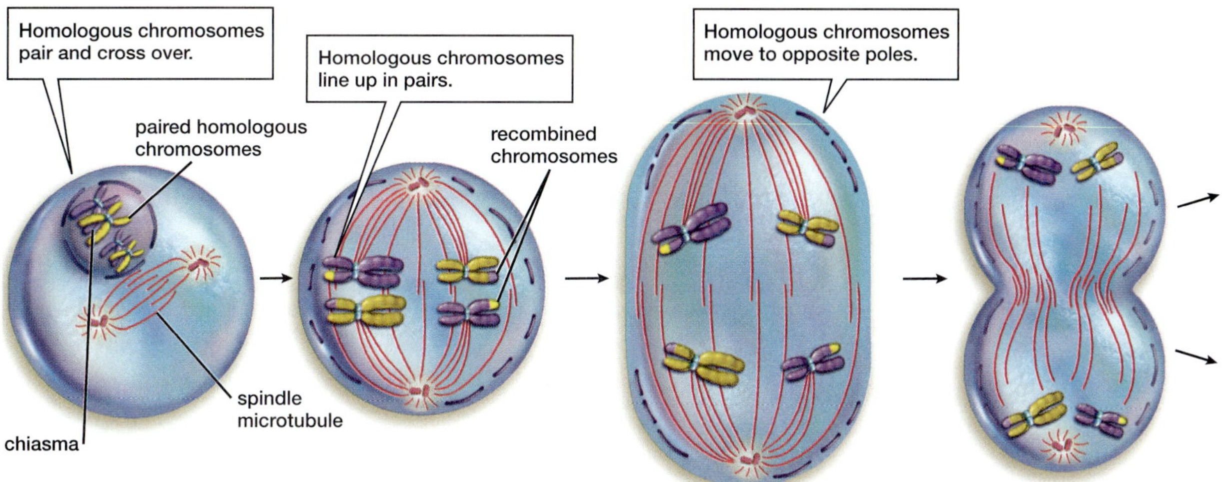

(a) Prophase I. Duplicated chromosomes condense. Homologous chromosomes pair up and chiasmata occur as chromatids of homologues exchange parts. The nuclear envelope disintegrates, and spindle microtubules form.

(b) Metaphase I. Paired homologous chromosomes line up along the equator of the cell. One homologue of each pair faces each pole of the cell and attaches to spindle microtubules via its kinetochore (blue).

(c) Anaphase I. Homologues separate, one member of each pair going to each pole of the cell. Sister chromatids do not separate.

(d) Telophase I. Spindle microtubules disappear. Two clusters of chromosomes have formed, each containing one member of each pair of homologues. The daughter nuclei are therefore haploid. Cytokinesis commonly occurs at this stage. There is little or no interphase between meiosis I and meiosis II.

MEIOSIS II

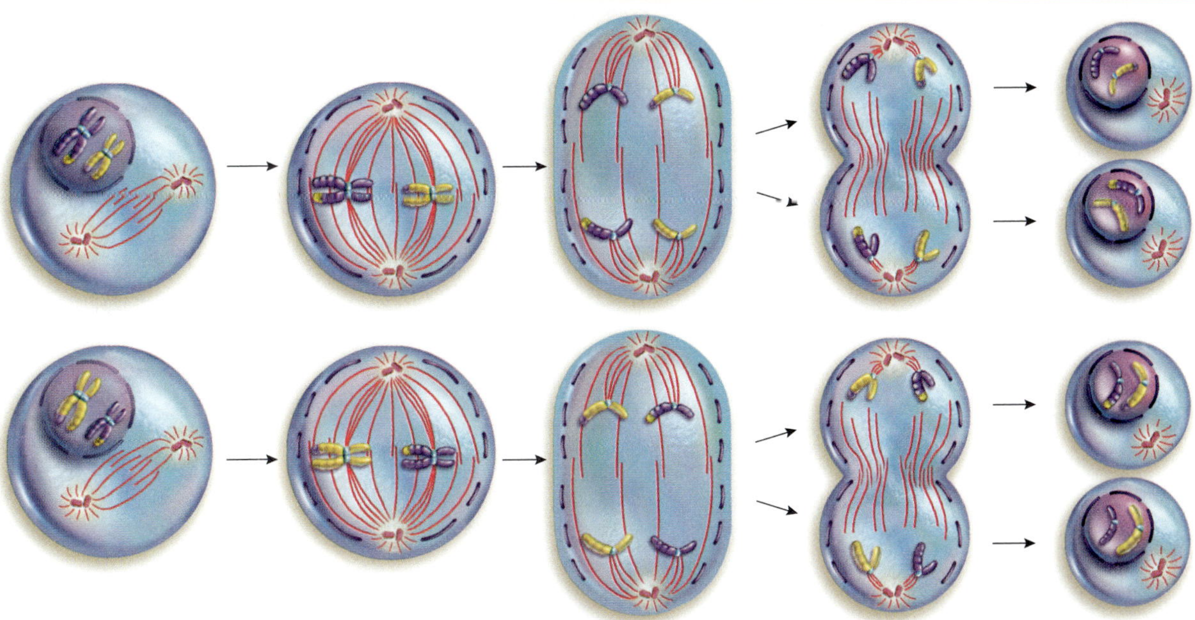

(e) Prophase II. If chromosomes have relaxed after telophase I, they recondense. Spindle microtubules re-form and attach to the sister chromatids.

(f) Metaphase II. Chromosomes line up along the equator, with sister chromatids of each chromosome attached to spindle microtubules that lead to opposite poles.

(g) Anaphase II. Chromatids separate into independent daughter chromosomes, one former chromatid moving toward each pole.

(h) Telophase II. Chromosomes finish moving to opposite poles. Nuclear envelopes re-form, and the chromosomes become extended again (not shown here).

(i) Four haploid cells. Cytokinesis results in four haploid cells, each containing one member of each pair of homologous chromosomes (shown here in condensed state).

Figure 11-21 Meiotic cell division

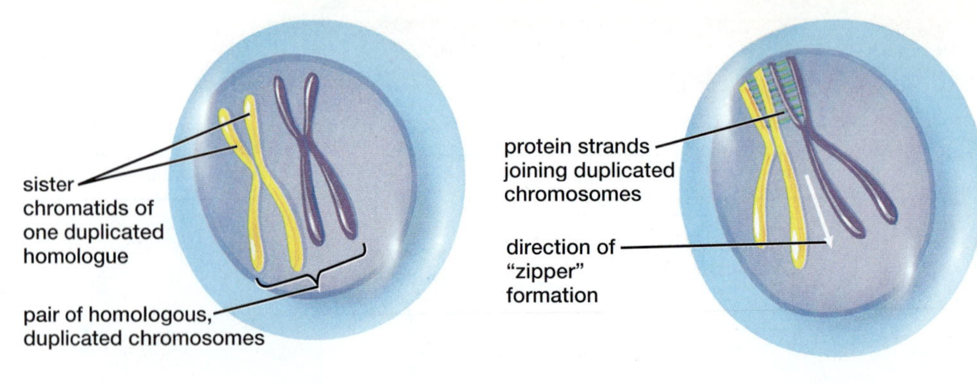

(a) **Duplicated homologous chromosomes pair up side by side.**

(b) **Protein strands "zip" the homologous chromosomes together.**

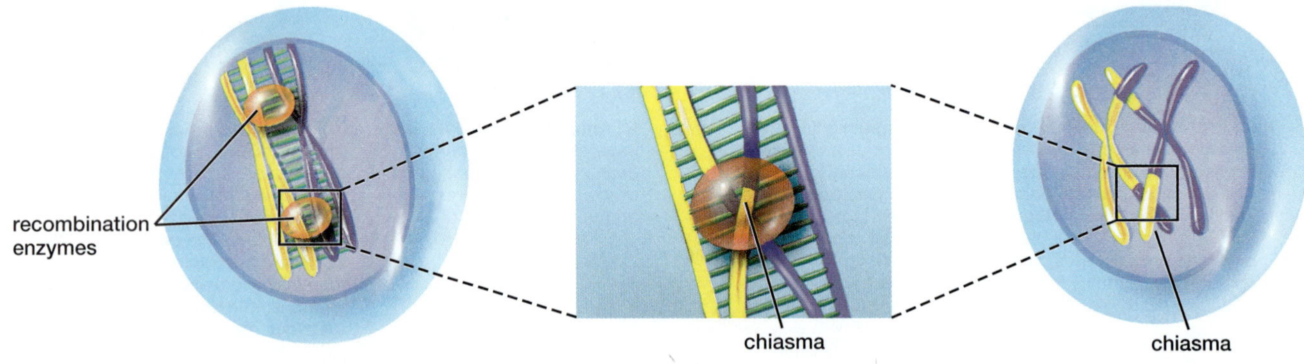

(c) **Recombination enzymes bind to the joined chromosomes.**

(d) **Recombination enzymes snip chromatids apart and reattach the free ends. Chiasmata (the sites of crossing over) form when one end of the paternal chromatid (yellow) attaches to the other end of a maternal chromatid (purple).**

(e) **Recombination enzymes and protein zippers leave. Chiasmata remain, helping to hold homologous chromosomes together.**

Figure 11-22 Mechanism of crossing over

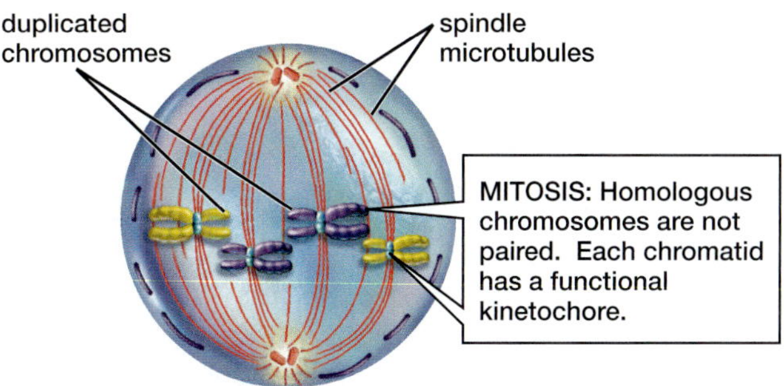

Figure 11-23 Spindle arrangement in mitosis

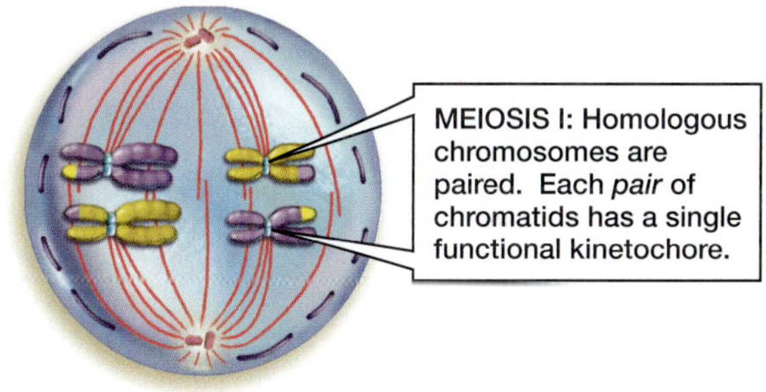

Figure 11-24 Spindle arrangement in meiosis I

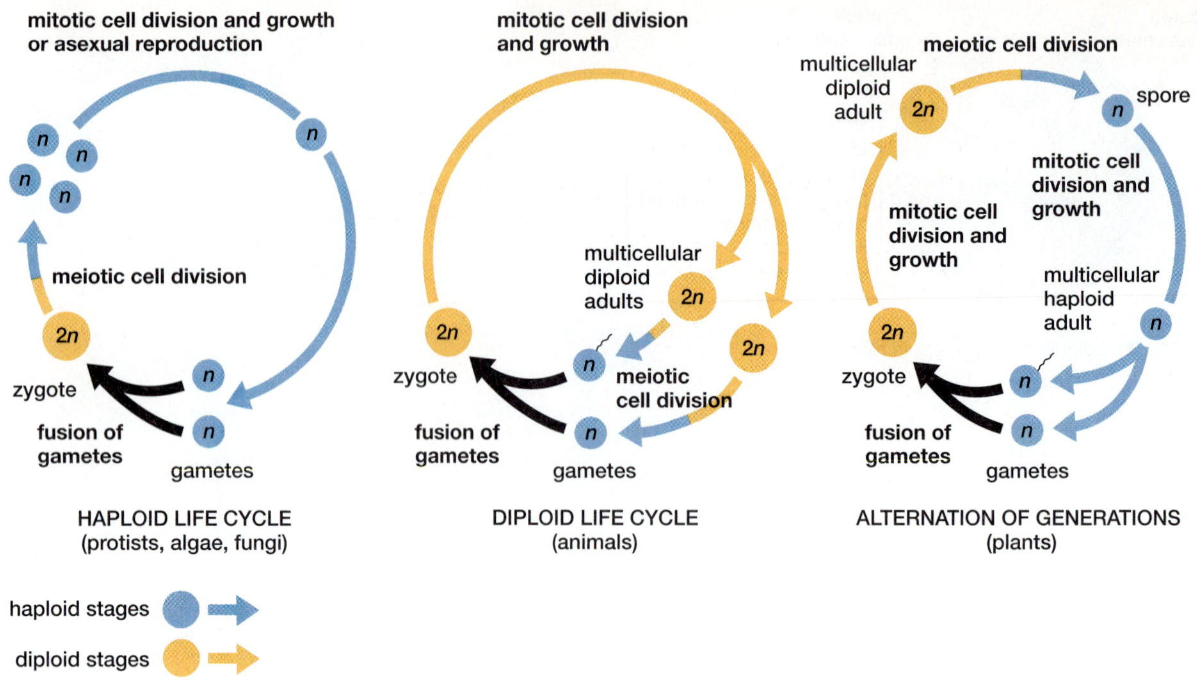

Figure 11-25 Types of eukaryotic life cycles

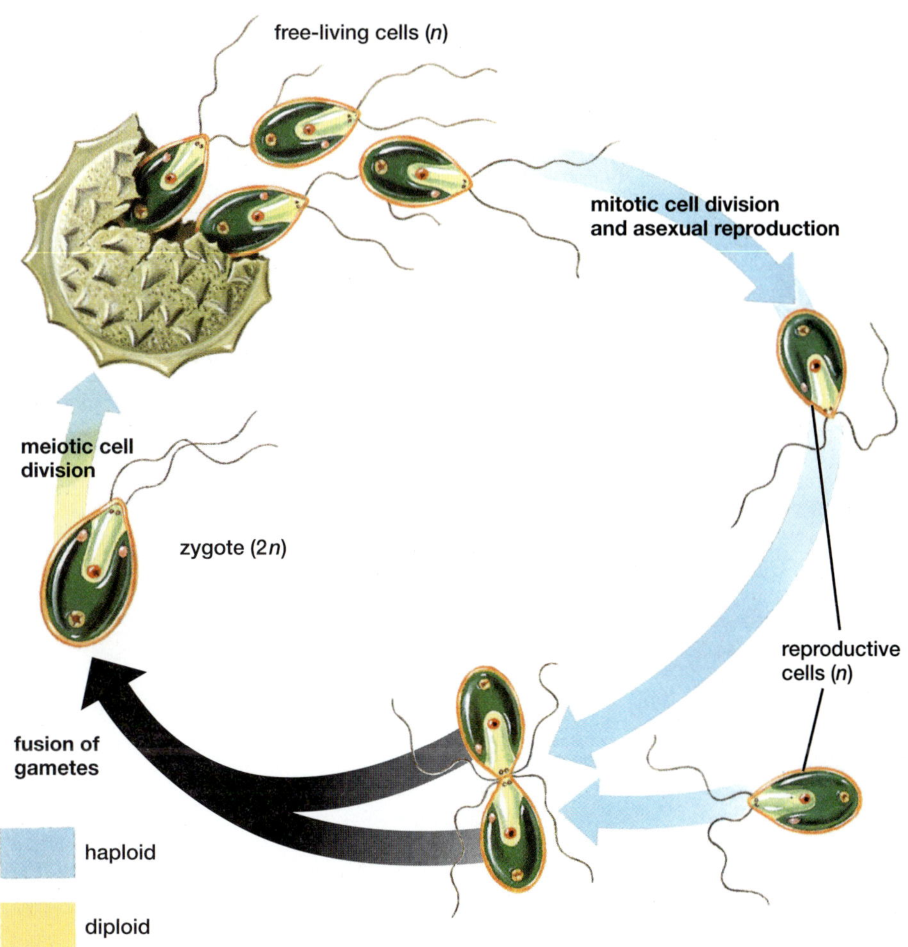

Figure 11-26 Life cycle of *Chlamydomonas*

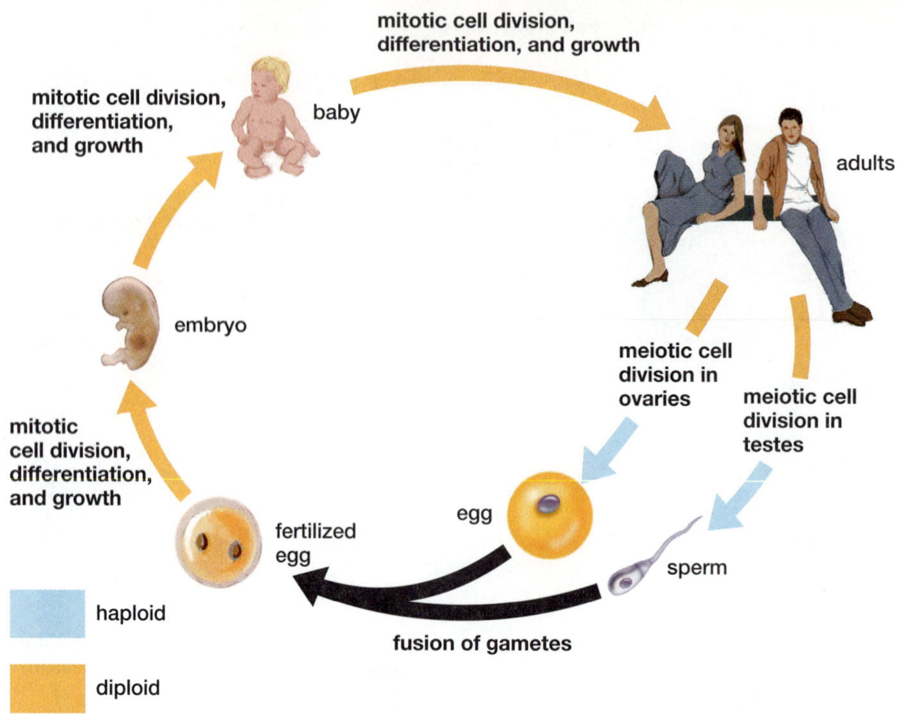

Figure 11-27 Human life cycle

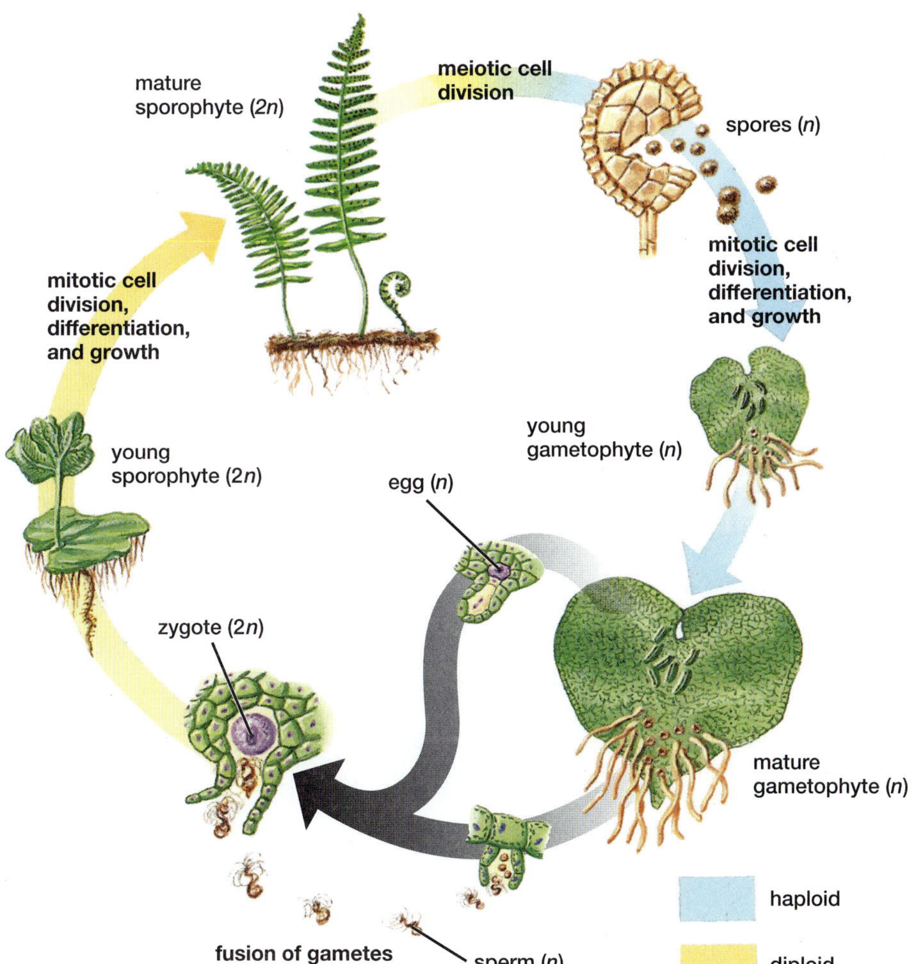

Figure 11-28 Fern life cycle

Figure 11-29 Possible chromosome arrangements at metaphase I

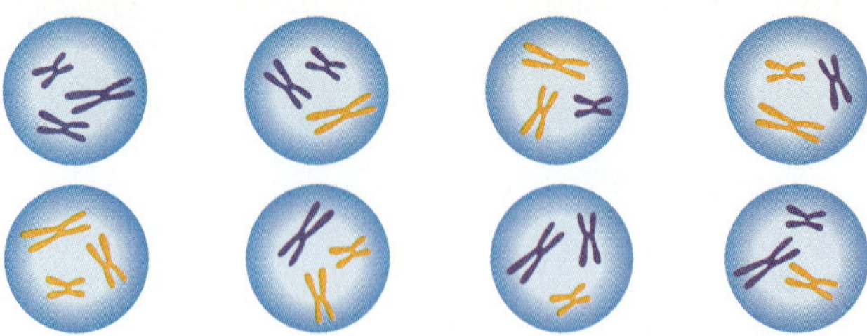

Figure 11-30 8 possible chromosome sets

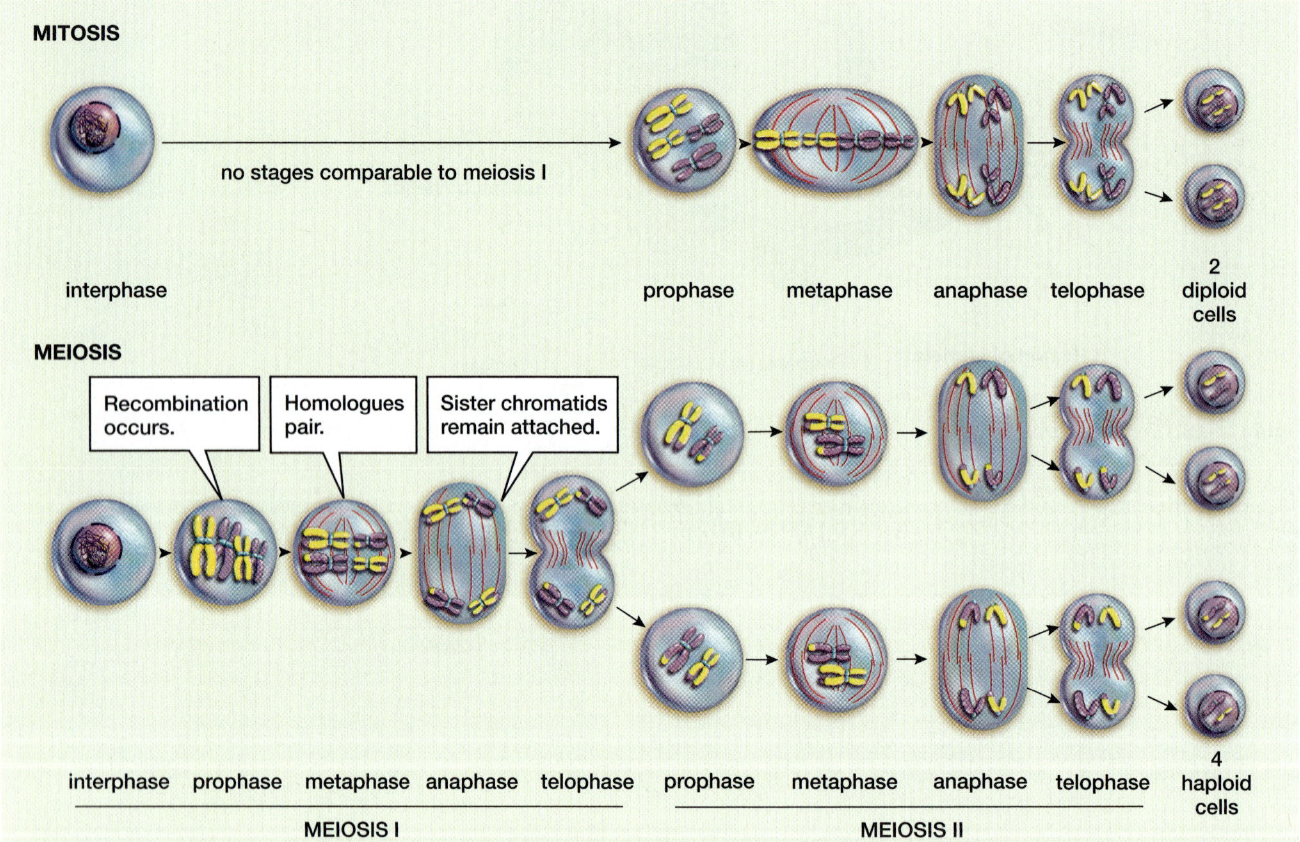

MITOSIS

no stages comparable to meiosis I

interphase | prophase | metaphase | anaphase | telophase | 2 diploid cells

MEIOSIS

Recombination occurs.

Homologues pair.

Sister chromatids remain attached.

interphase | prophase | metaphase | anaphase | telophase | prophase | metaphase | anaphase | telophase | 4 haploid cells

MEIOSIS I

MEIOSIS II

In these diagrams, comparable phases are aligned. In both mitosis and meiosis, chromosomes are replicated during interphase. Meiosis I, with the pairing of homologous chromosomes, formation of chiasmata, exchange of chromosome parts, and separation of homologues to form haploid daughter nuclei, has no counterpart in mitosis. Meiosis II, however, is similar to mitosis.

Table 11-1 Comparison of mitosis and meiosis

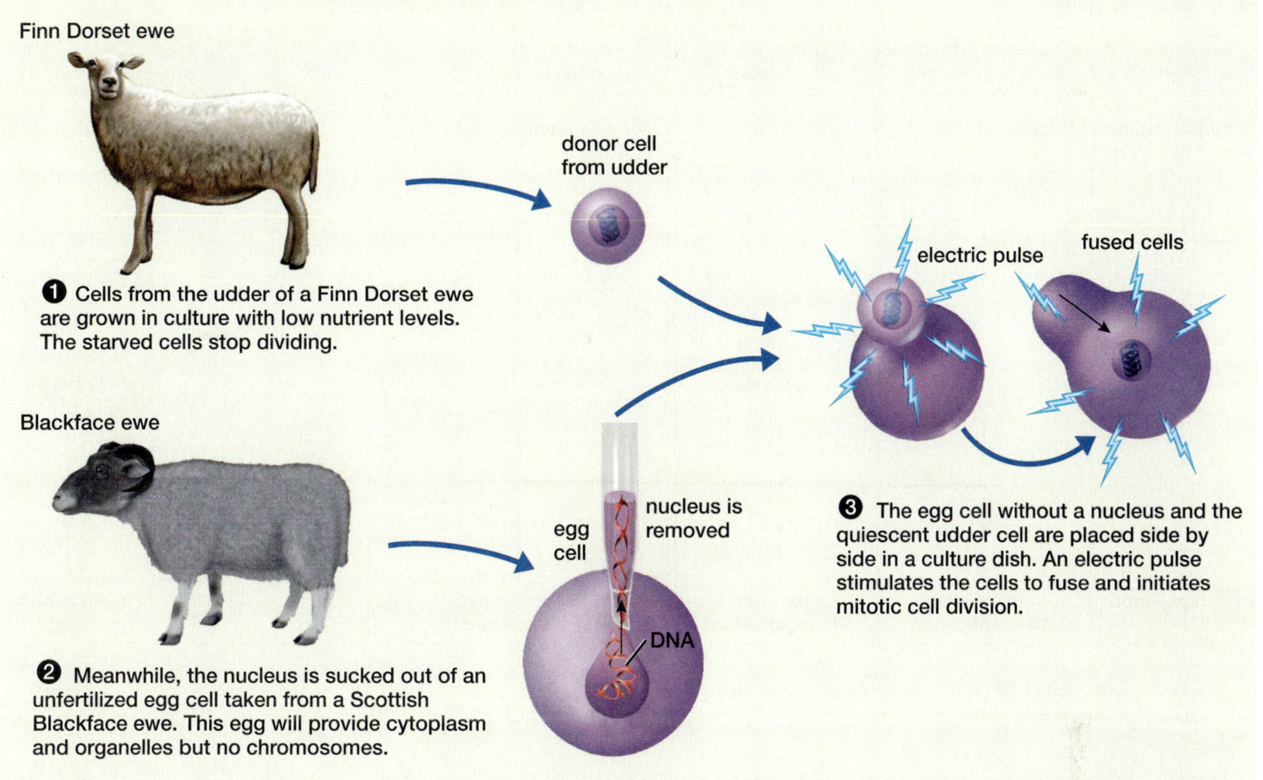

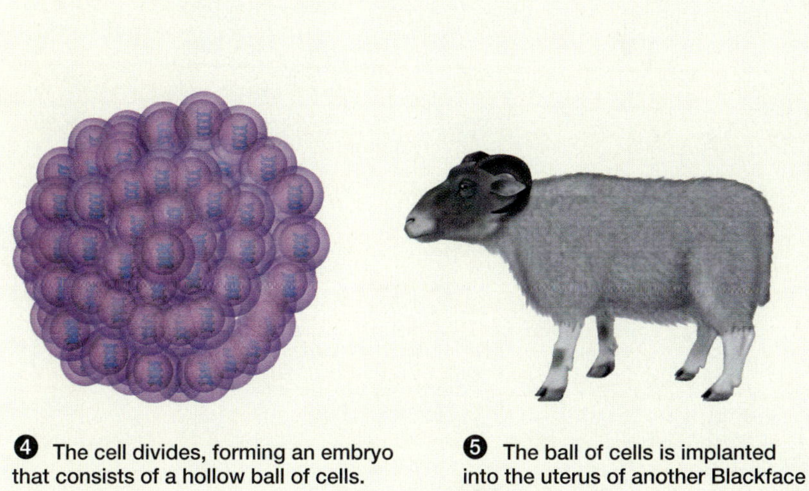

Figure E11-1 Cloning Dolly the sheep

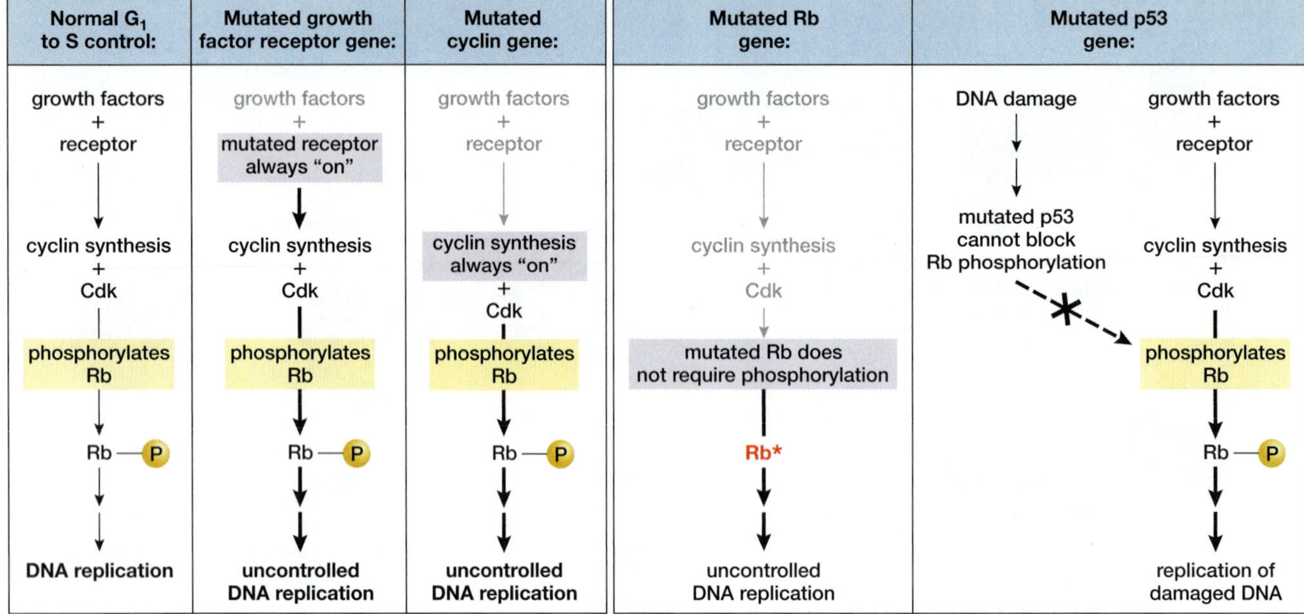

Figure E11-3 Oncogenes and tumor suppressor gene action

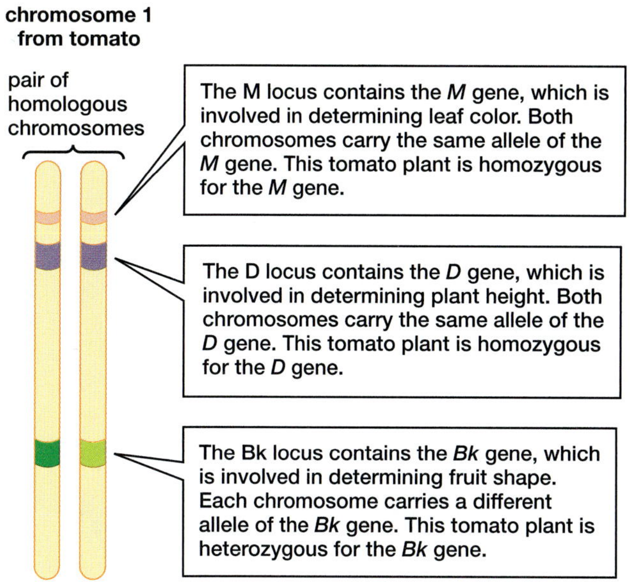

chromosome 1
from tomato
pair of
homologous
chromosomes
The M locus contains the *M* gene, which is involved in determining leaf color. Both chromosomes carry the same allele of the *M* gene. This tomato plant is homozygous for the *M* gene.
The D locus contains the *D* gene, which is involved in determining plant height. Both chromosomes carry the same allele of the *D* gene. This tomato plant is homozygous for the *D* gene.
The Bk locus contains the *Bk* gene, which is involved in determining fruit shape. Each chromosome carries a different allele of the *Bk* gene. This tomato plant is heterozygous for the *Bk* gene.

Figure 12-1 Chromosomes and alleles

pollen
pollen
Parental
generation (P)
cross-fertilize
true-breeding,
purple-flowered
plant
true-breeding,
white-flowered
plant
First-generation
offspring (F_1)
all purple-flowered
plants

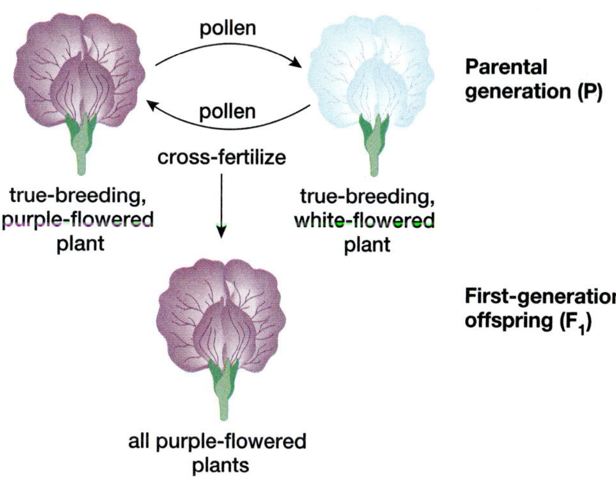

Figure 12-4 Cross of white and purple pea flowers

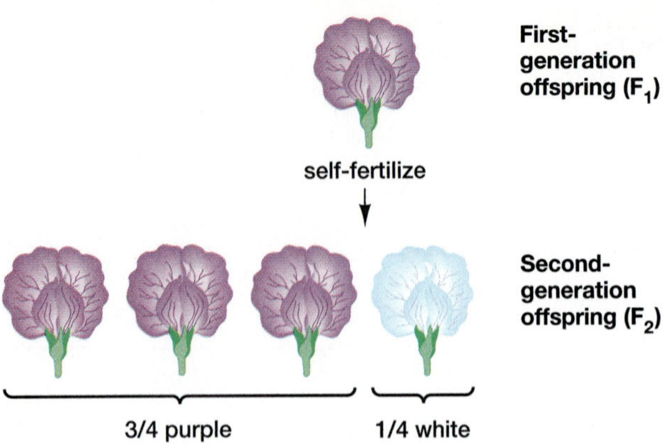

Figure 12-5 Cross of F_1 purple pea flowers

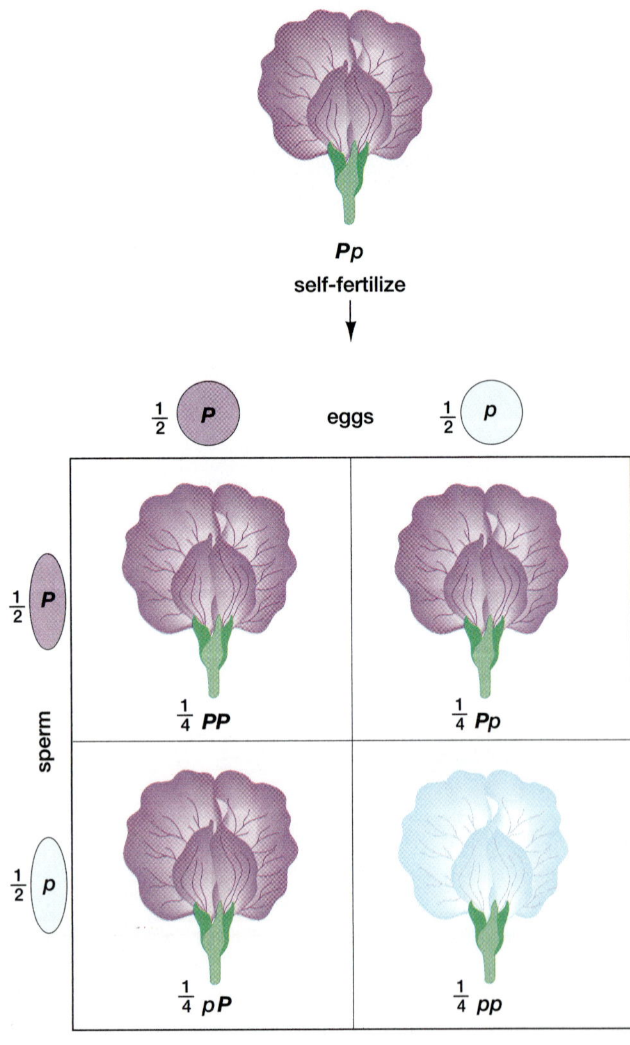

Figure 12-11a Punnett square of monohybrid

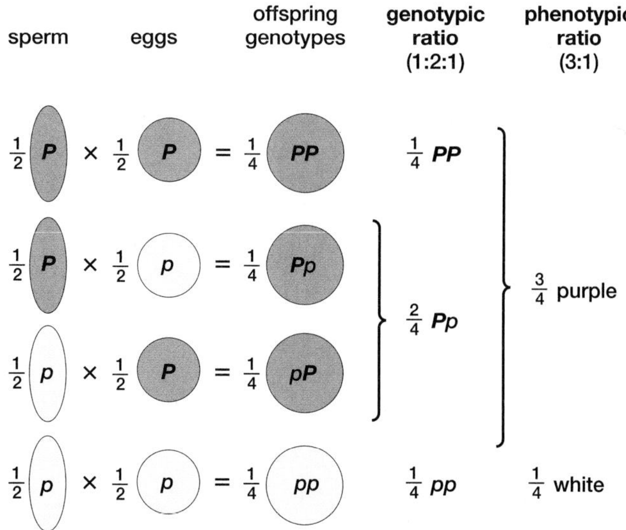

Figure 12-11b Probability method of calculating monohybrid cross

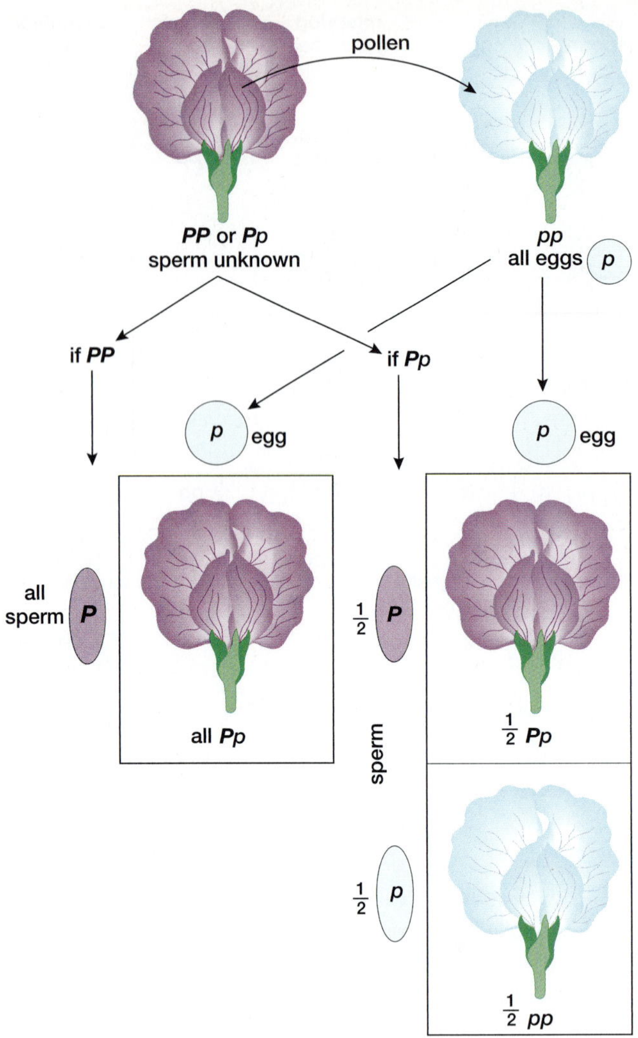

Figure 12-12 Test cross

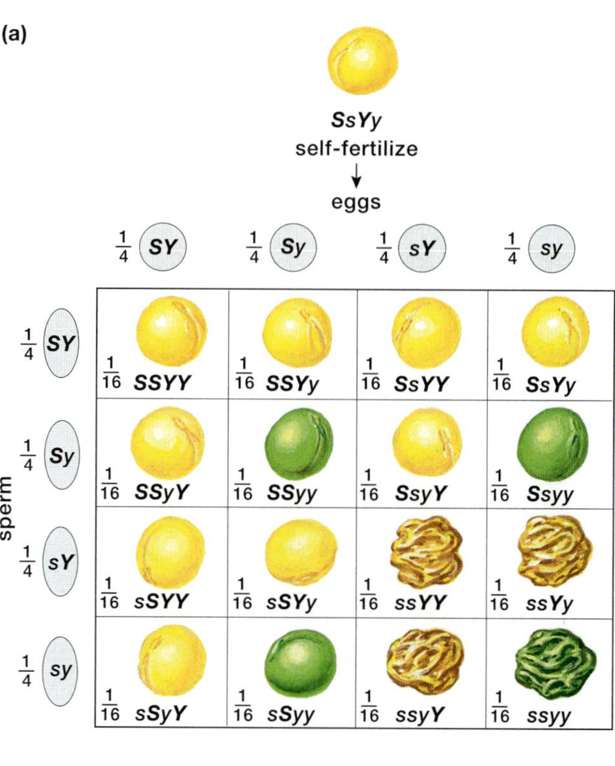

(b)

seed shape		seed color		phenotypic ratio (9:3:3:1)
$\frac{3}{4}$ smooth	×	$\frac{3}{4}$ yellow	=	$\frac{9}{16}$ smooth yellow
$\frac{3}{4}$ smooth	×	$\frac{1}{4}$ green	=	$\frac{3}{16}$ smooth green
$\frac{1}{4}$ wrinkled	×	$\frac{3}{4}$ yellow	=	$\frac{3}{16}$ wrinkled yellow
$\frac{1}{4}$ wrinkled	×	$\frac{1}{4}$ green	=	$\frac{1}{16}$ wrinkled green

Figure 12-14 Punnett square, dihybrid cross

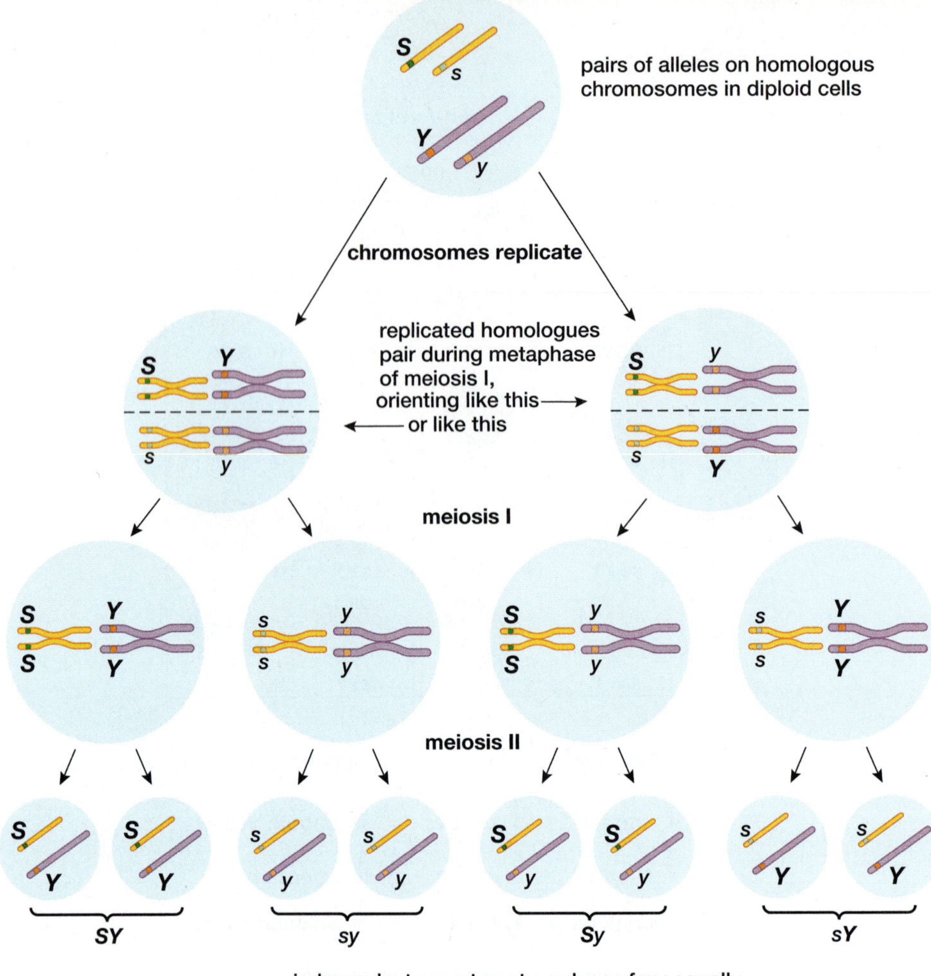

Figure 12-15 Independent assortment of chromosomes and alleles

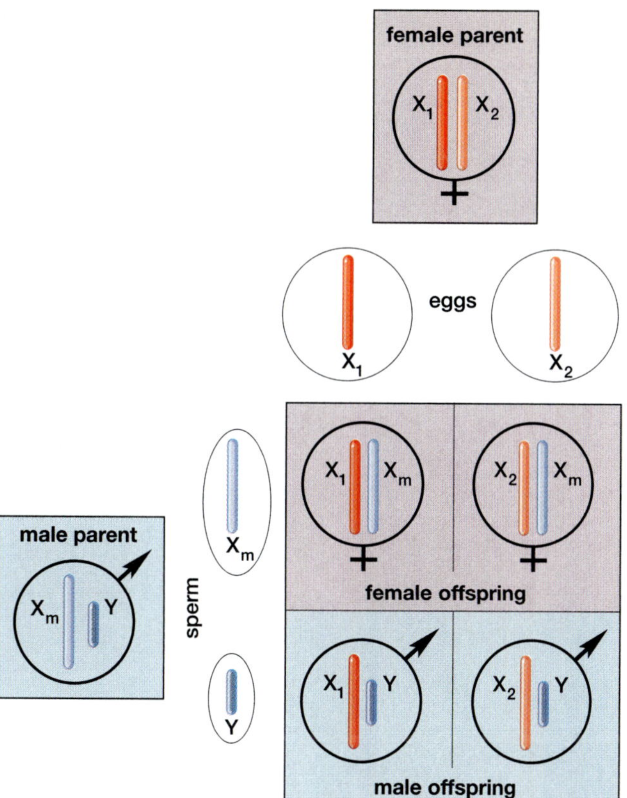

Figure 12-22 Punnett square of sex determination in mammals

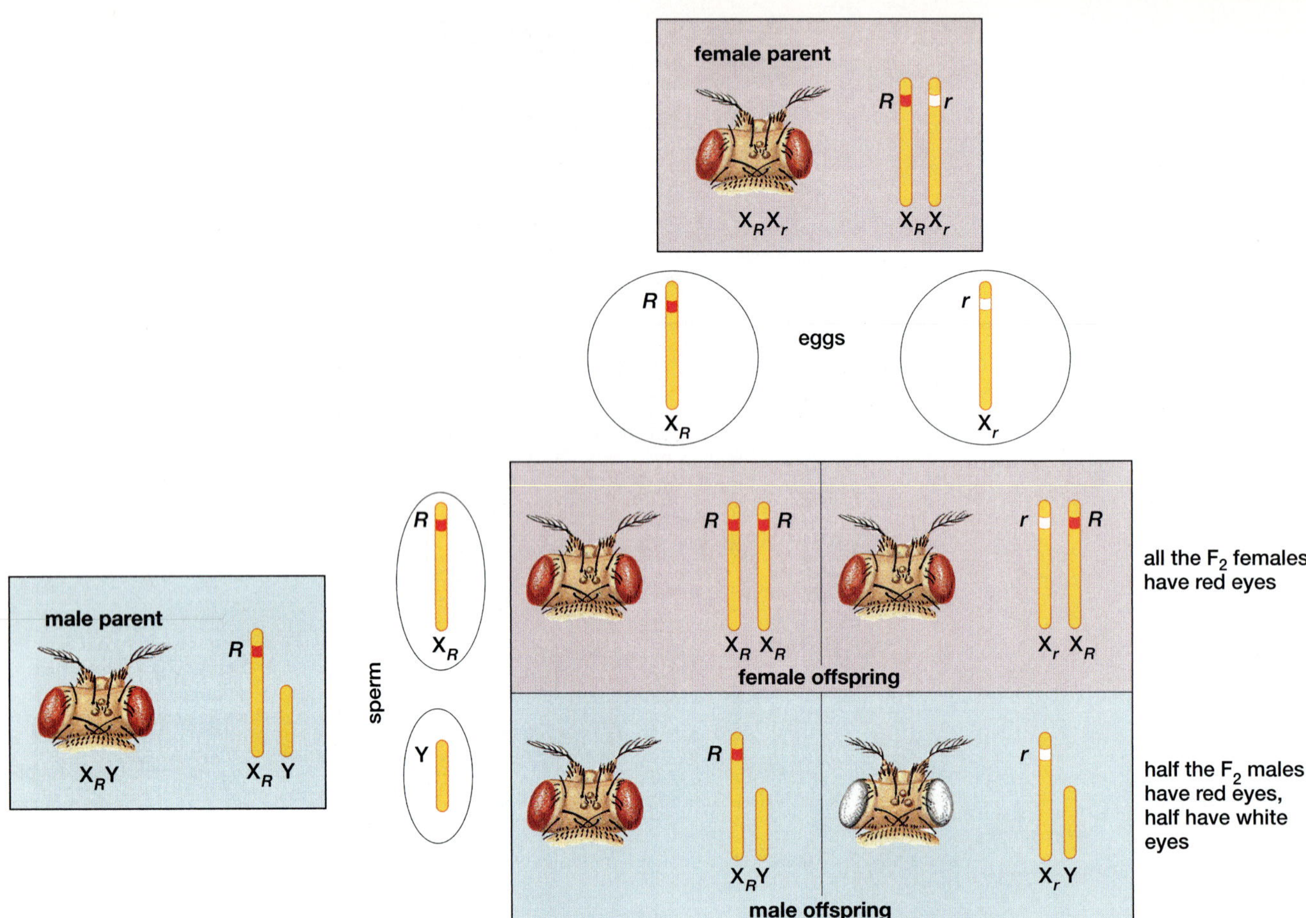

Figure 12-23 Punnett square of sex-linked inheritance in flies

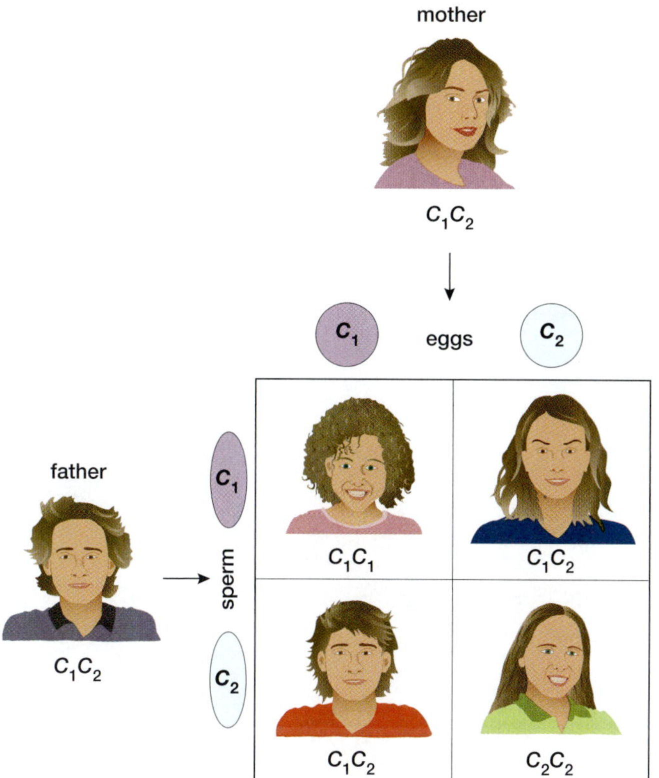

Figure 12-24 Punnett square of incomplete dominance in hair curliness

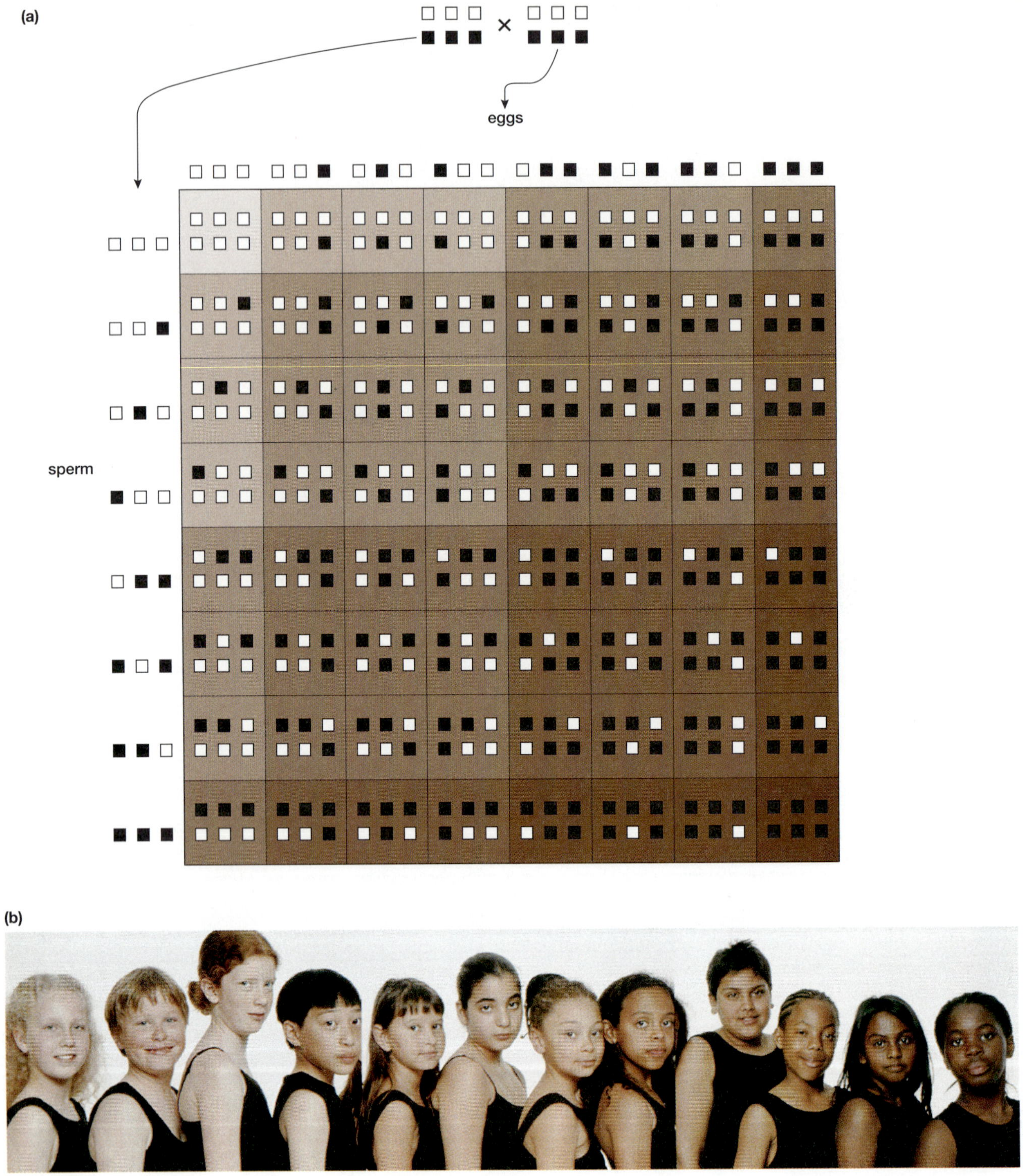

(a)

×

eggs

sperm

(b)

Figure 12-25 Punnett square of polygenic inheritance of skin color

(a) A pedigree for a dominant trait

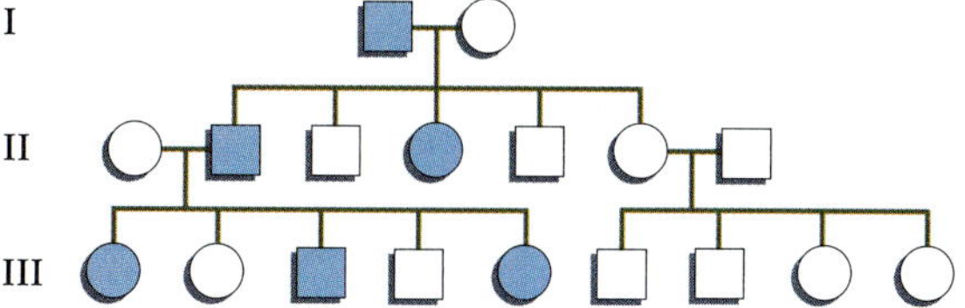

(b) A pedigree for a recessive trait

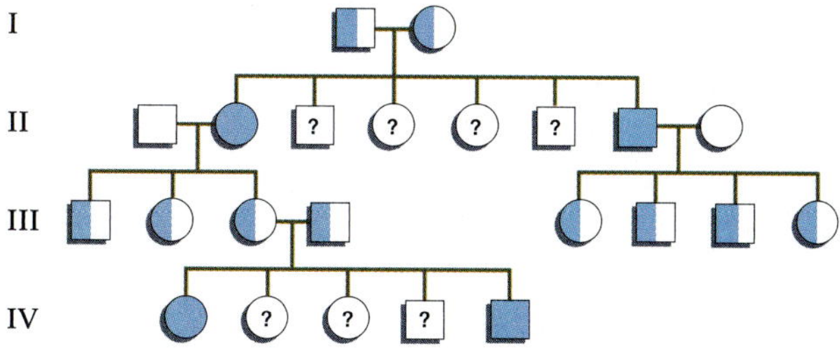

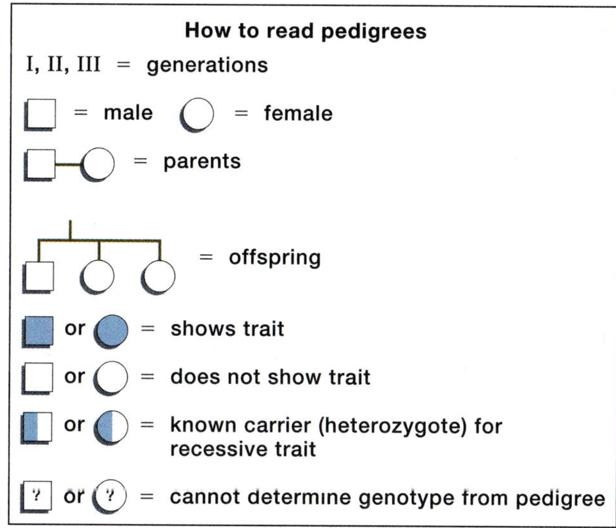

Figure 12-27 Family pedigrees

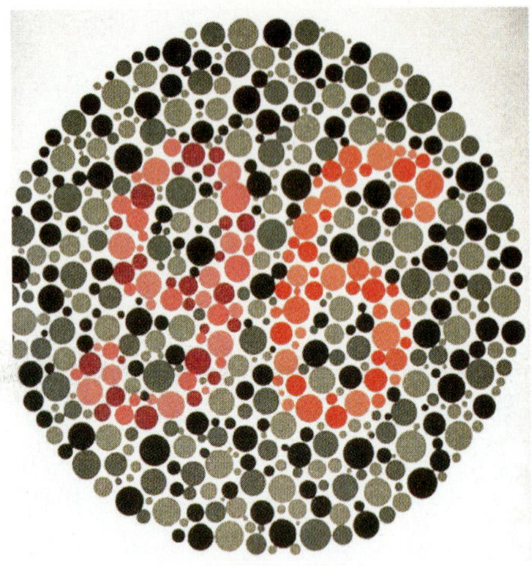

Figure 12-30a Ishihara test for colorblindness

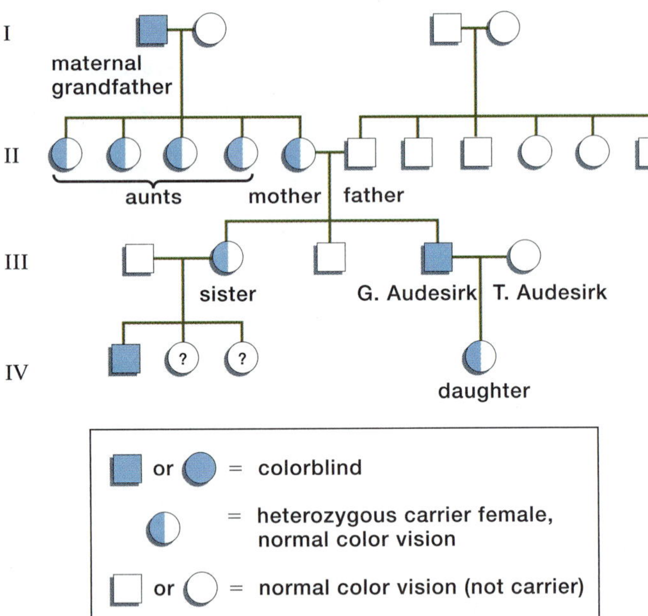

Figure 12-30b Family pedigree for colorblindness

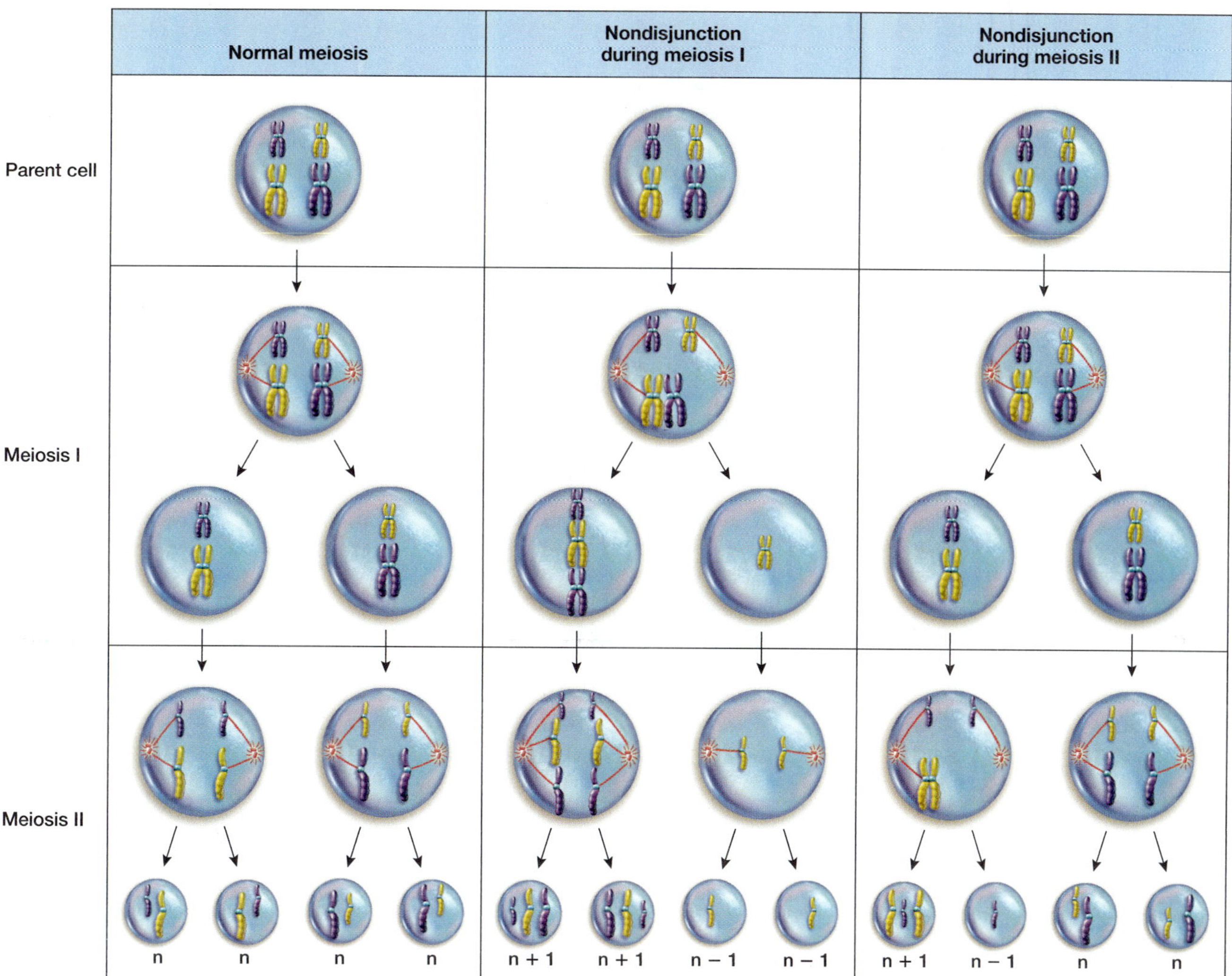

Figure 12-32 Nondisjunction during meiosis

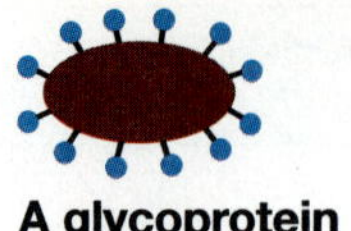

A glycoprotein

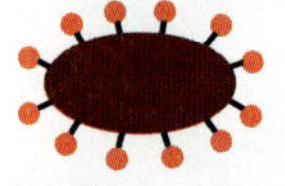

B glycoprotein

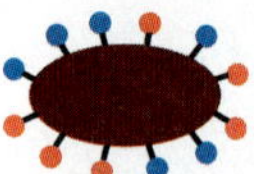

Both A and B glycoprotein

Neither A nor B glycoprotein

Table 12-1 Antigens on red blood cells

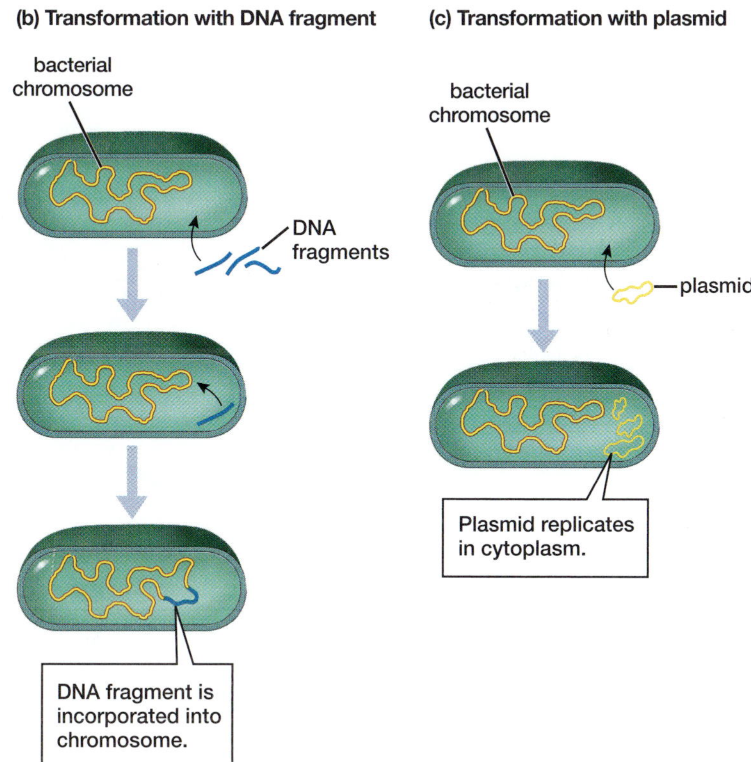

Figure 13-1 Bacterial recombination

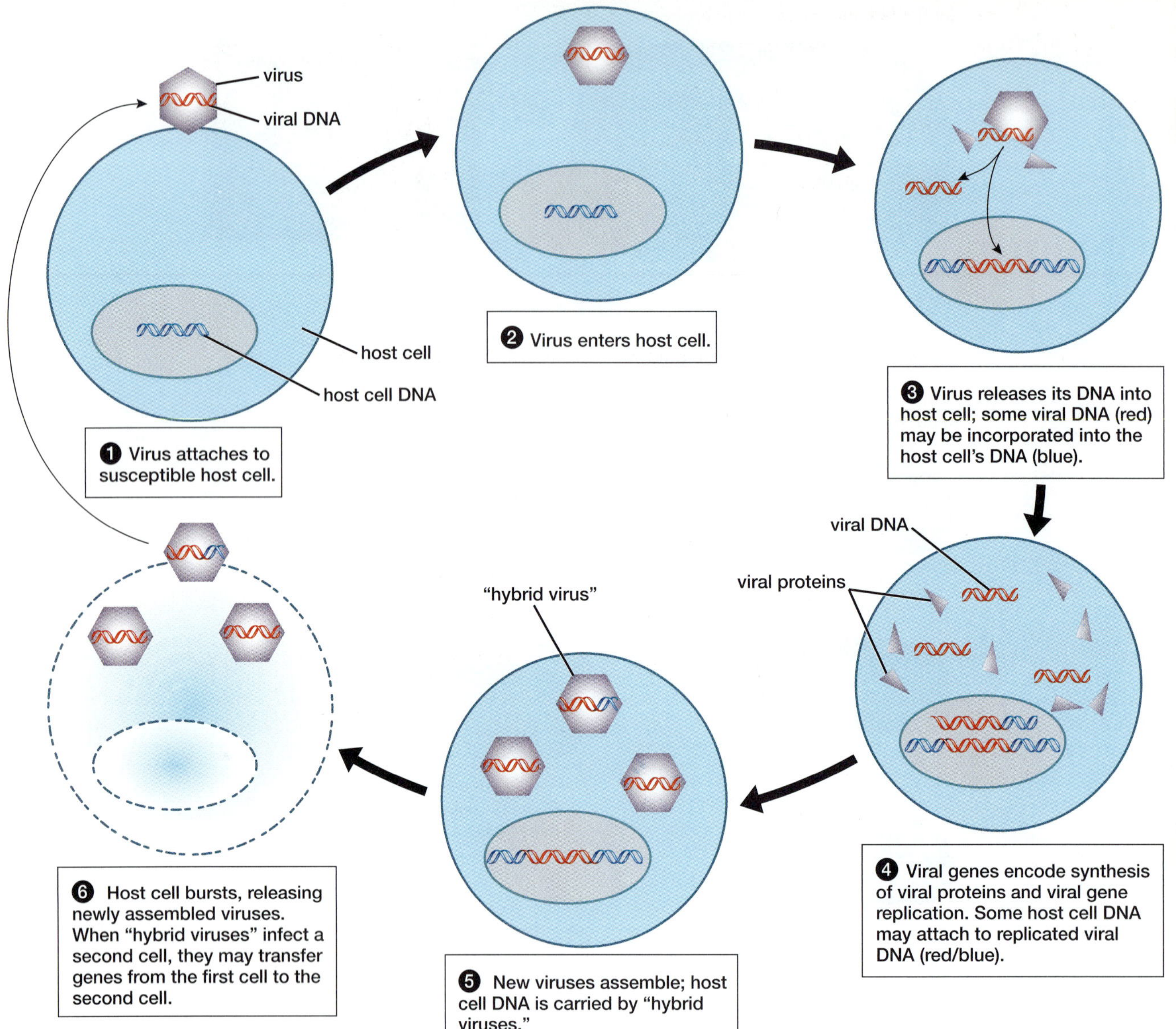

Figure 13-2 Viral transfer of genes

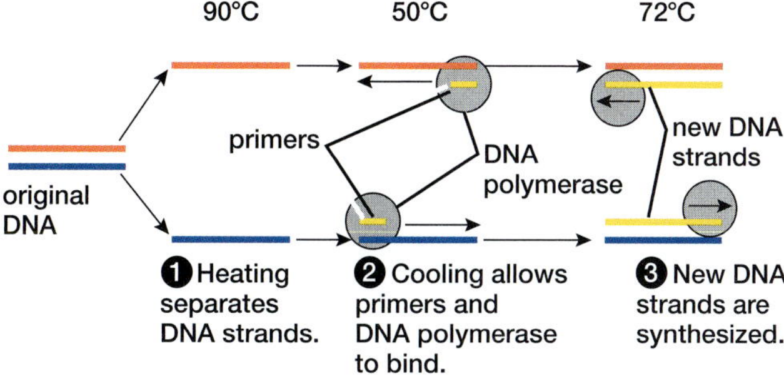

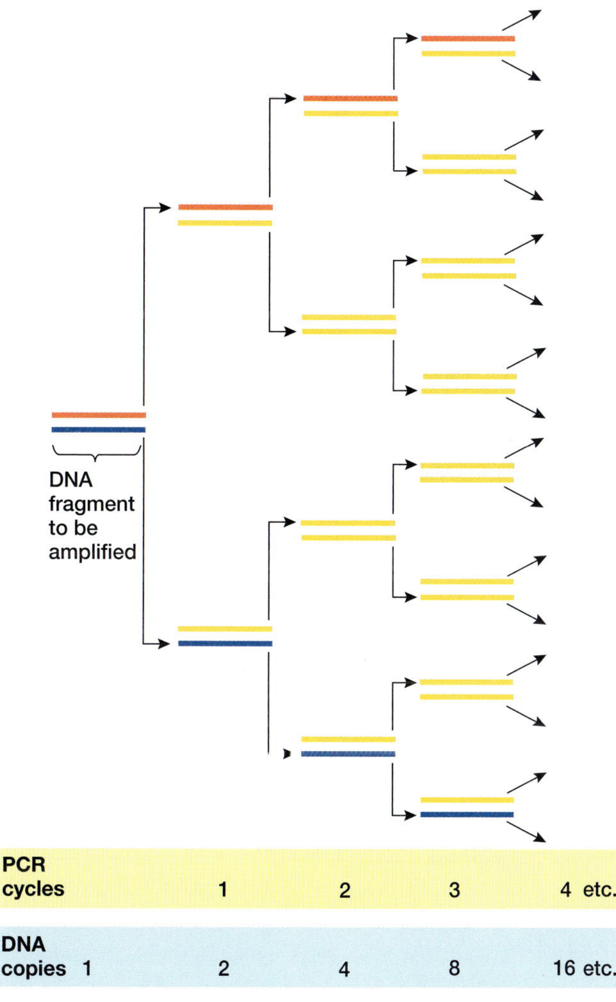

PCR cycles		1	2	3	4 etc.
DNA copies	1	2	4	8	16 etc.

Figure 13-3 Polymerase chain reaction

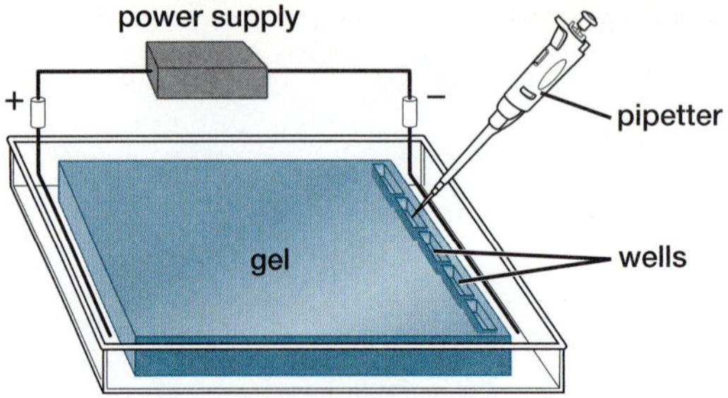

(a) DNA samples are pipetted into wells (shallow slots) in the gel. Electrical current is sent through the gel (negative at end with wells, positive at opposite end).

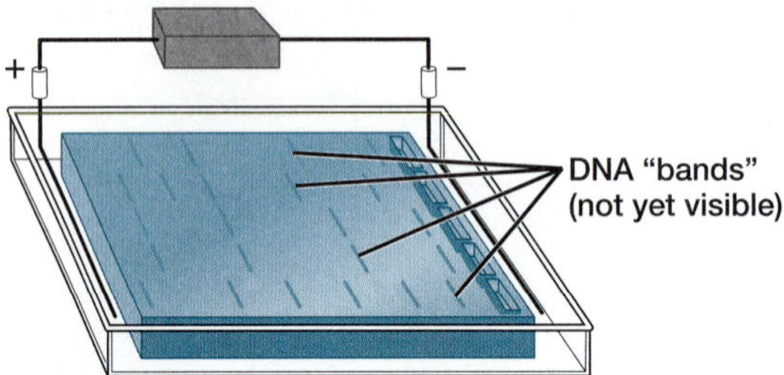

(b) Electrical current moves DNA segments through the gel. Smaller pieces of DNA move farther toward the positive electrode.

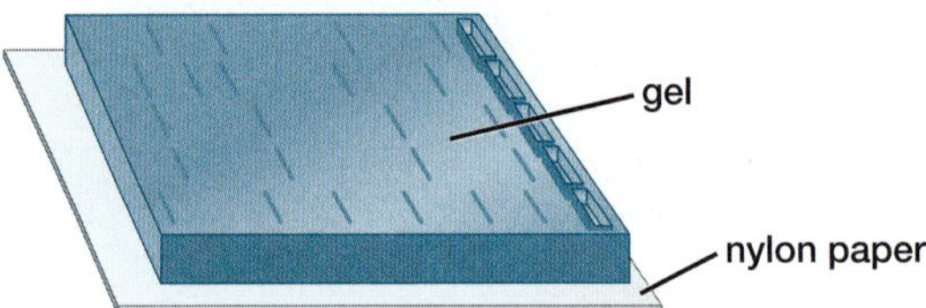

(c) Gel is placed on special nylon "paper." Electrical current drives DNA out of gel onto nylon.

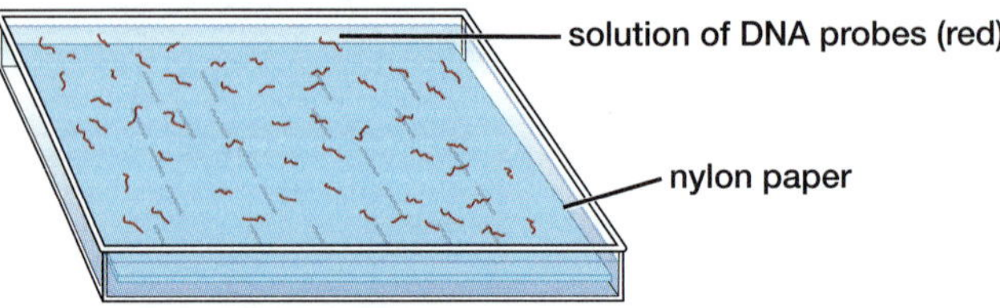

(d) Nylon paper with DNA is bathed in a solution of labeled DNA probes (red) that are complementary to specific DNA segments in the original DNA sample.

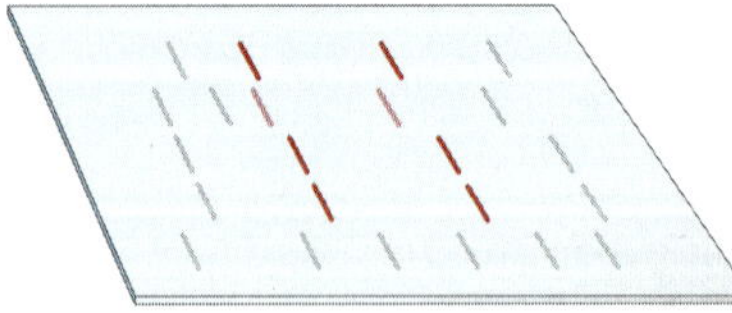

(e) Complementary DNA segments are labeled by probes (red bands).

Figure 13-5 Principles of gel electrophoresis

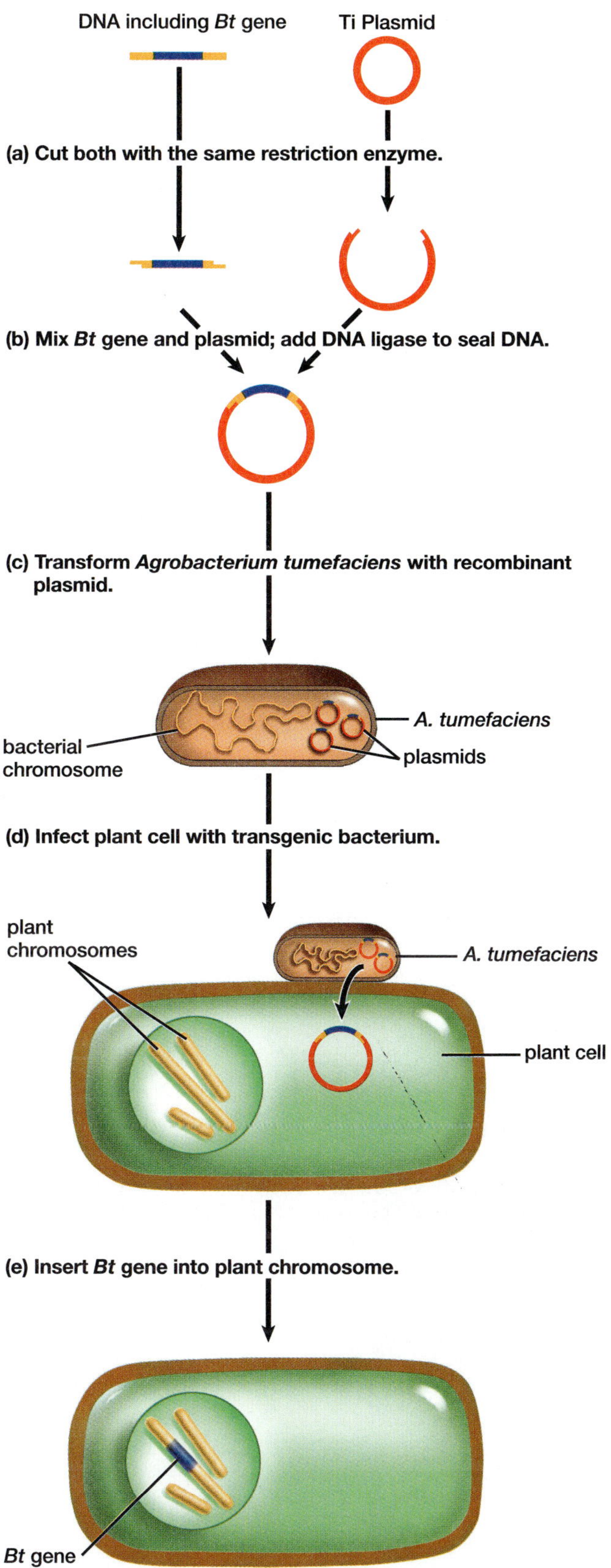

Figure 13-10 Using bacteria to transfect plant cells

(a) **Mst II cuts a normal globin allele in 2 places, but cuts the sickle-cell allele in 1 place.**

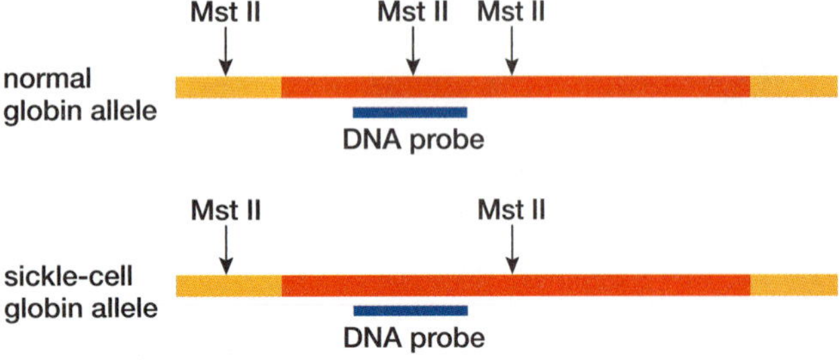

(b) **Gel electrophoresis of globin alleles**

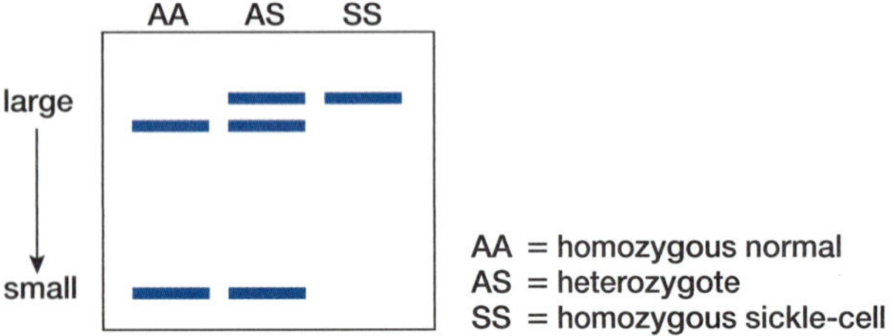

Figure 13-11 Using restriction enzymes to diagnose sickle-cell anemia

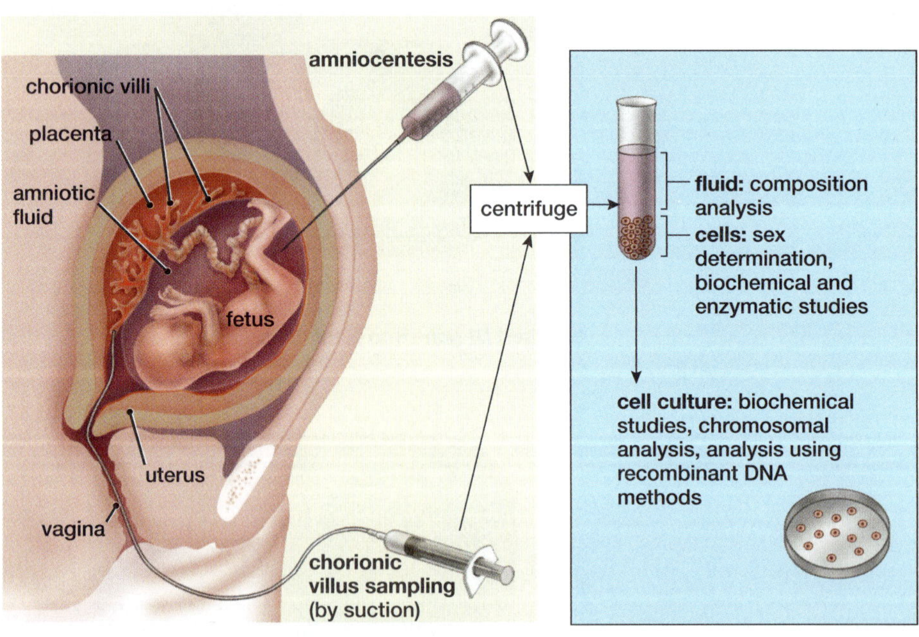

Figure E13-5 Amniocentesis and chorionic villus sampling

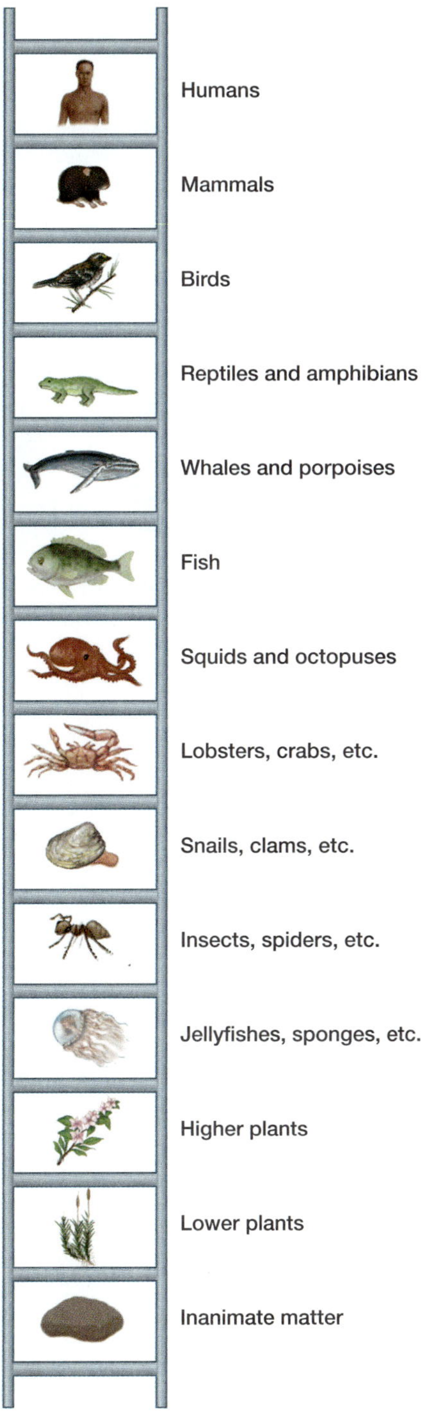

Figure 14-2 Aristotle's ladder of nature

Figure 14-3 Fossils

Figure 14-4 Rock layers with fossils

(a) Large ground finch, beak suited to large seeds

(b) Small ground finch, beak suited to small seeds

(c) Warbler finch, beak suited to insects

(d) Vegetarian tree finch, beak suited to leaves

Figure 14-5 Darwin's finches

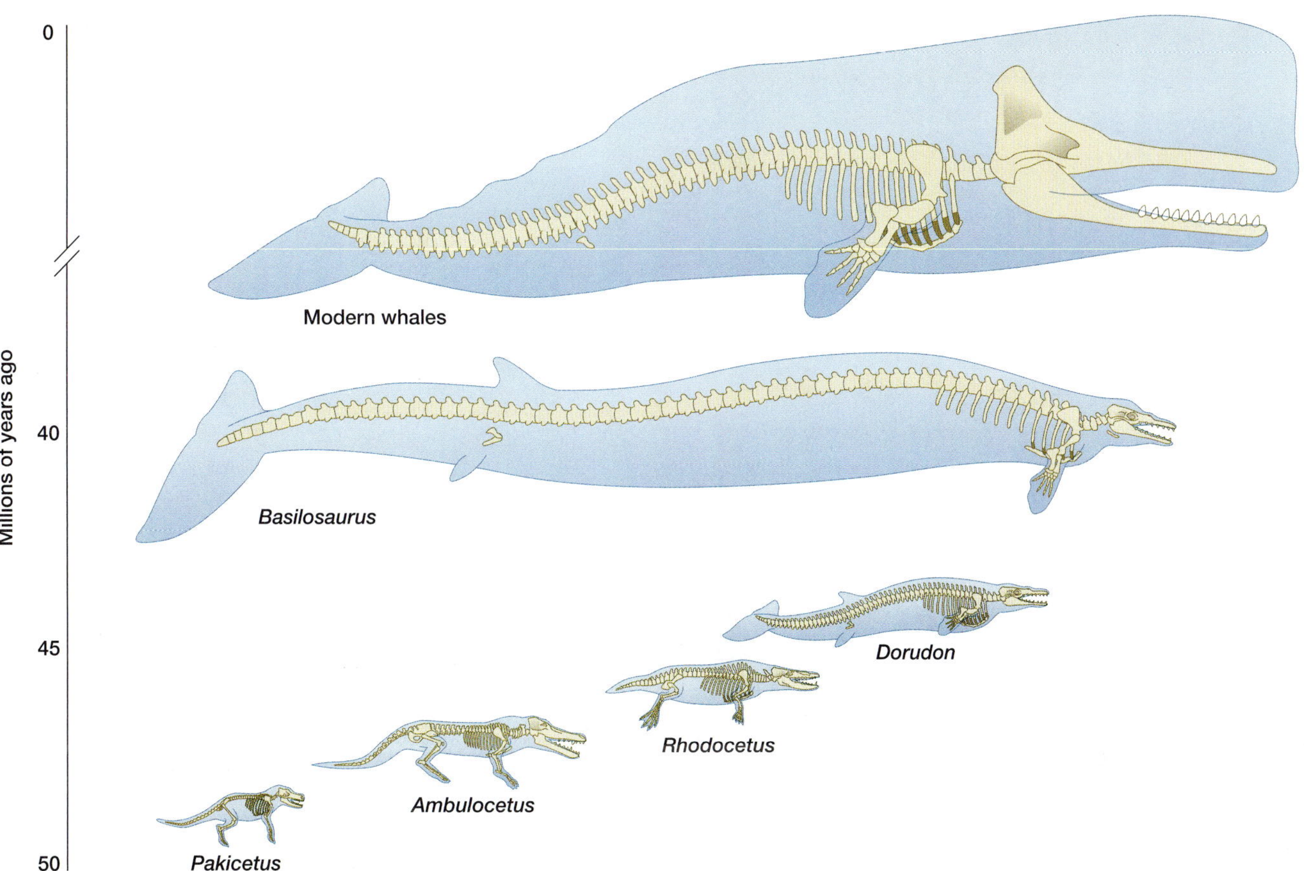

Figure 14-6 Whale evolution

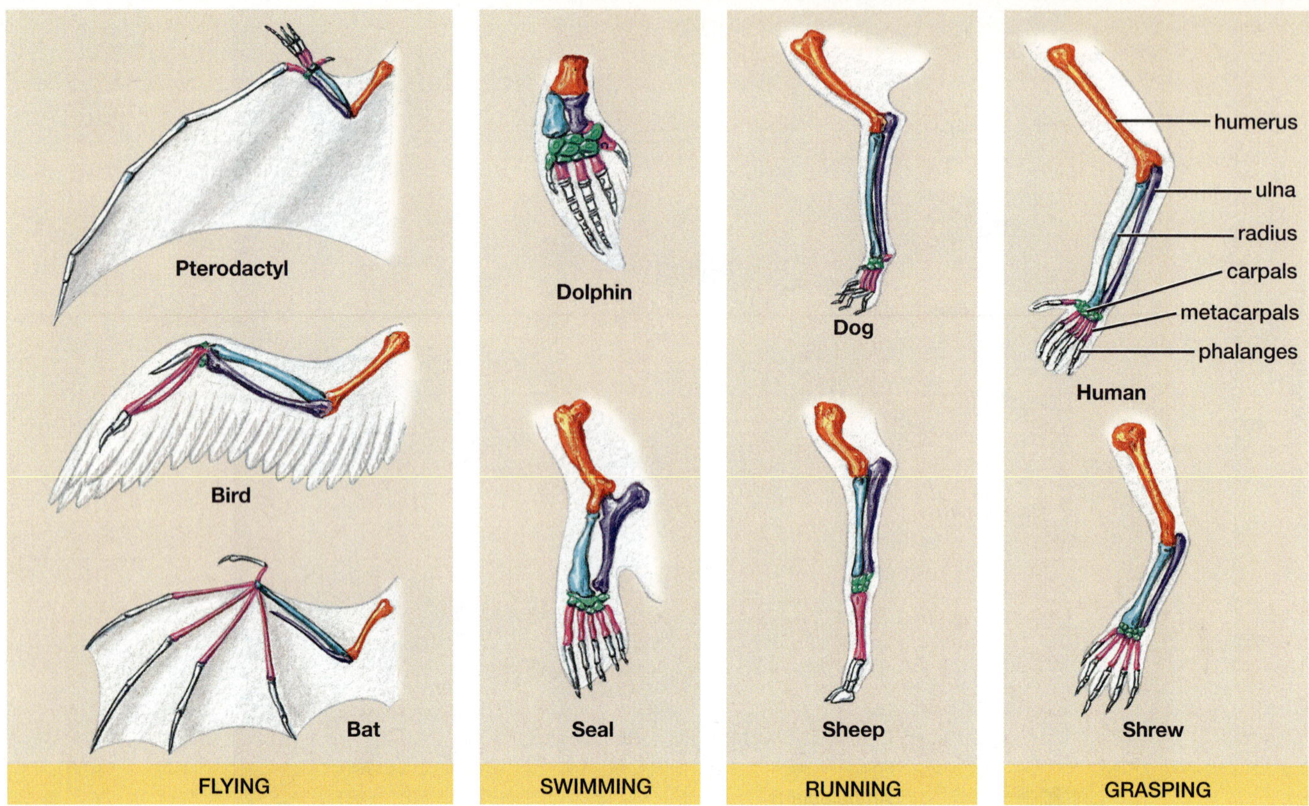

Figure 14-7 Homologous structures

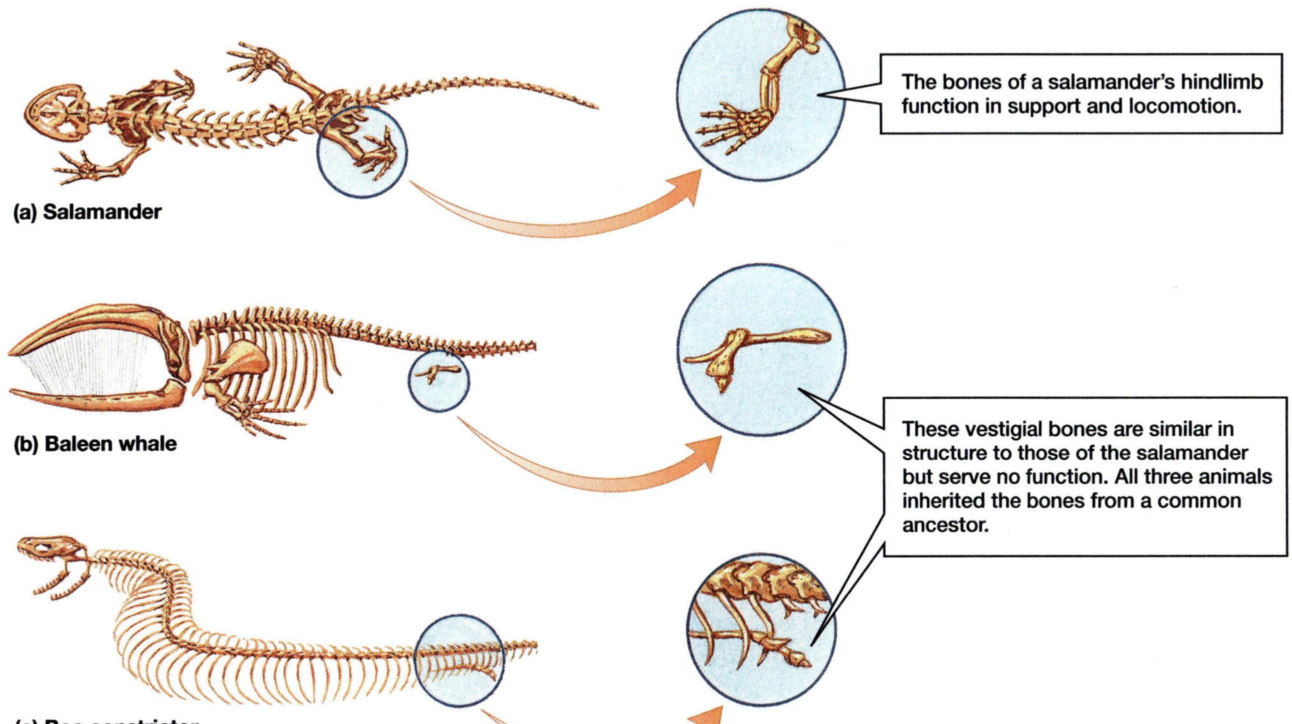

Figure 14-8 Vestigial structures

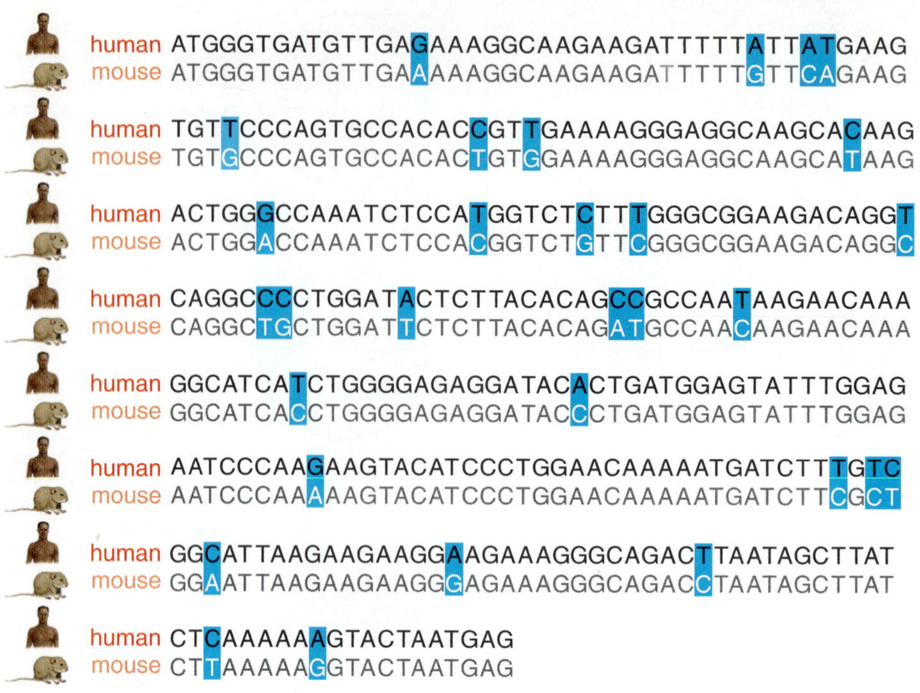

Figure 14-11 Cytochrome c sequences

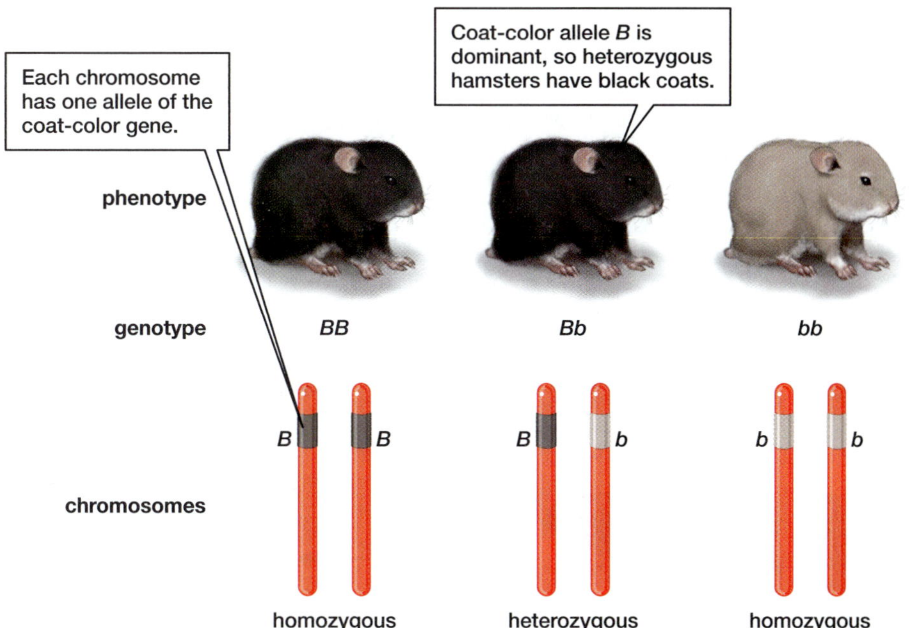

Figure 15-1 Genotype and phenotype

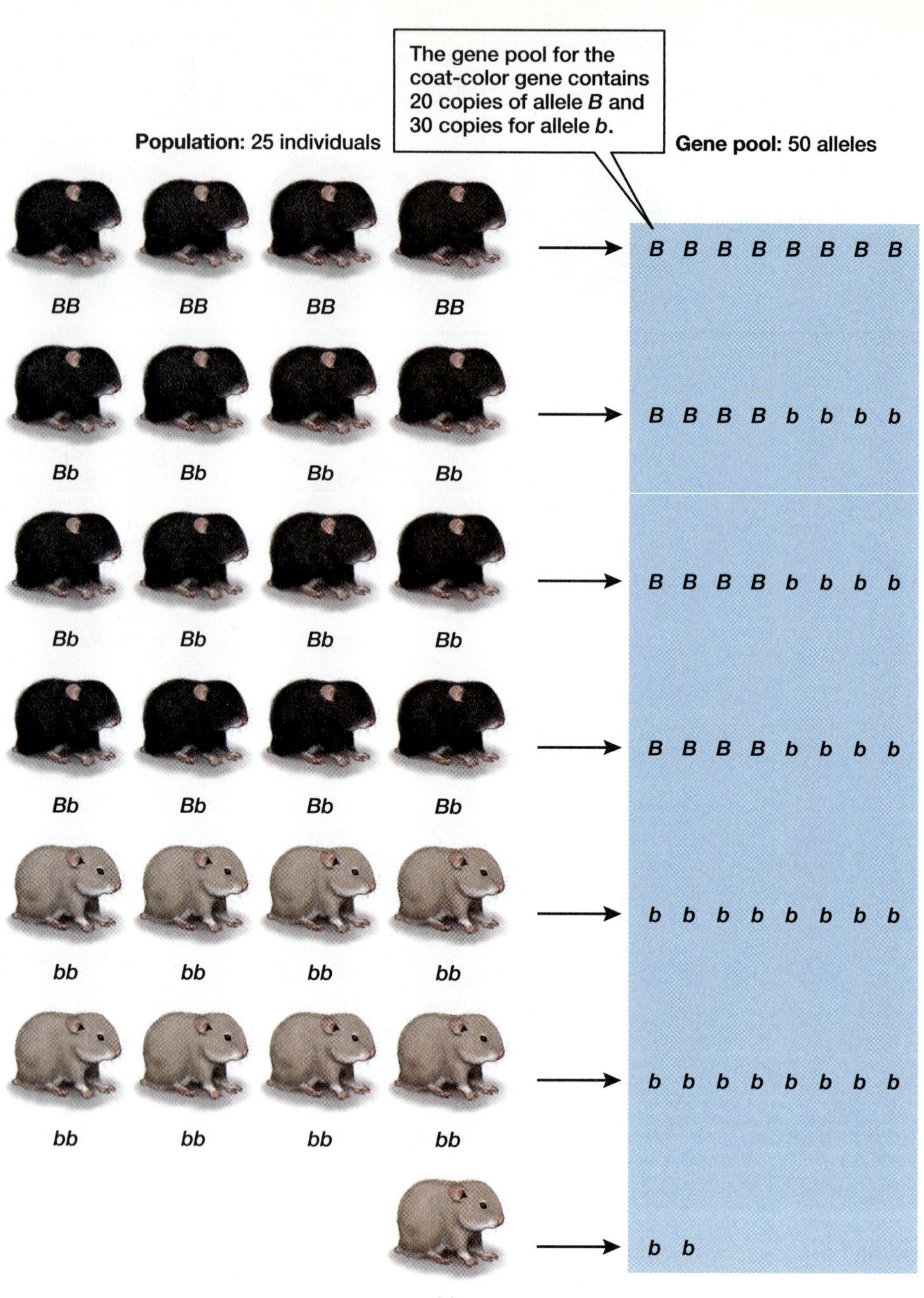

Figure 15-2 Gene pool

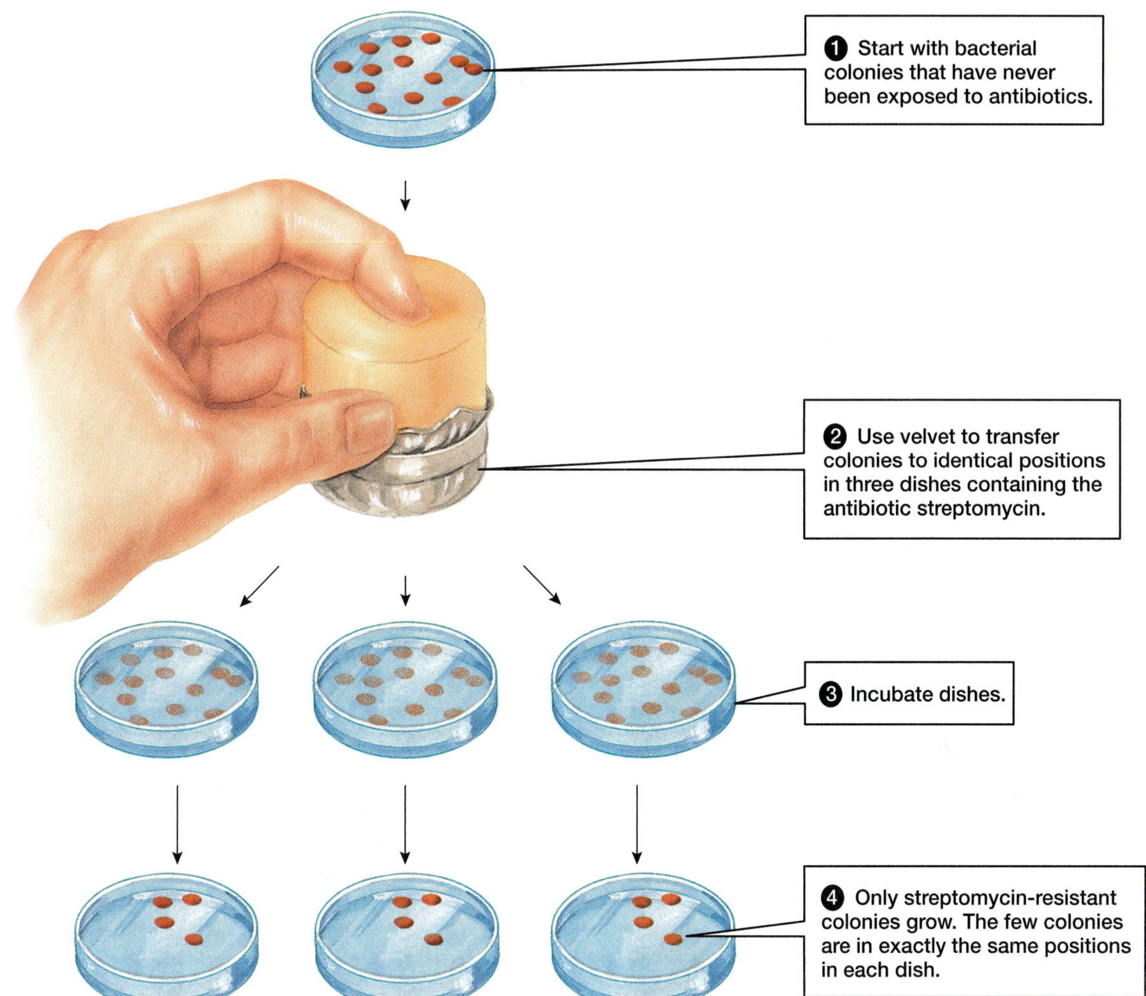

Figure 15-3 Mutation experiment

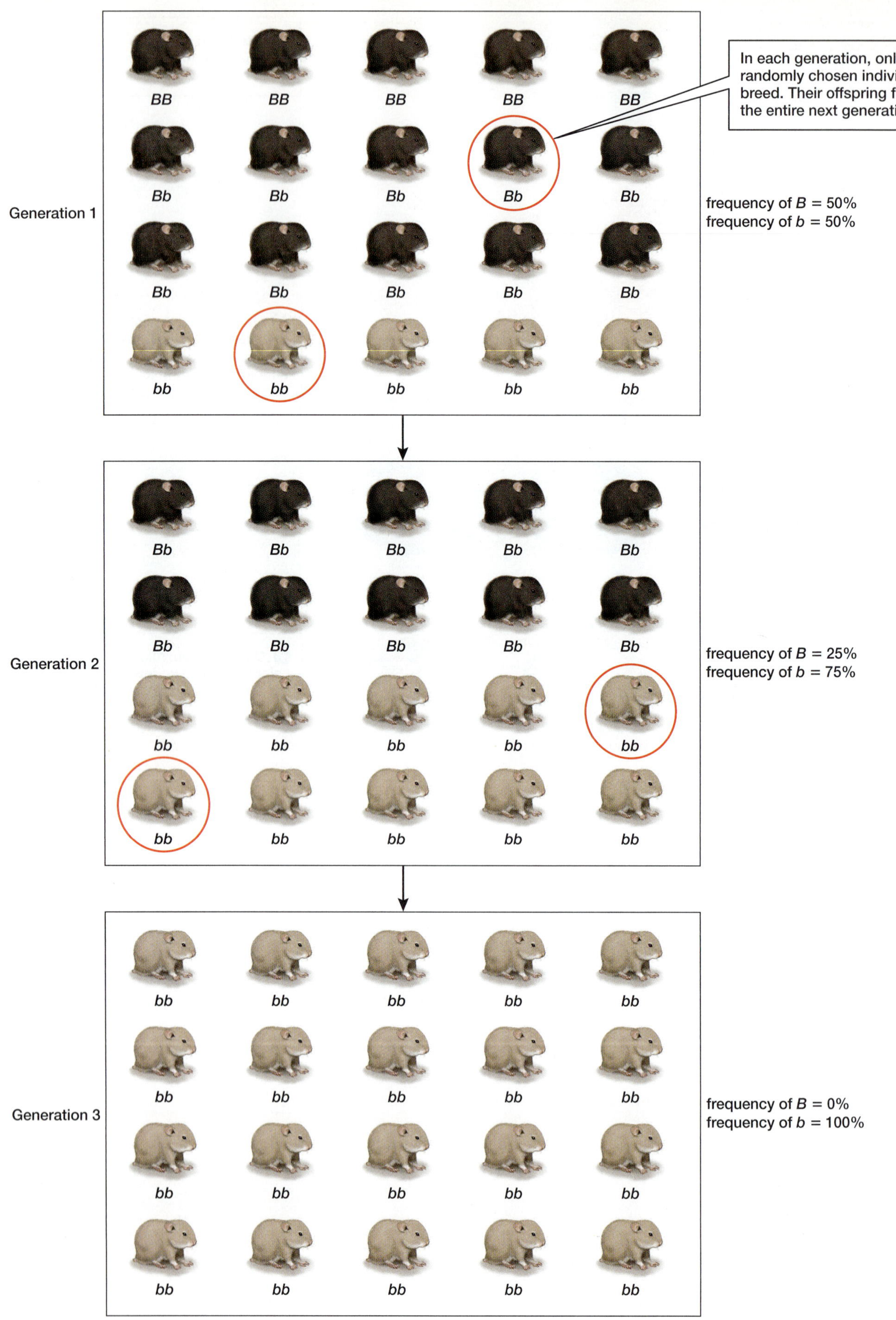

Figure 15-5 Genetic drift

(a) Population size = 10,000

1.0
0.9
0.8
0.7
0.6
0.5
0.4
0.3
0.2
0.1
0.0

frequency of allele *A*

In the large population, allele frequencies remain relatively constant.

0 1 2 3 4 5 6

generation

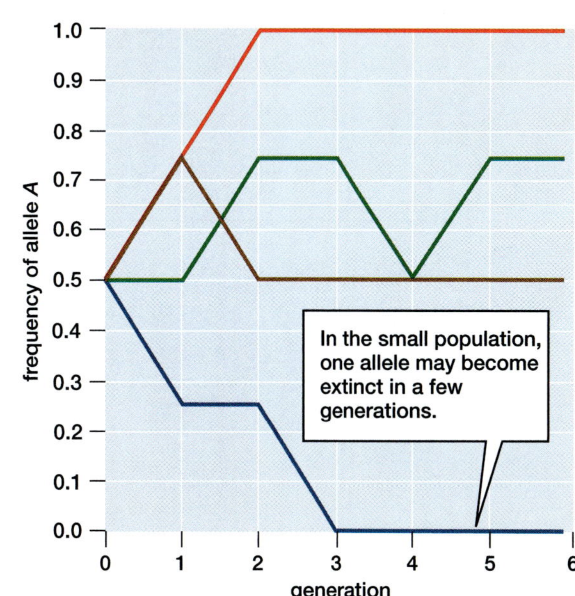

Figure 15-6 Genetic drift graphs

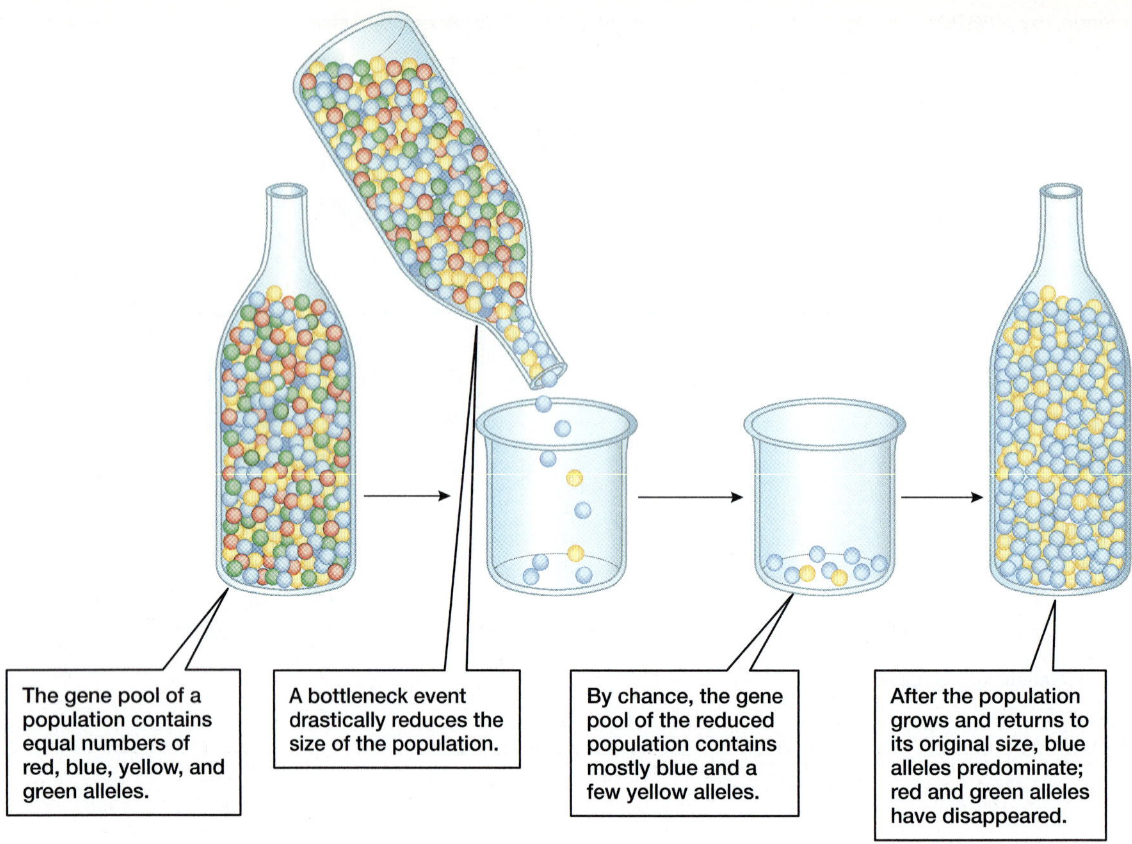

Figure 15-7a Bottleneck

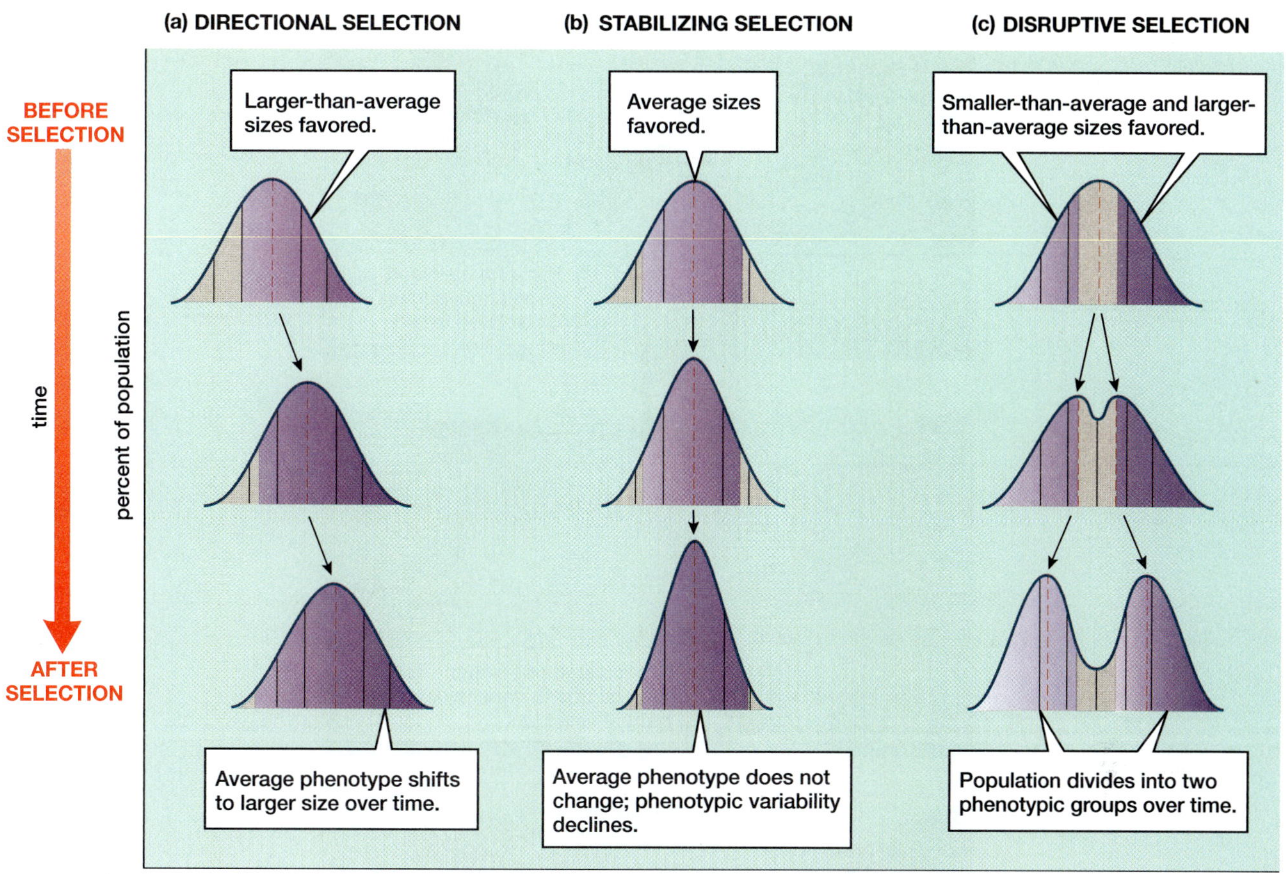

Figure 15-13 Modes of selection

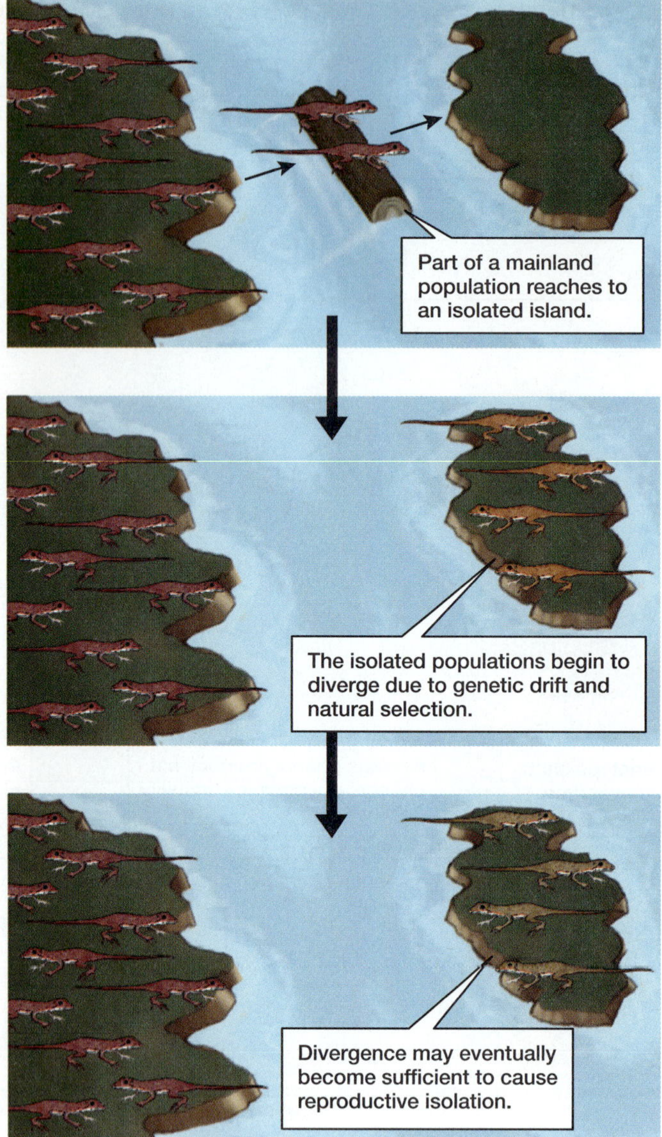

Figure 16-9 Allopatric speciation

Figure 16-10 Sympatric speciation

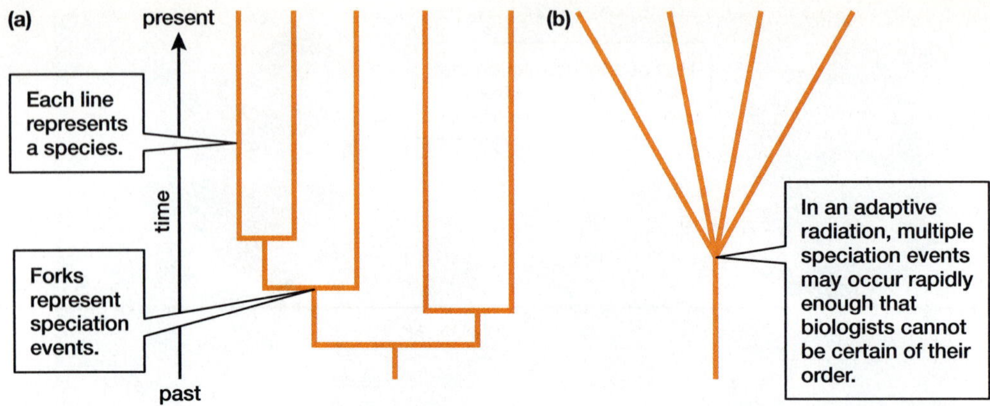

Figure 16-11 Evolutionary trees

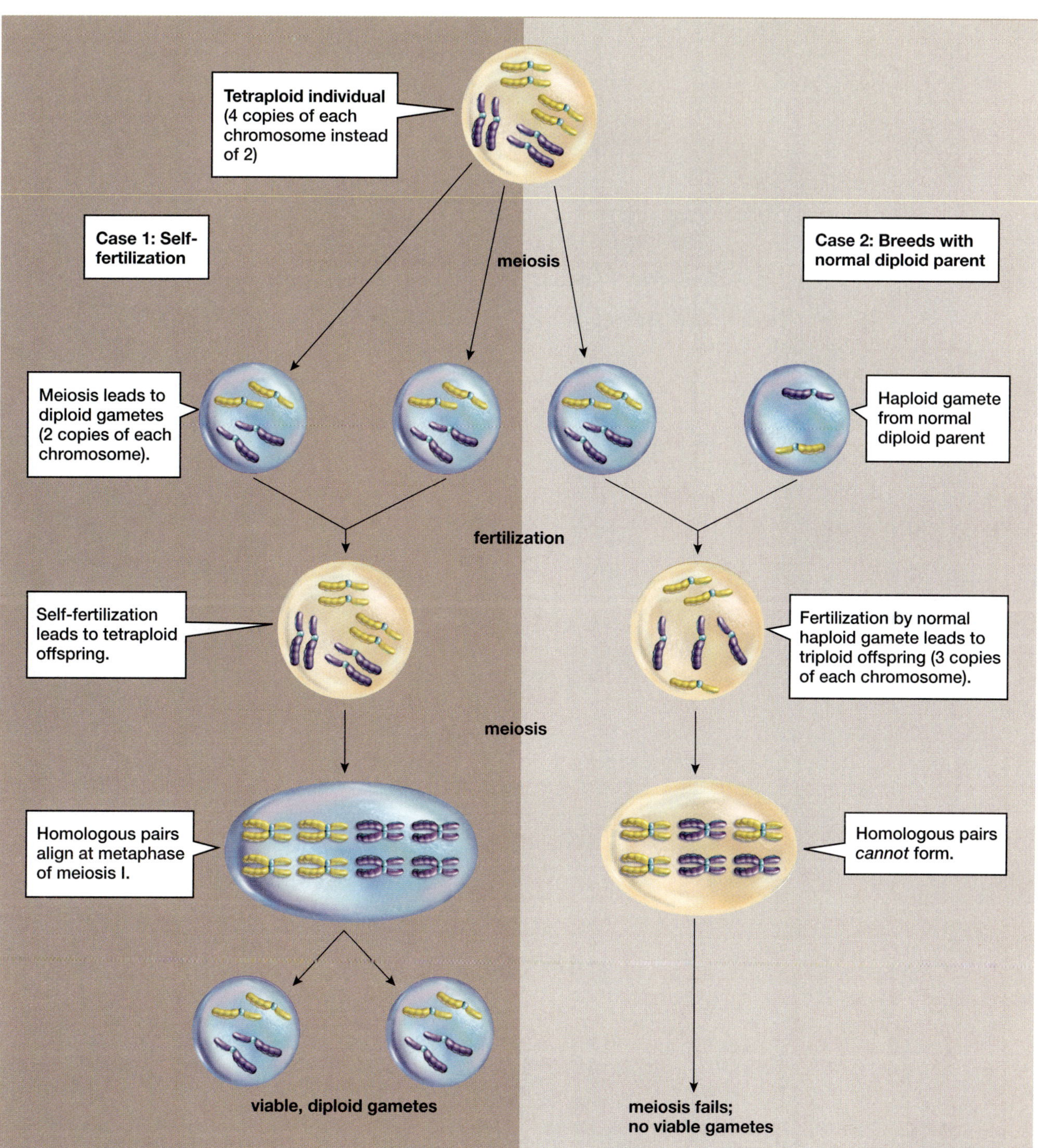

Figure E16-2 Polyploidy

Figure E16-3 Golden Palace monkey

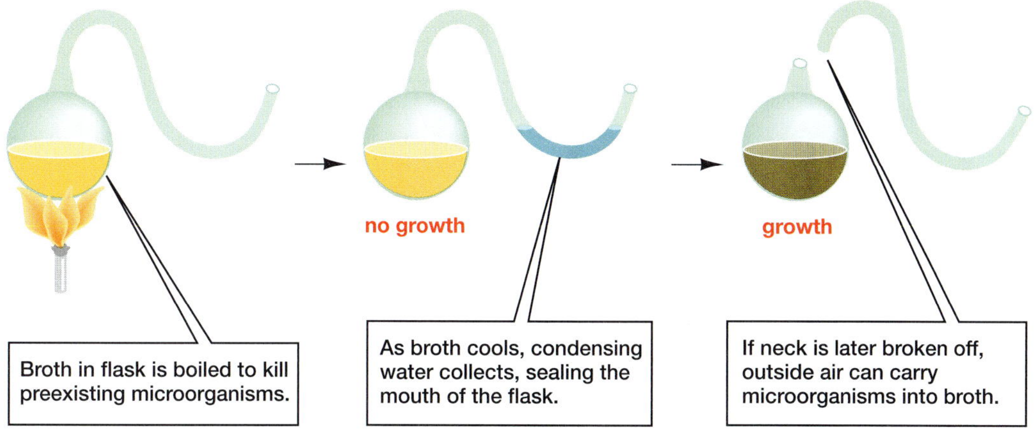

Figure 17-1 Spontaneous generation refuted

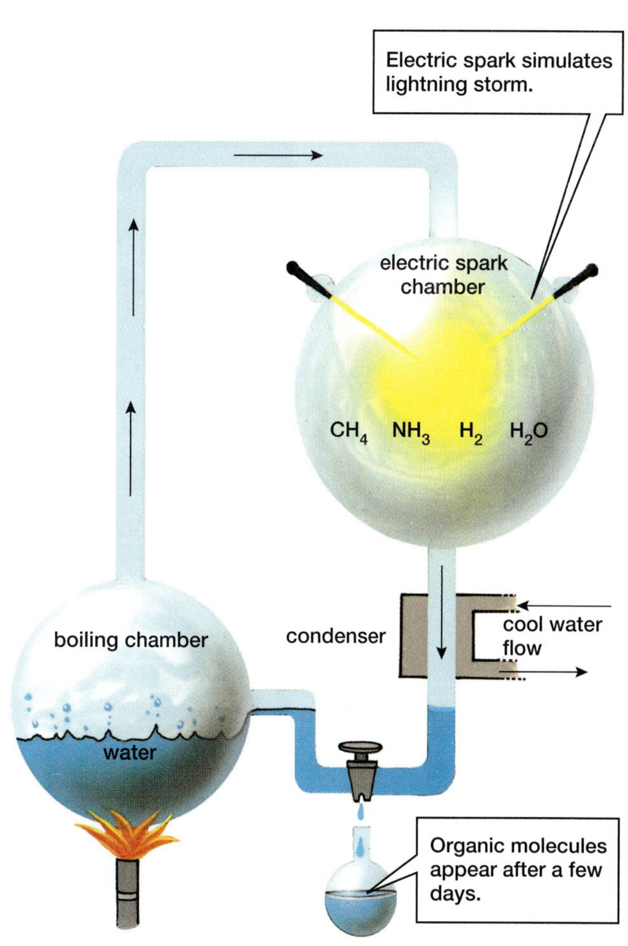

Figure 17-2 Miller-Urey experiment

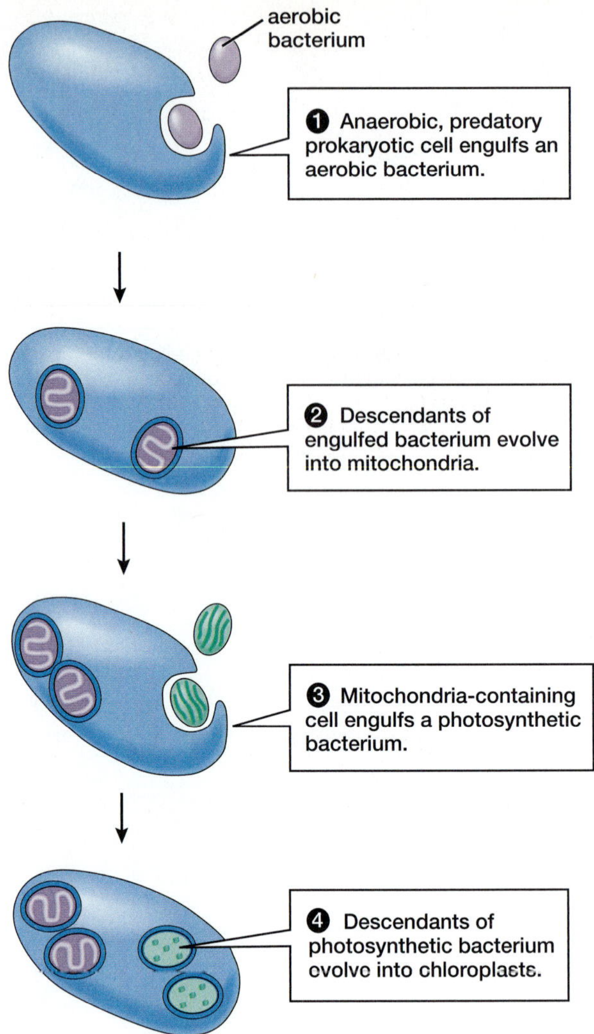

Figure 17-4 Endosymbiotic hypothesis

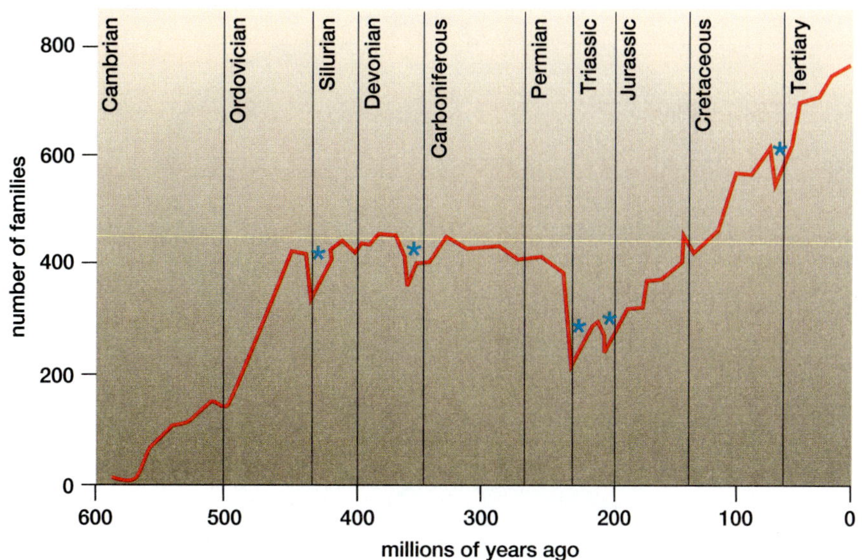

Figure 17-10 Mass extinctions

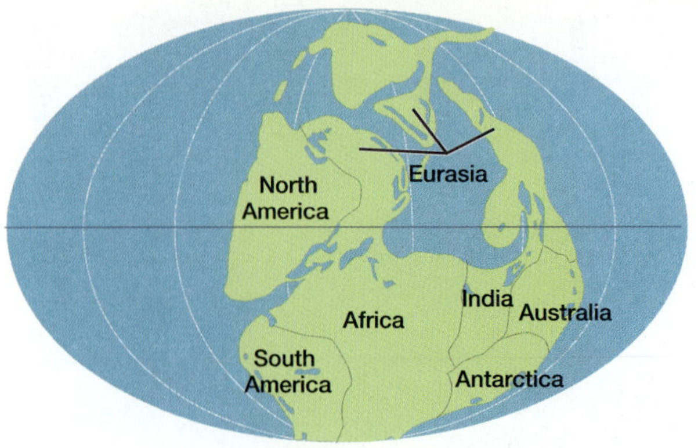

(a) 340 million years ago

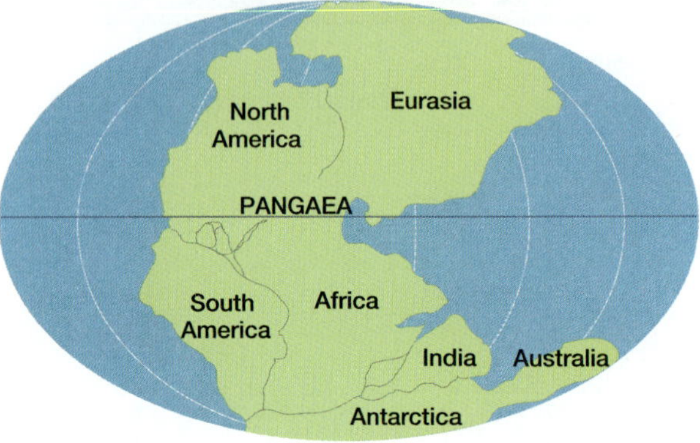

(b) 225 million years ago

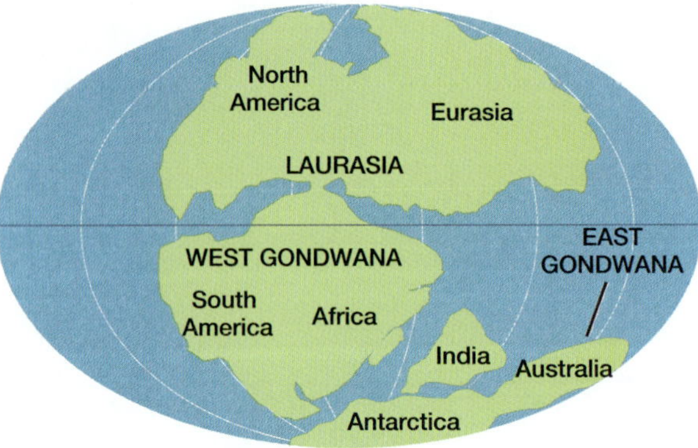

(c) 135 million years ago

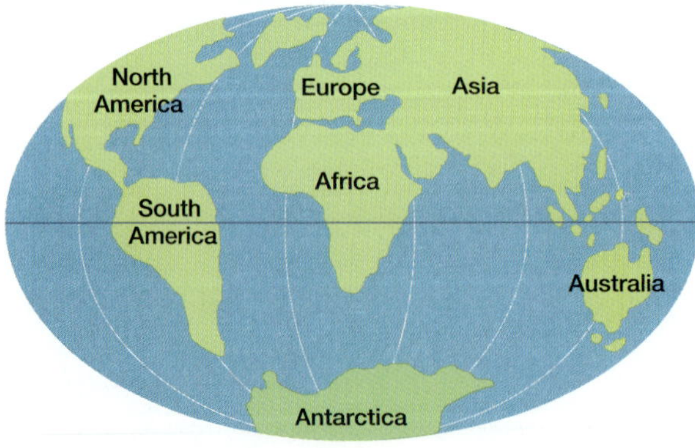

(d) Present

Figure 17-11 Plate tectonics

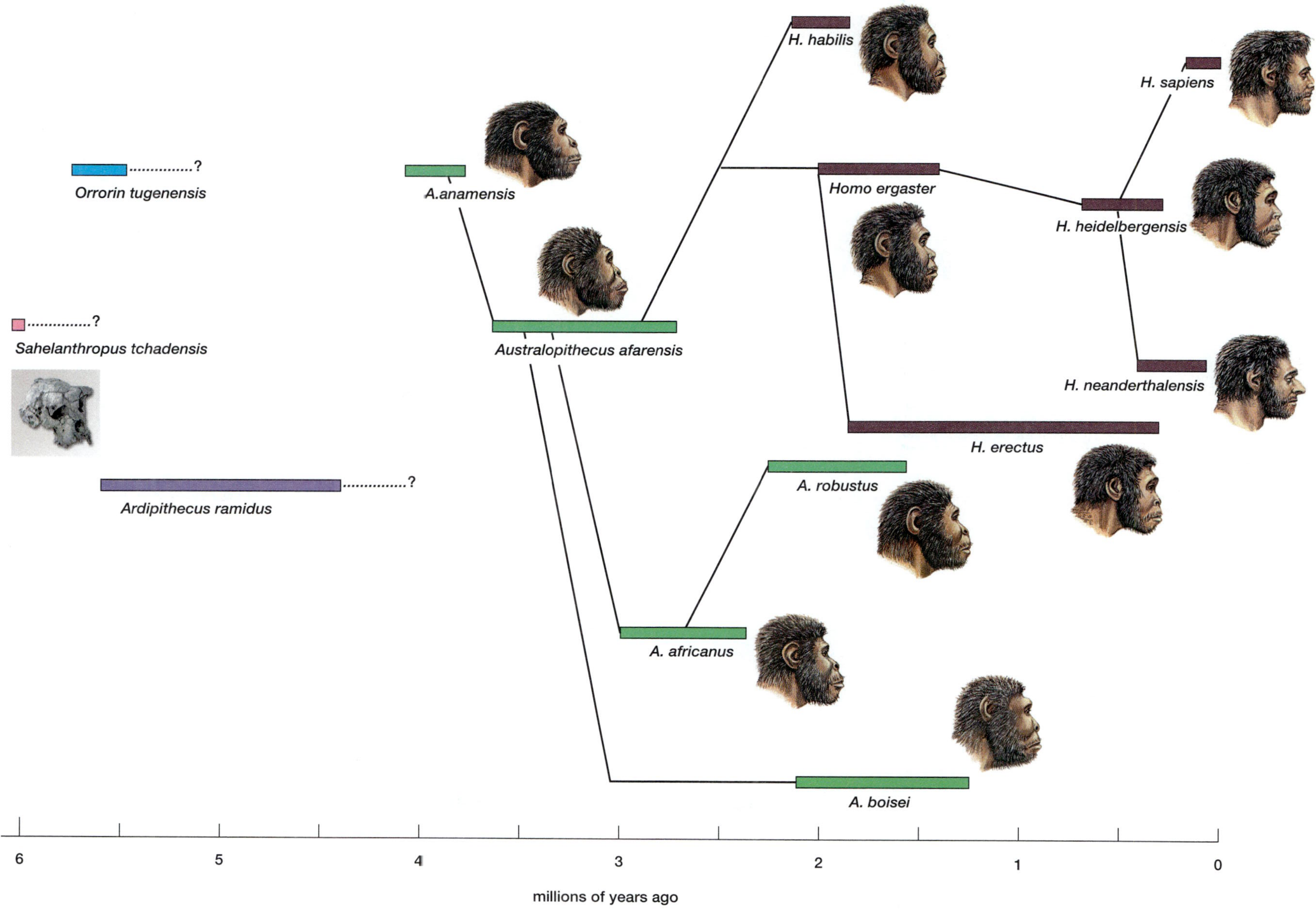

Figure 17-14 Human evolutionary tree

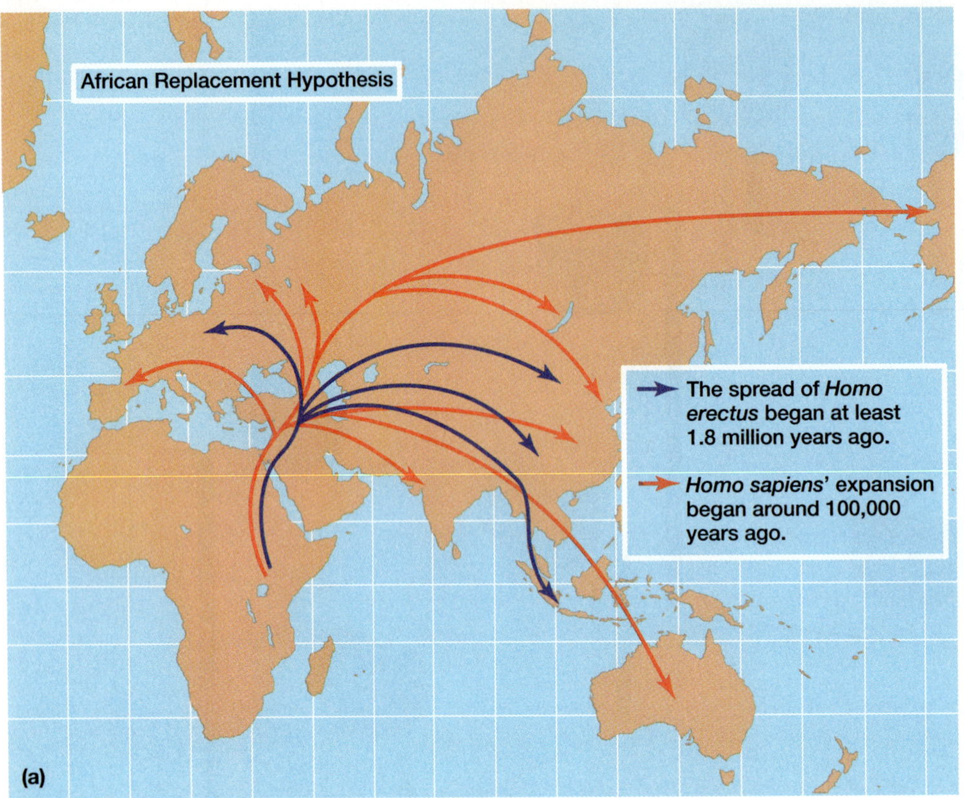

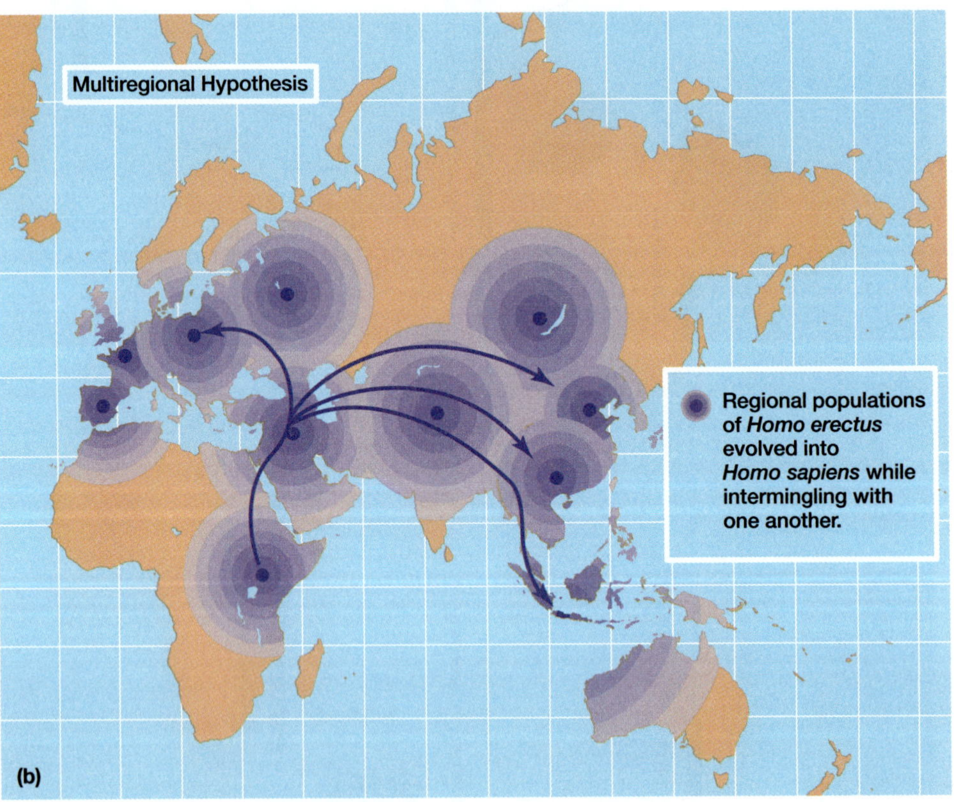

Figure 17-18 Human origin hypothesis maps

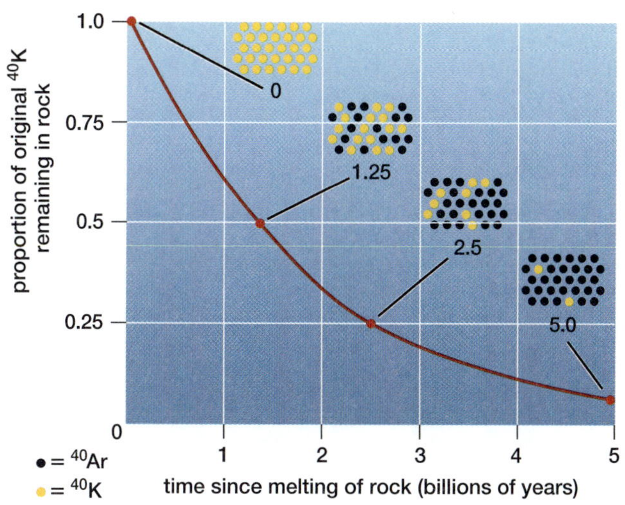

Figure E17-1 Time *vs* decay graph

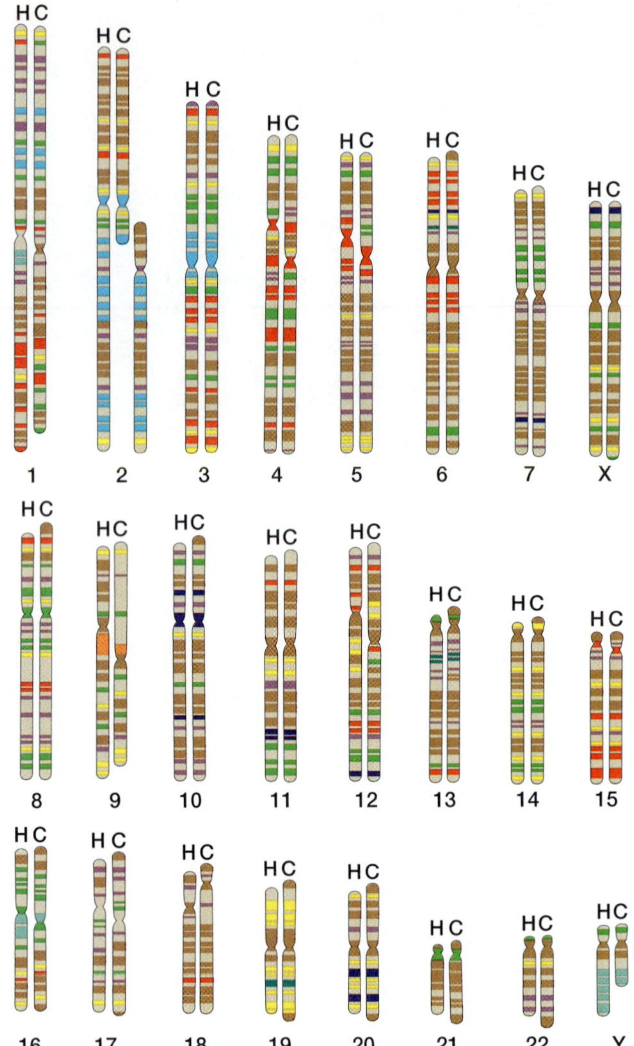

Figure 18-3 Human and chimp banded chromosomes

BACTERIA
ARCHAEA
EUKARYA
animals
fungi
plants
protists

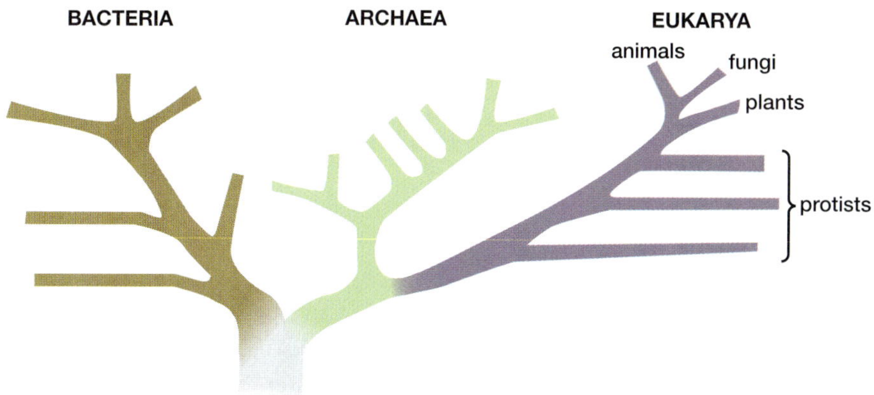

Figure 18-5 Three-domains tree

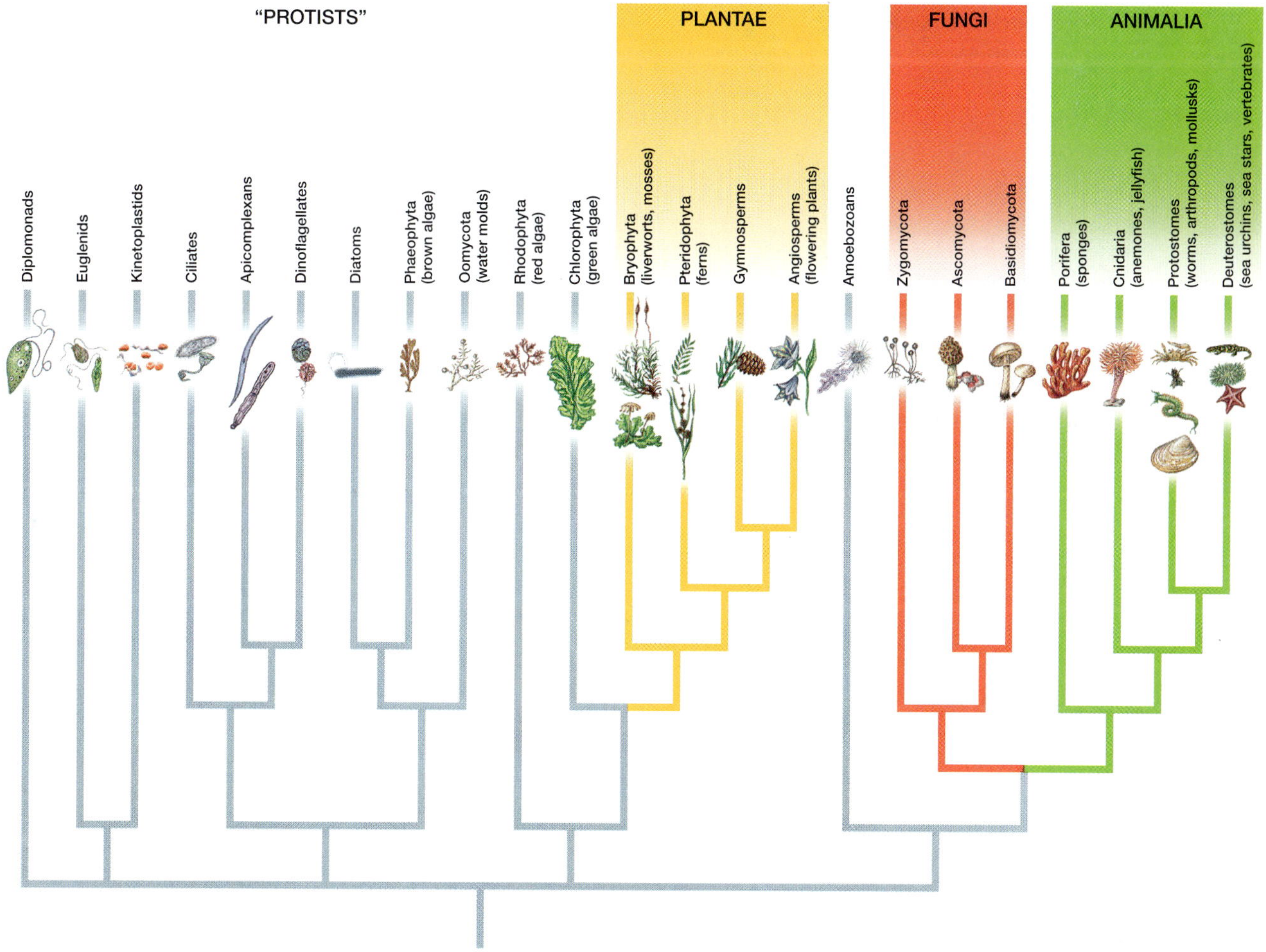

"PROTISTS"
PLANTAE
FUNGI
ANIMALIA
Diplomonads
Euglenids
Kinetoplastids
Ciliates
Apicomplexans
Dinoflagellates
Diatoms
Phaeophyta (brown algae)
Oomycota (water molds)
Rhodophyta (red algae)
Chlorophyta (green algae)
Bryophyta (liverworts, mosses)
Pteridophyta (ferns)
Gymnosperms
Angiosperms (flowering plants)
Amoebozoans
Zygomycota
Ascomycota
Basidiomycota
Porifera (sponges)
Cnidaria (anemones, jellyfish)
Protostomes (worms, arthropods, mollusks)
Deuterostomes (sea urchins, sea stars, vertebrates)

Figure 18-6 Eukaryote tree

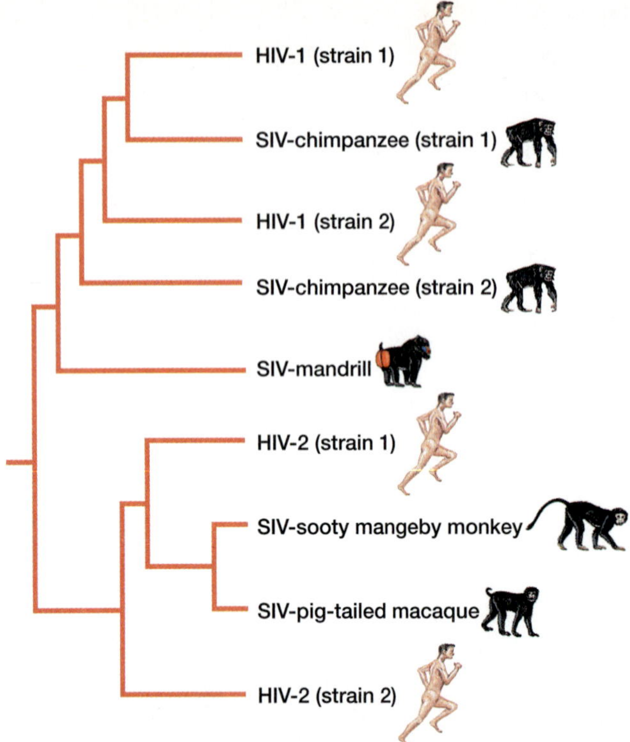

Figure 18-8 HIV tree

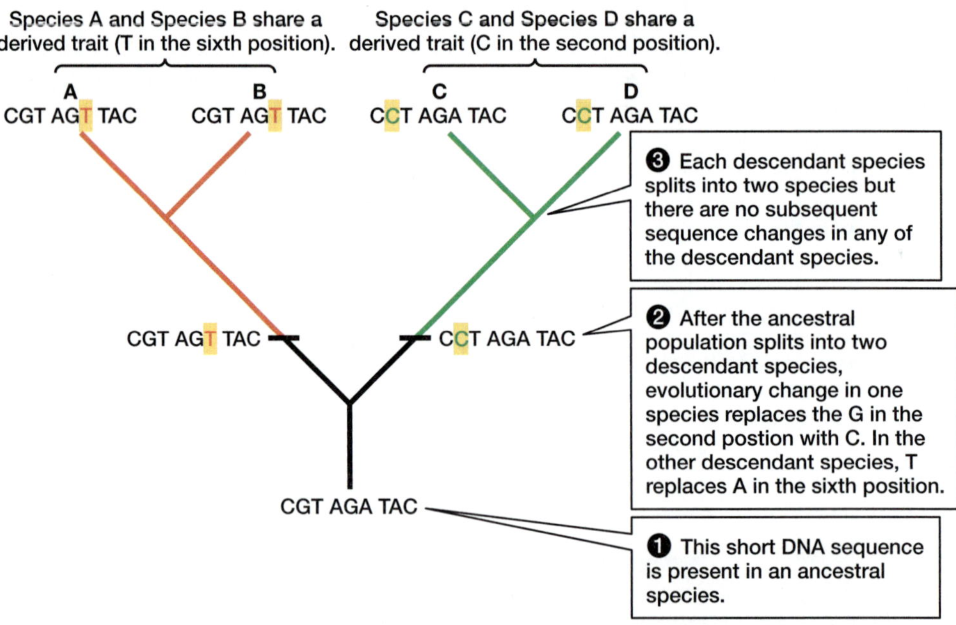

Figure E18-1 Synapomorphies

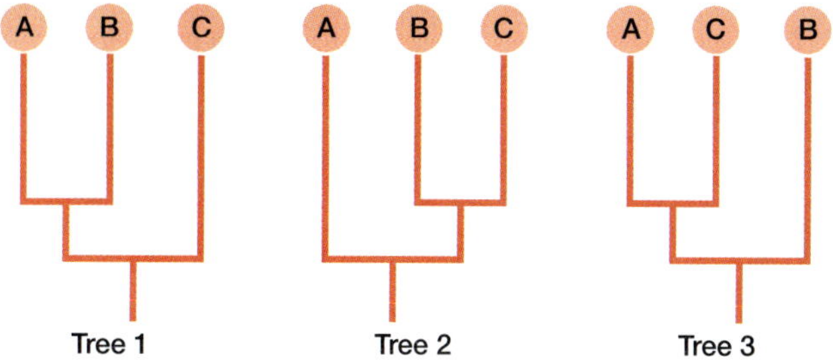

Figure E18-2 Monophyletic groups

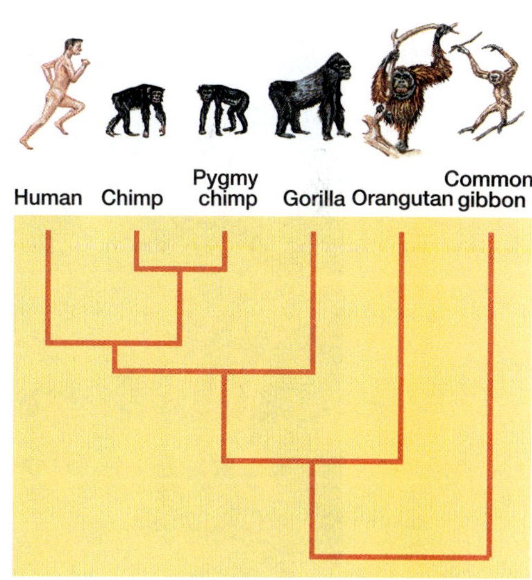

Figure E18-4 Human/ape tree

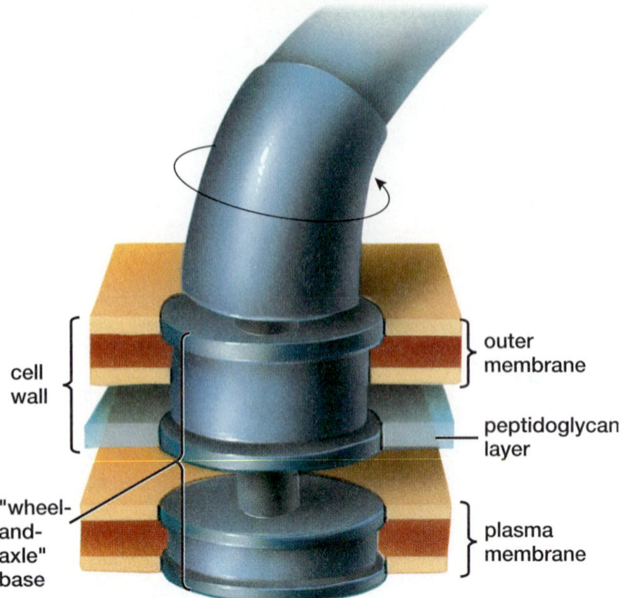

Figure 19-2b Bacterial flagellum

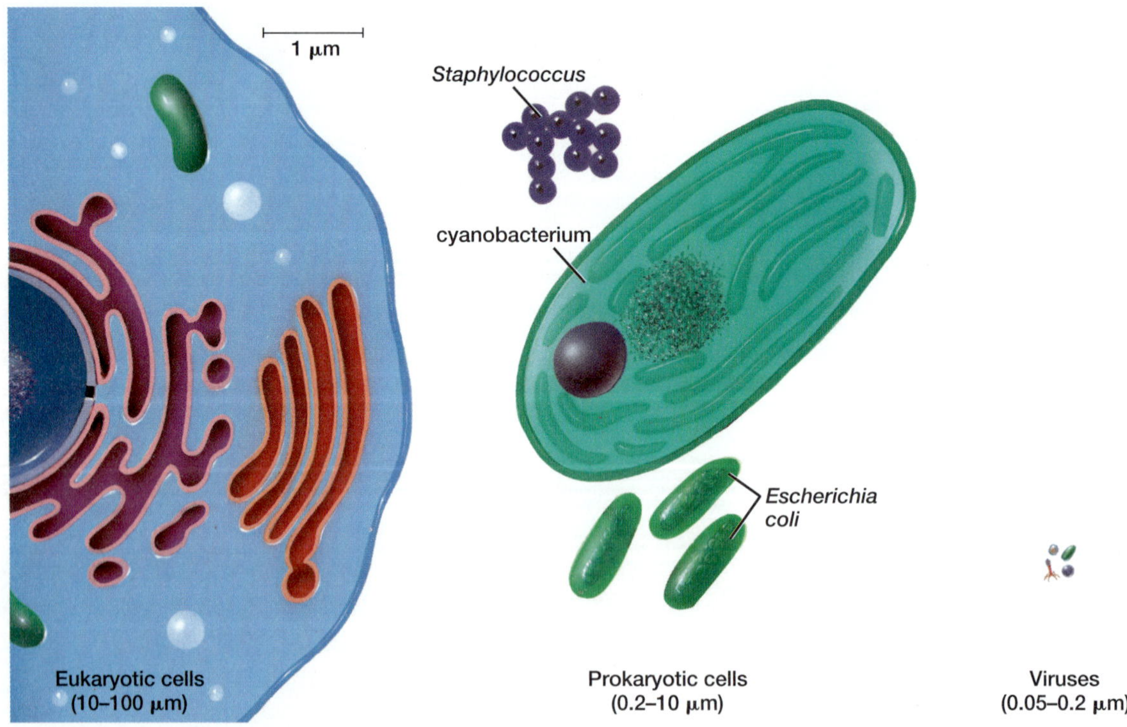

Figure 19-10 Sizes of microorganisms

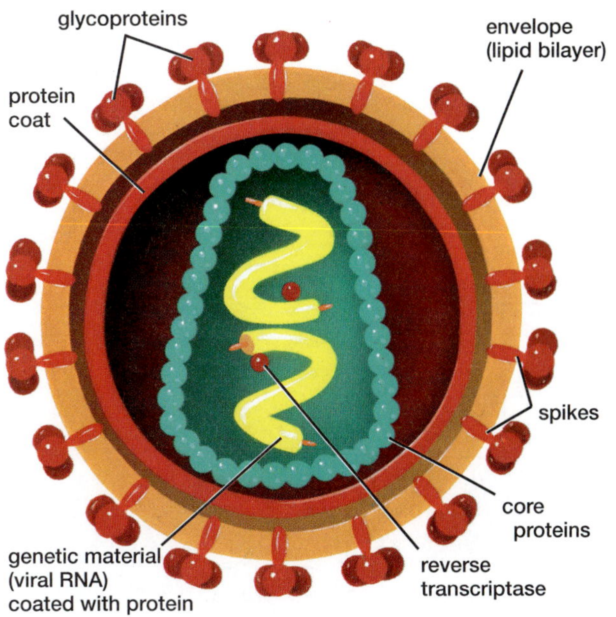

Figure 19-12 Structure of HIV

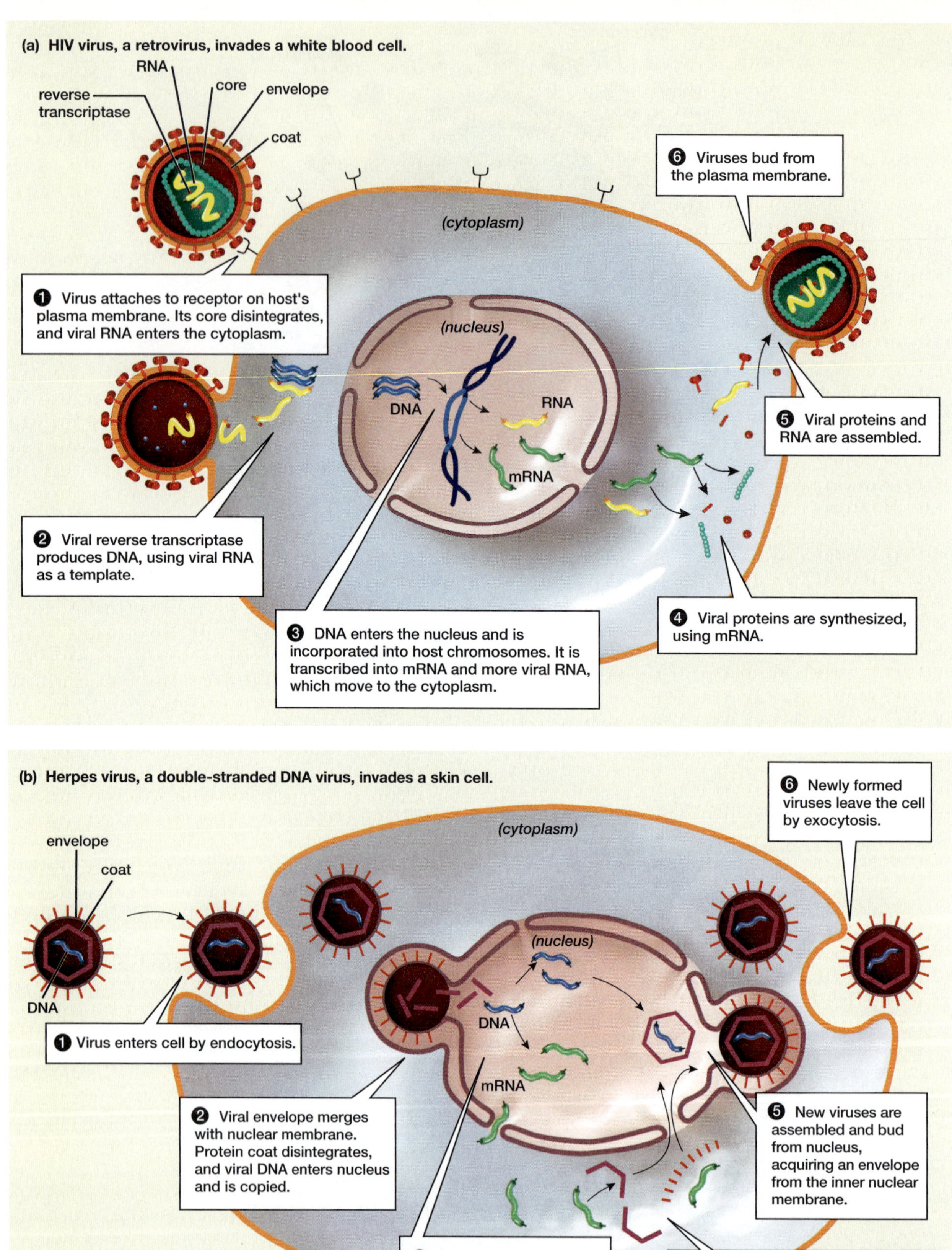

Figure E19-2 Viral replication

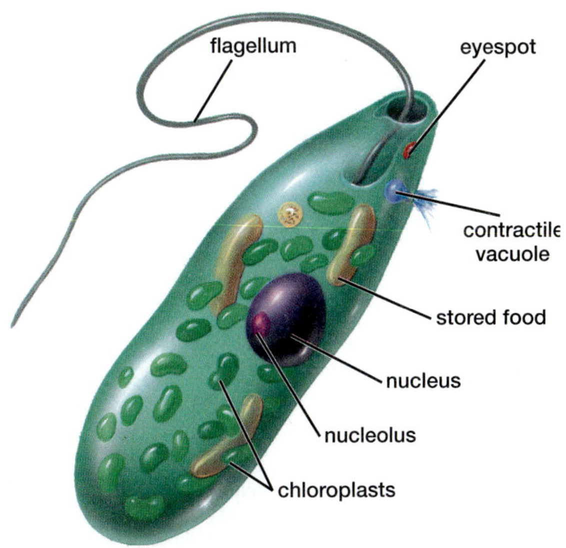

Figure 20-5 *Euglena*

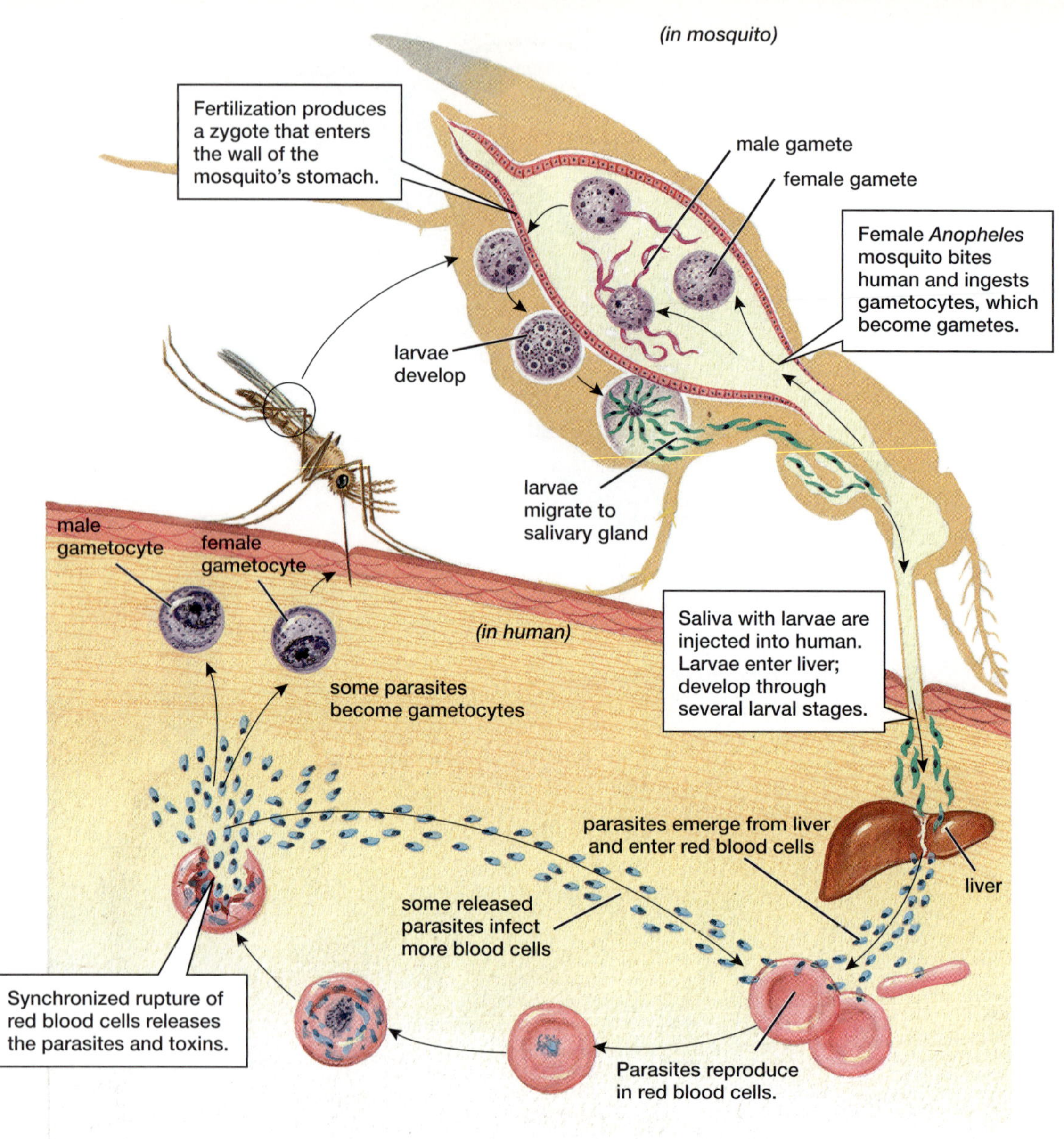

Figure 20-12 Malaria life cycle

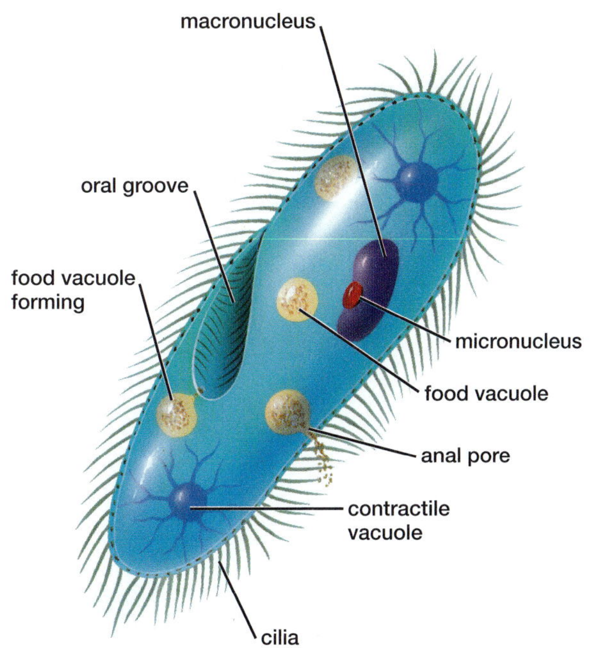

Figure 20-13 *Paramecium*

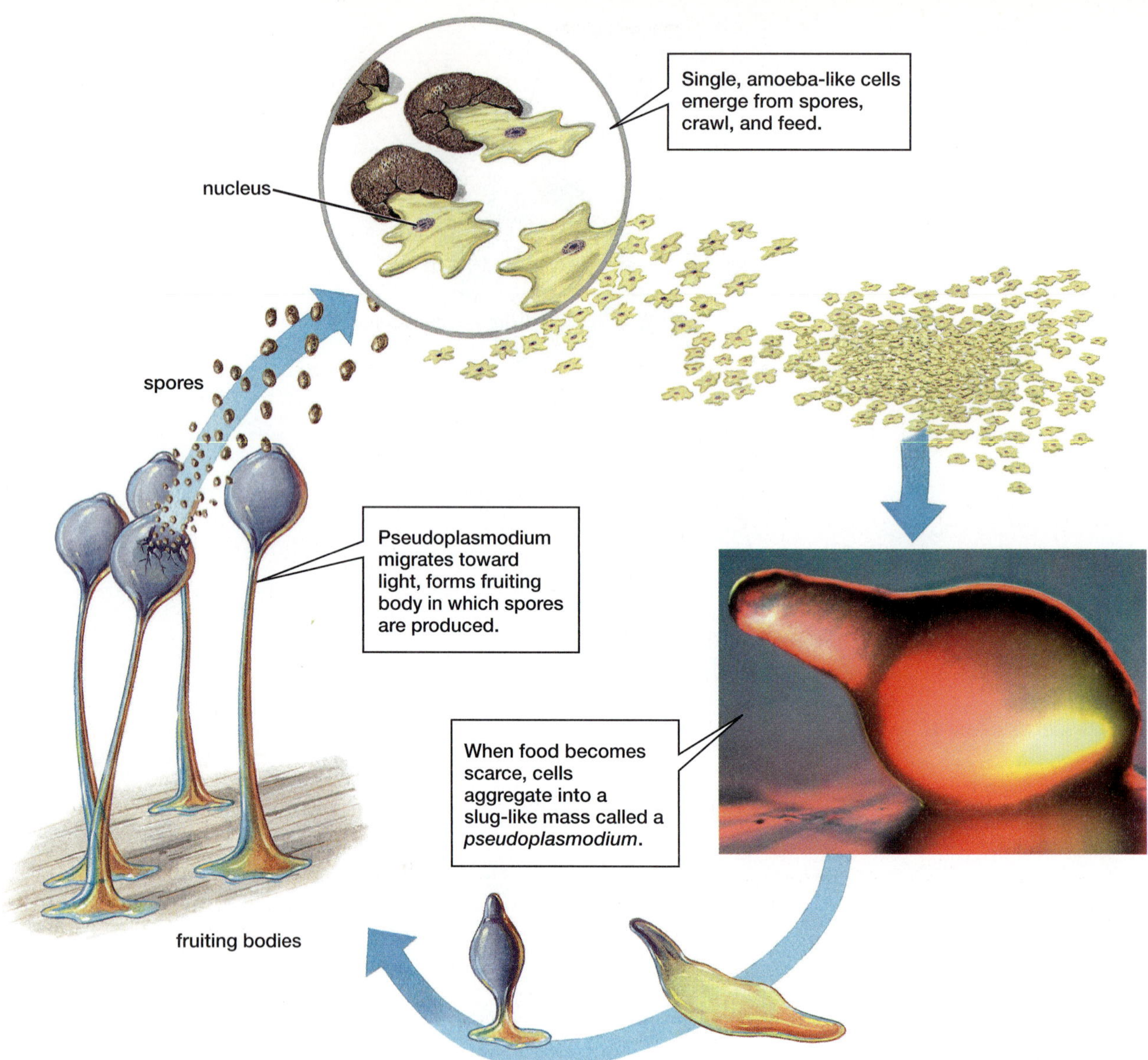

Figure 20-18 Slime mold life cycle

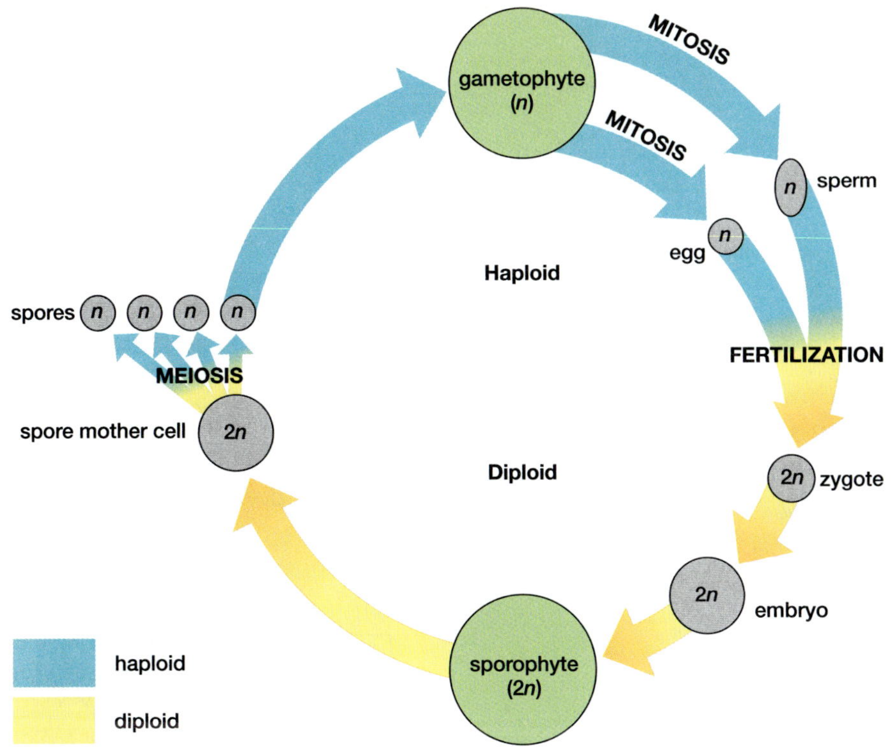

Figure 21-1 Generalized plant life cycle

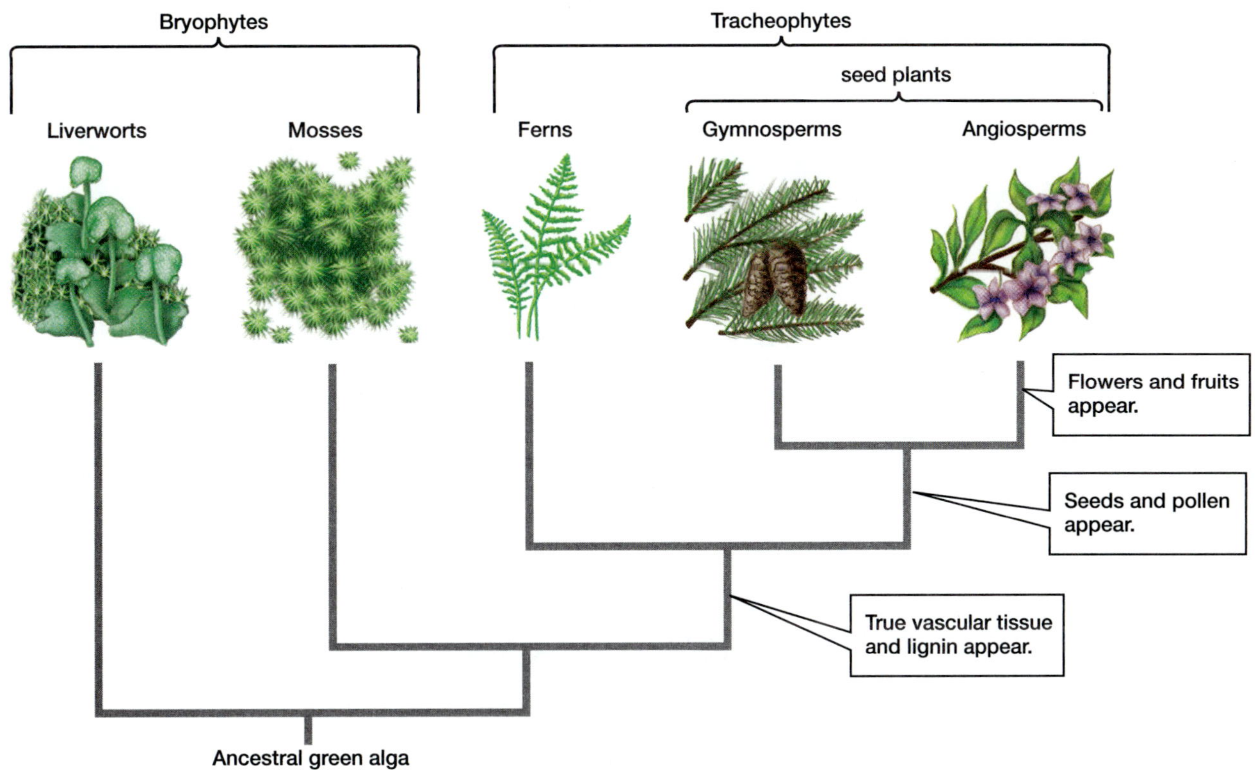

Figure 21-2 Family tree of plants

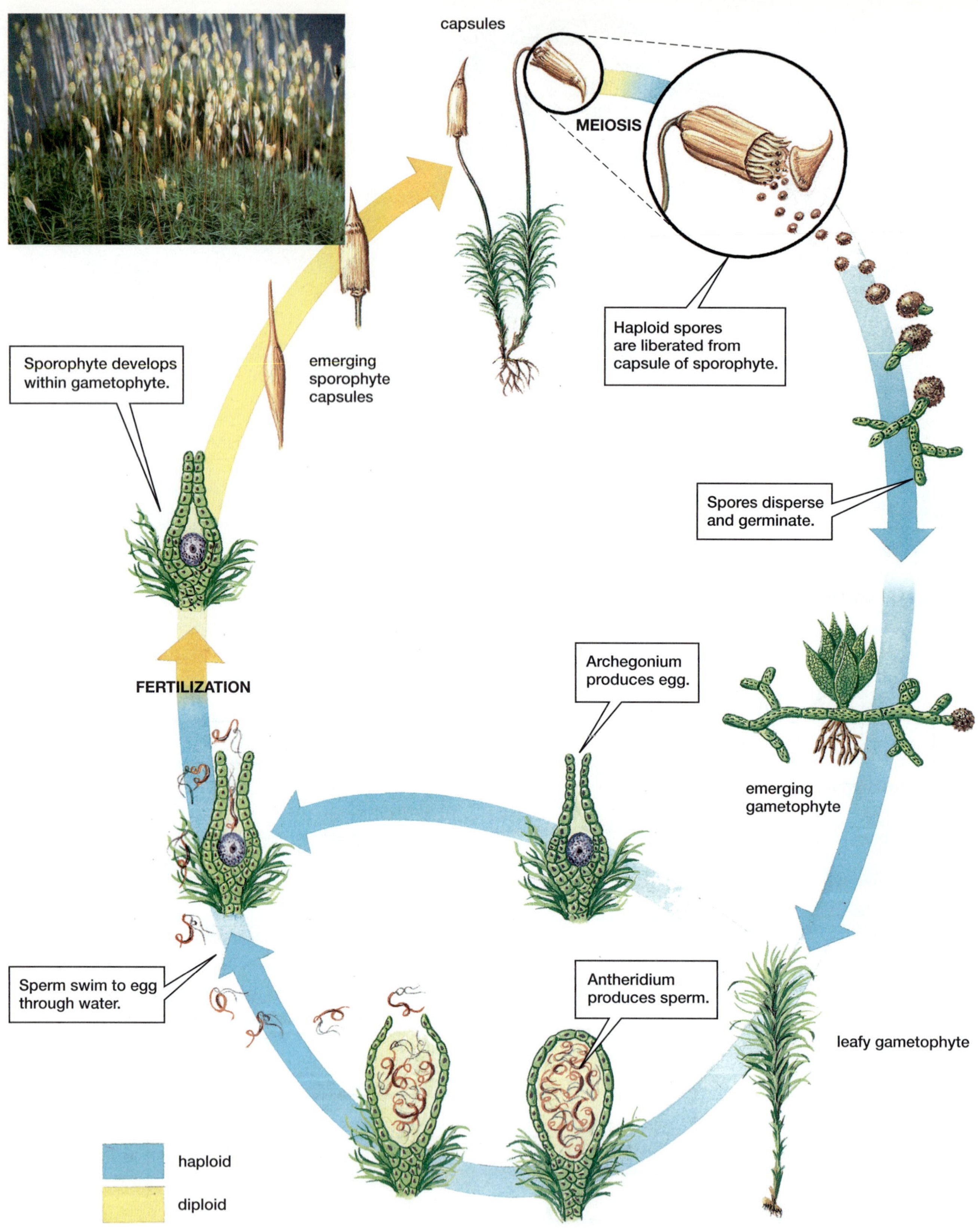

Figure 21-4 Moss life cycle

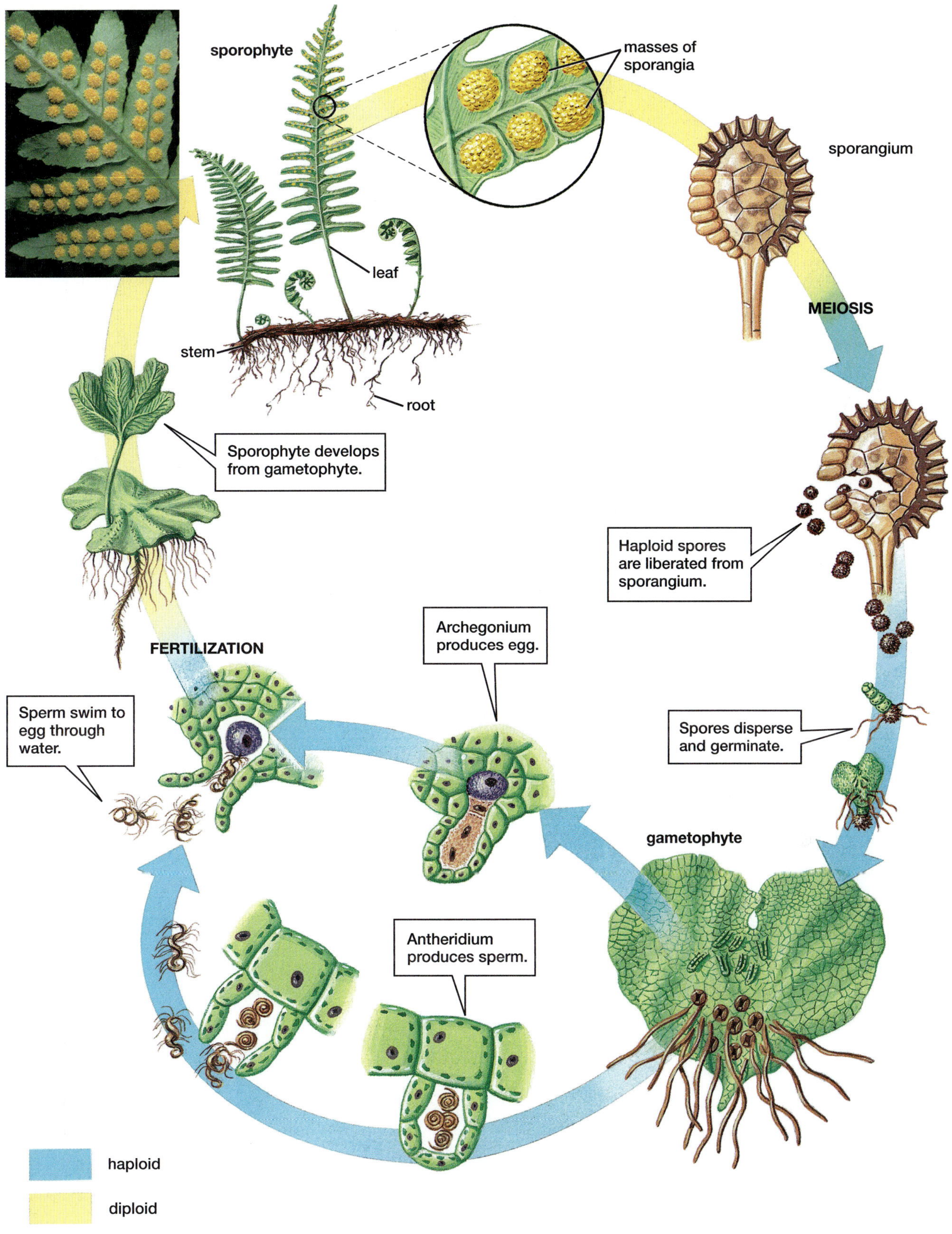

Figure 21-6 Fern life cycle

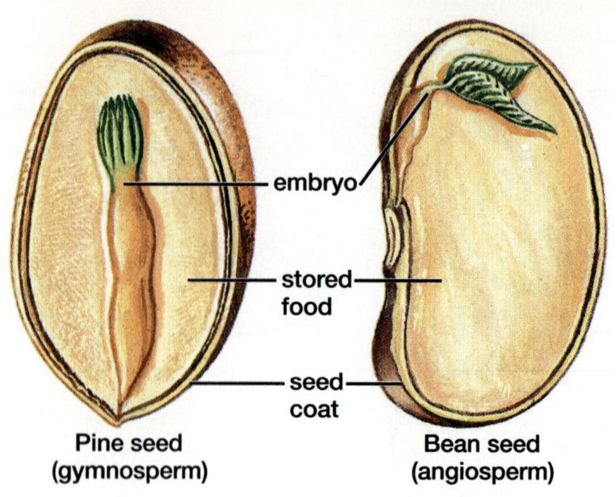

Figure 21-7 Pine seed, bean seed

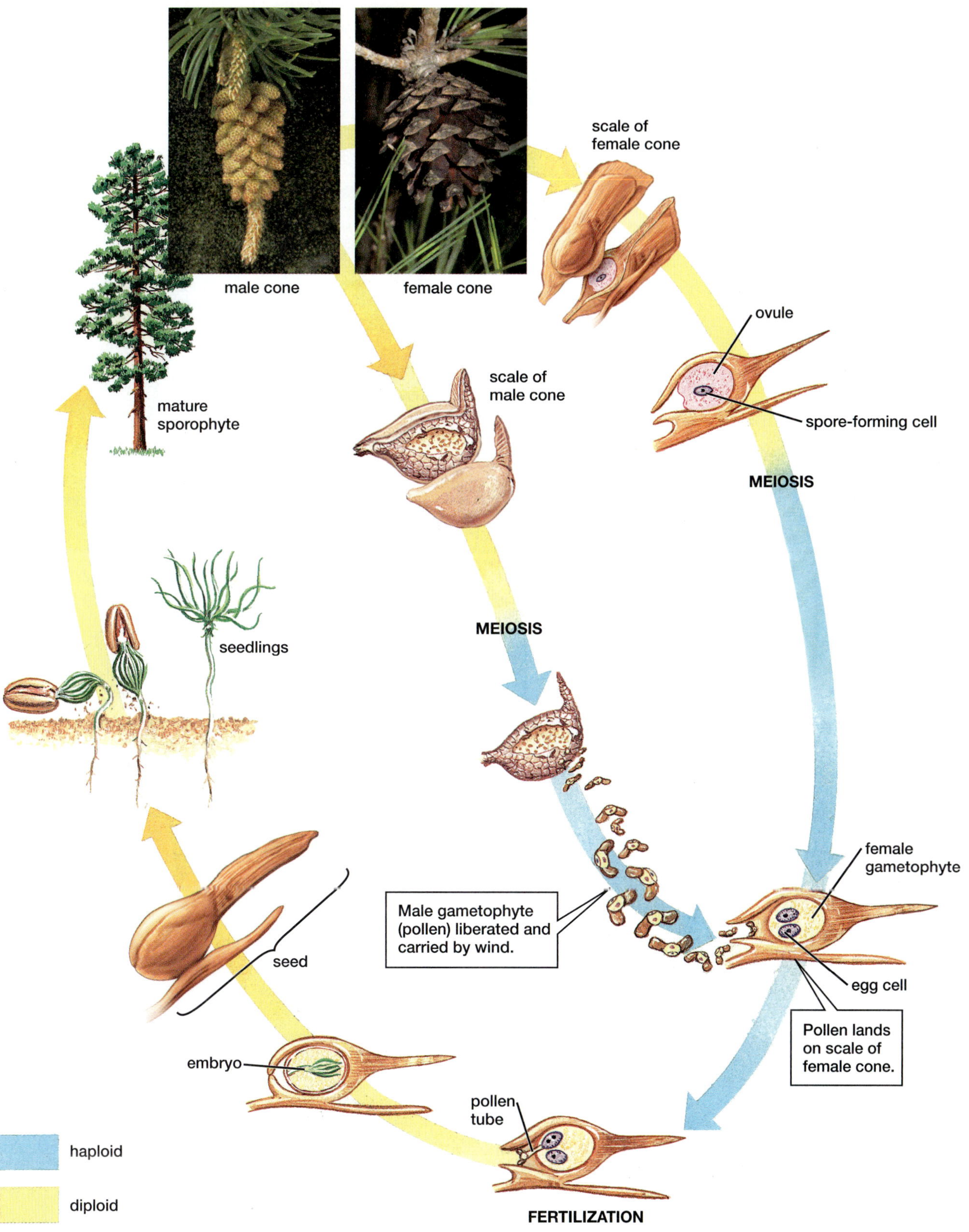

Figure 21-9 Pine life cycle

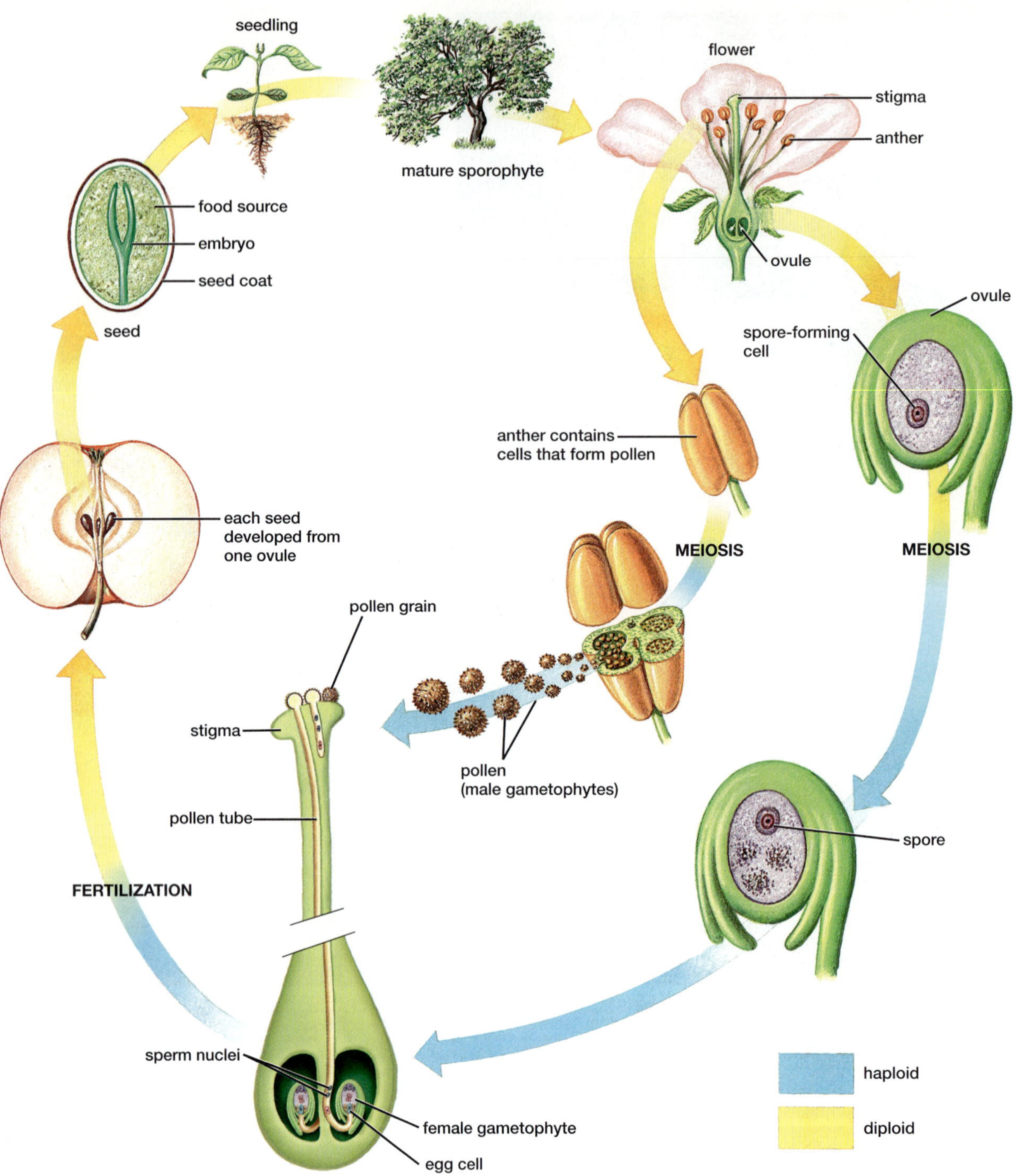

Figure 21-11 Flowering plant life cycle

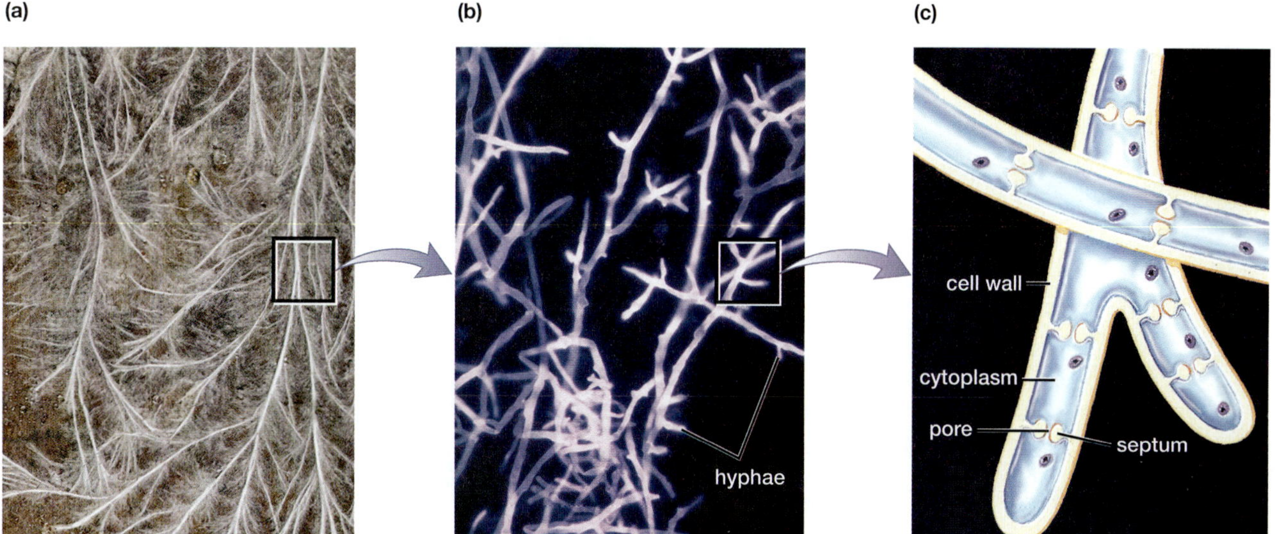

Figure 22-1 Fungal mycelia

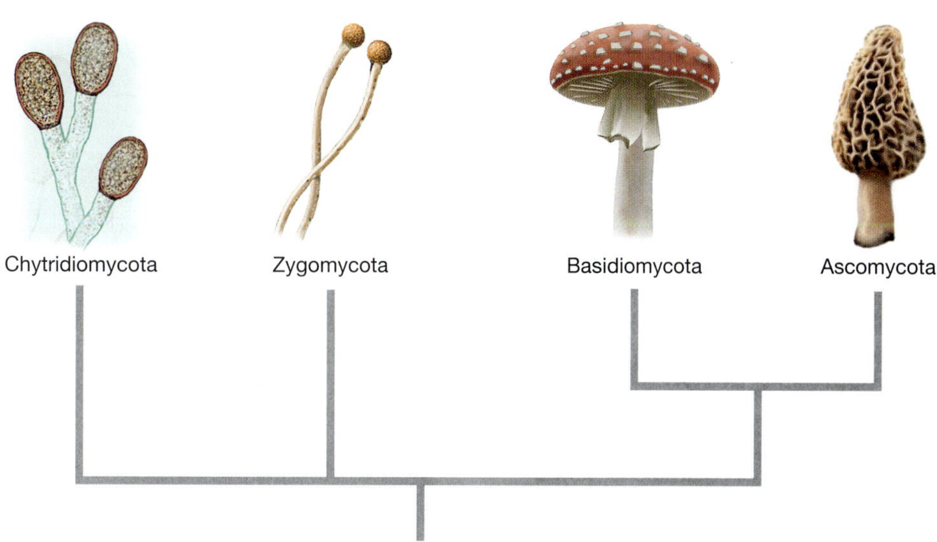

Figure 22-3 Fungi evolutionary tree

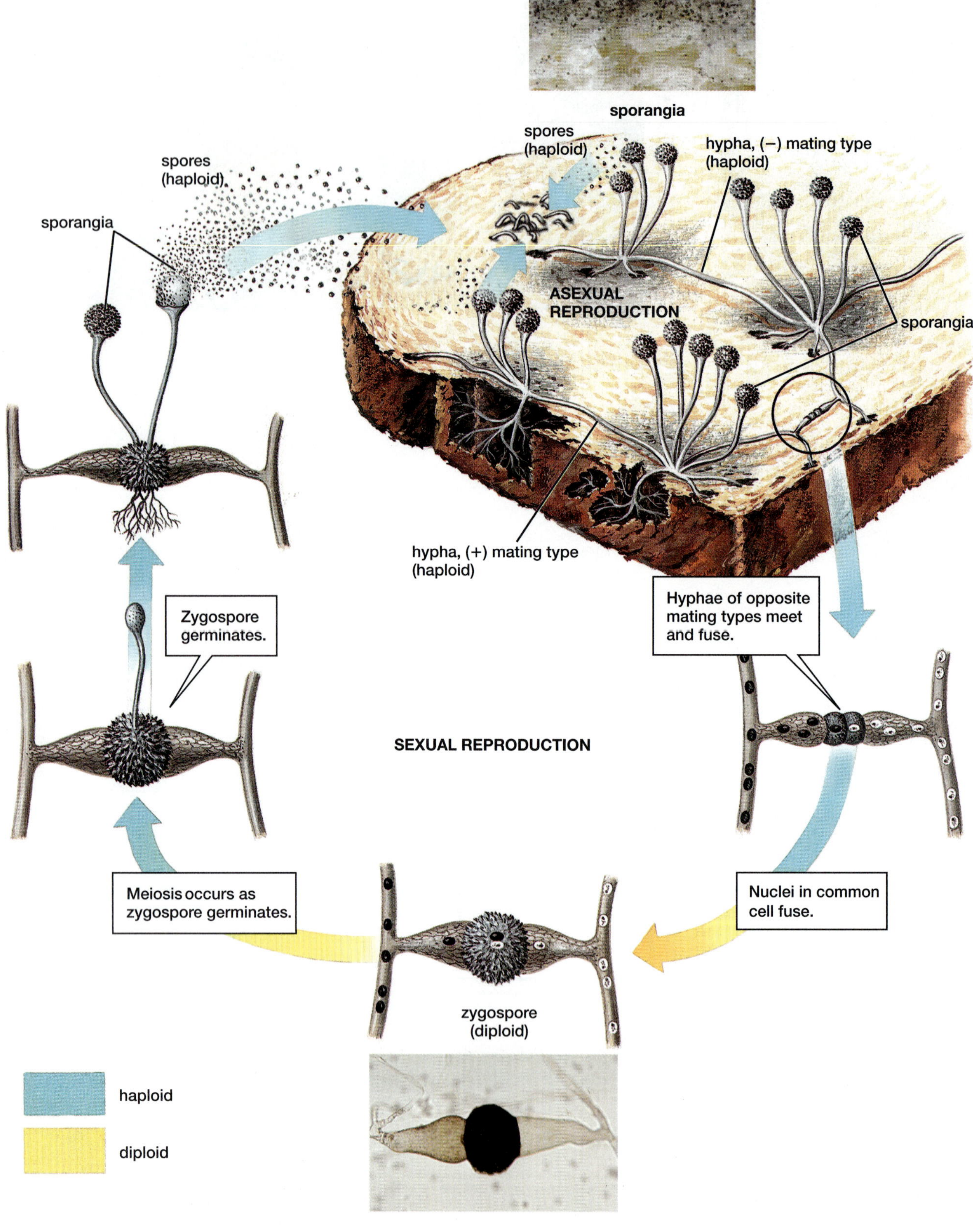

Figure 22-5 Bread mold life cycle

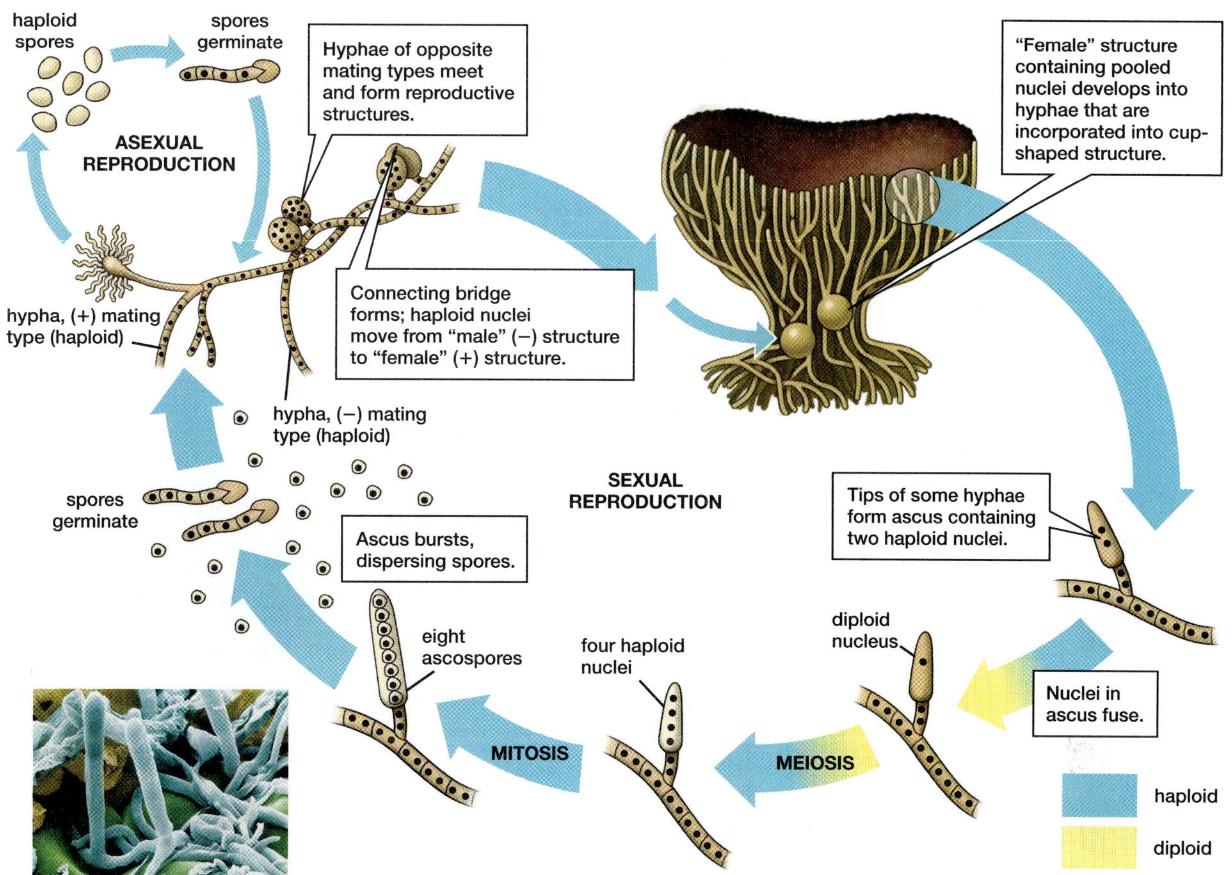

Figure 22-6 Ascomycete life cycle

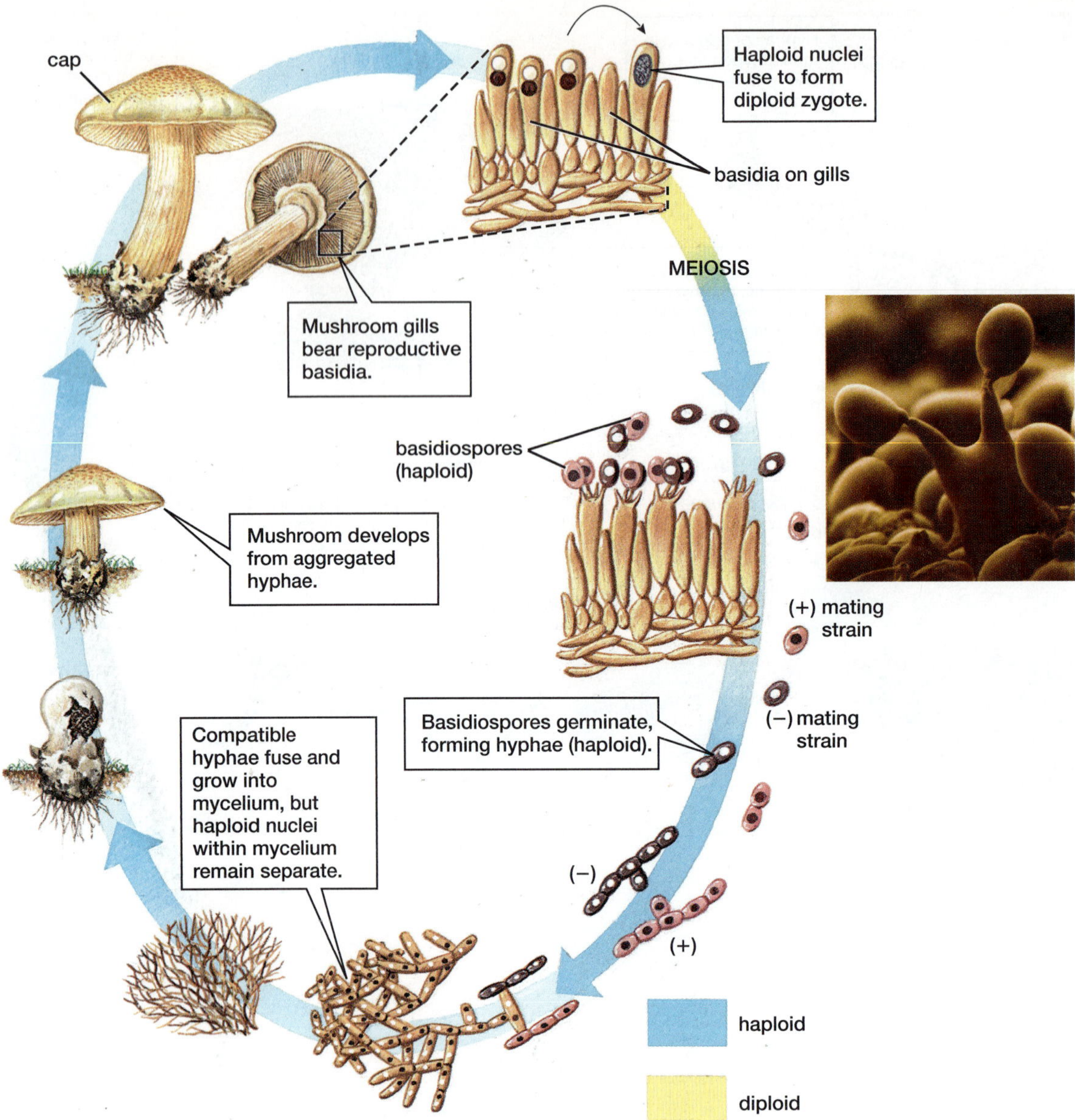

Figure 22-8 Basidiomycete life cycle

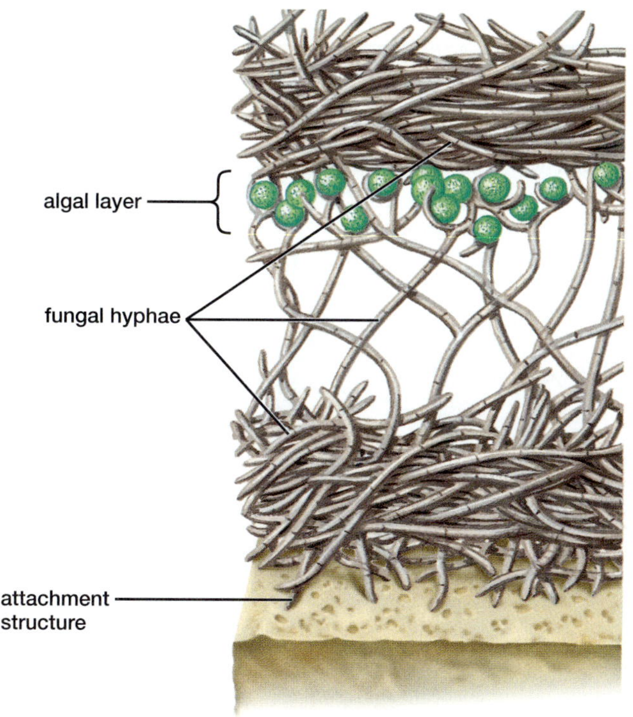

Figure 22-11 Lichen structure

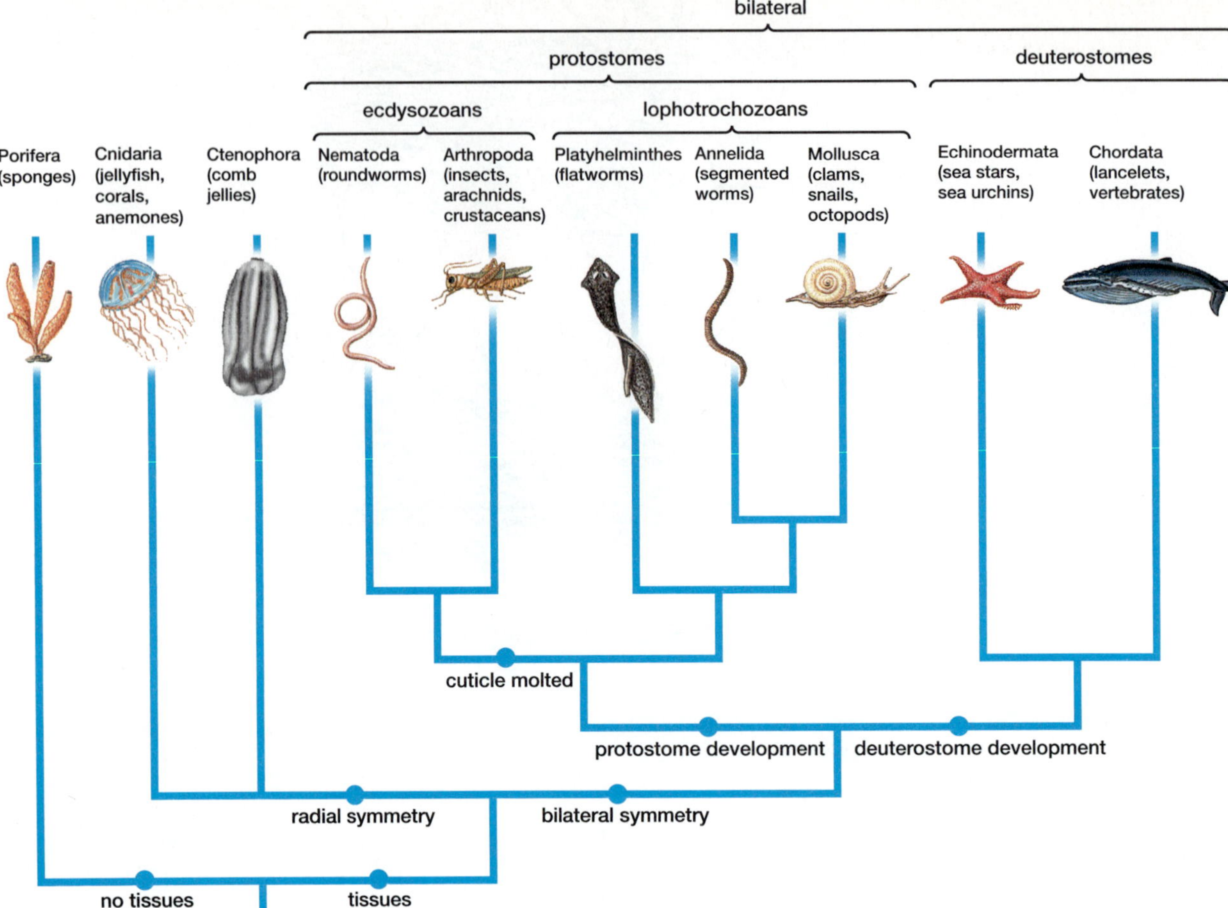

Figure 23-1 Animal evolutionary tree

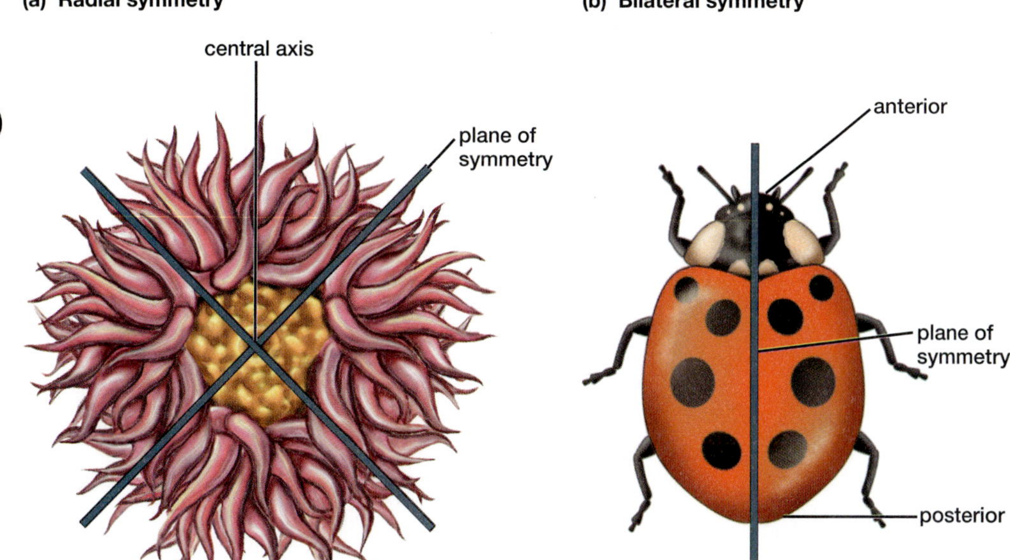

Figure 23-2 Radial and bilateral symmetry

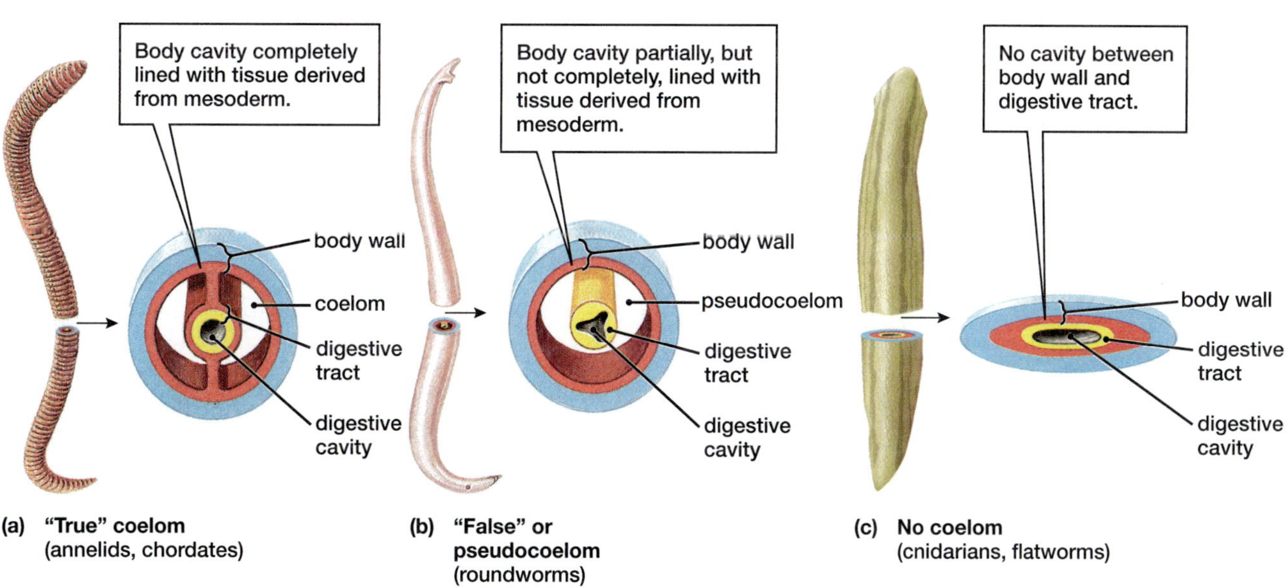

Figure 23-3 Body cavities

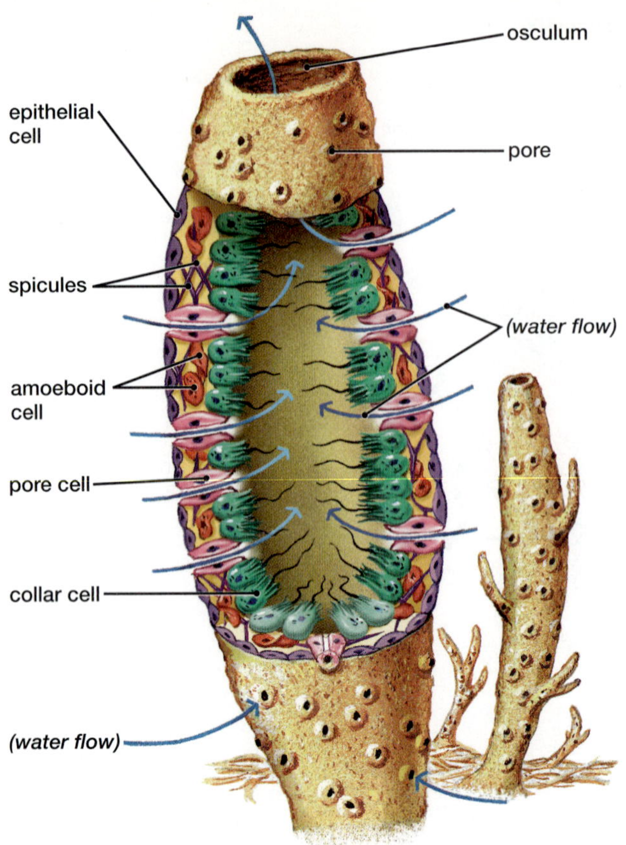

Figure 23-5 Sponge structure

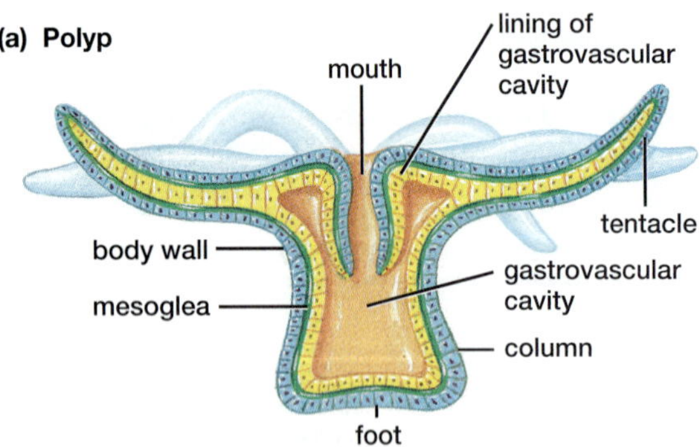

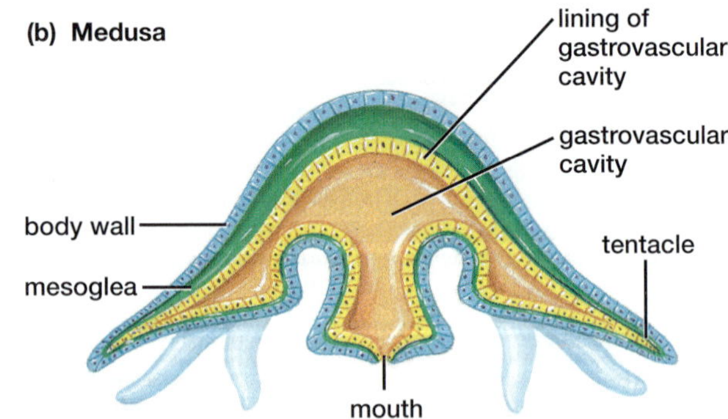

Figure 23-7 Medusa and polyp

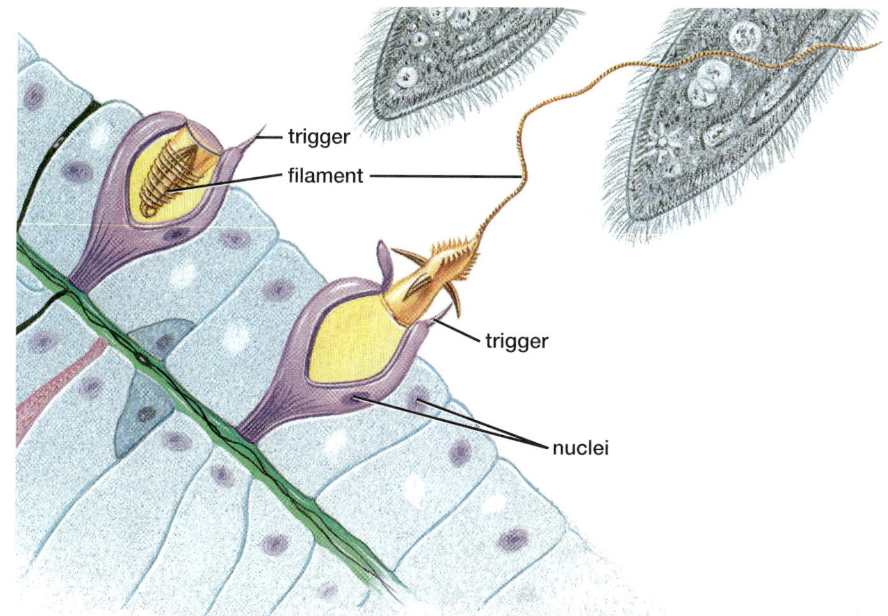

Figure 23-8 Cnidocyte

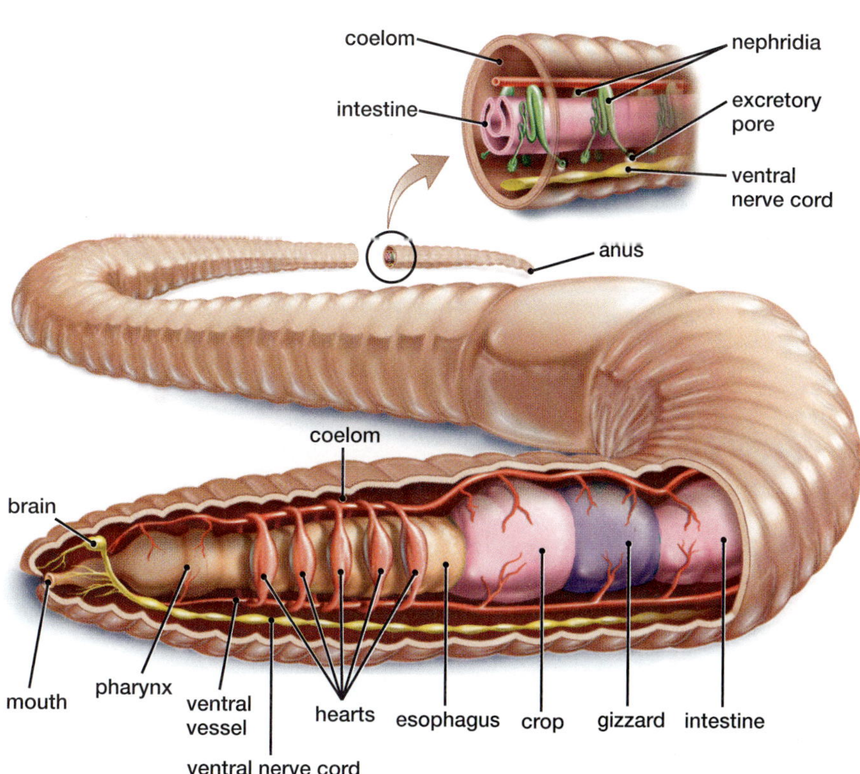

Figure 23-11 Earthworm

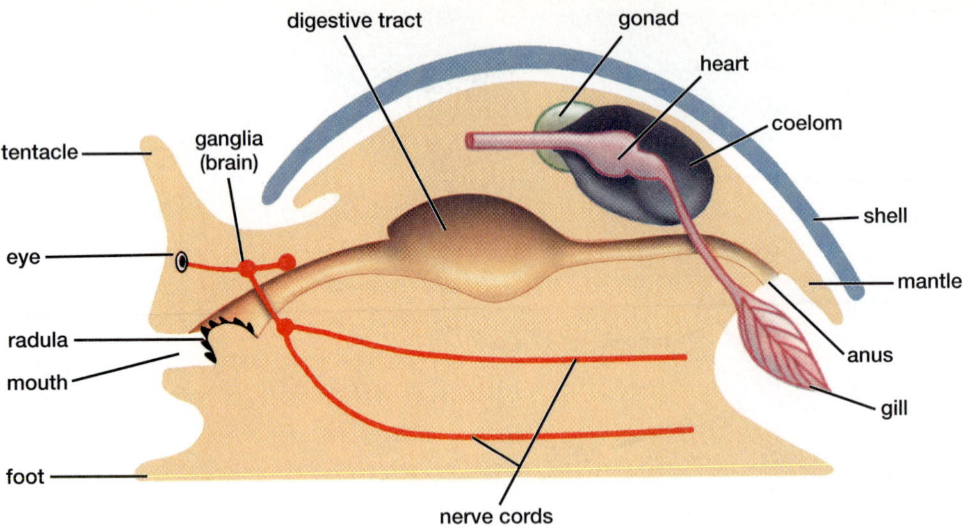

Figure 23-13 Mollusk

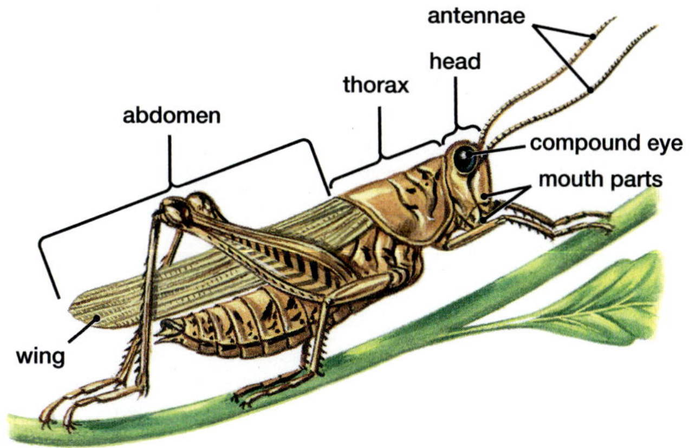

Figure 23-19 Grasshopper

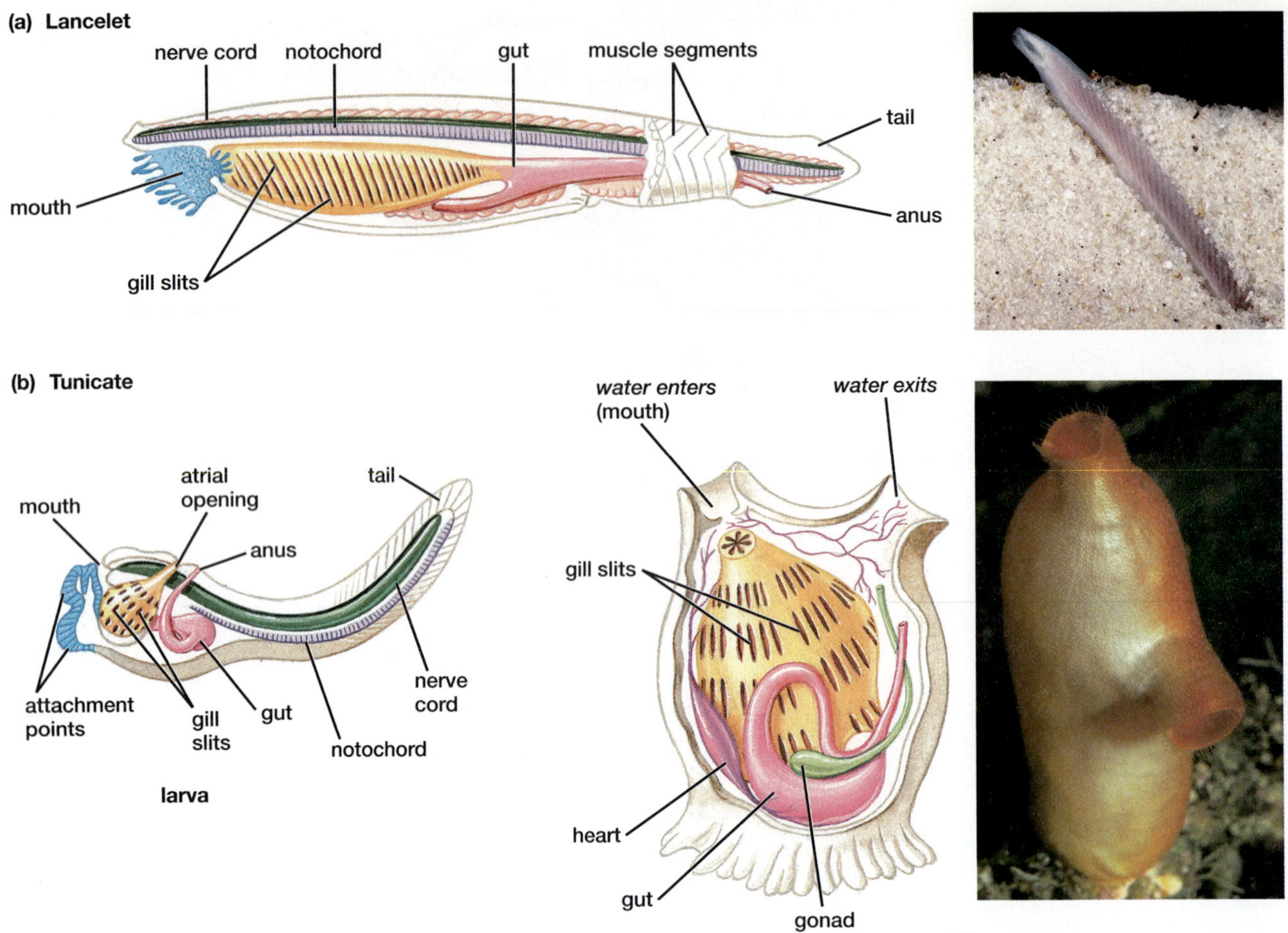

Figure 24-3 Invertebrate chordates

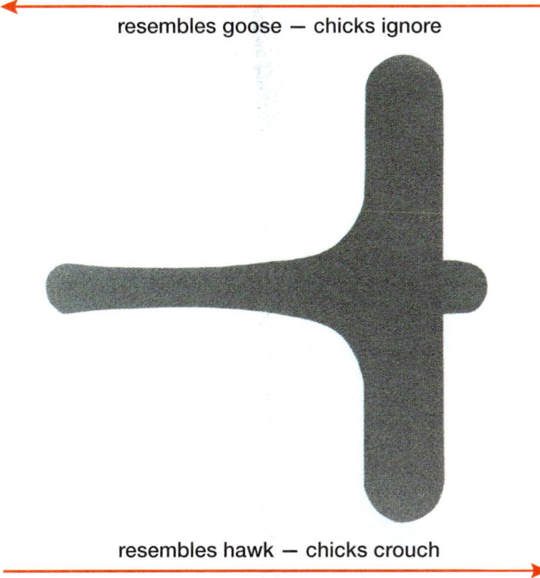

Figure 25-5 Bird silhouette model

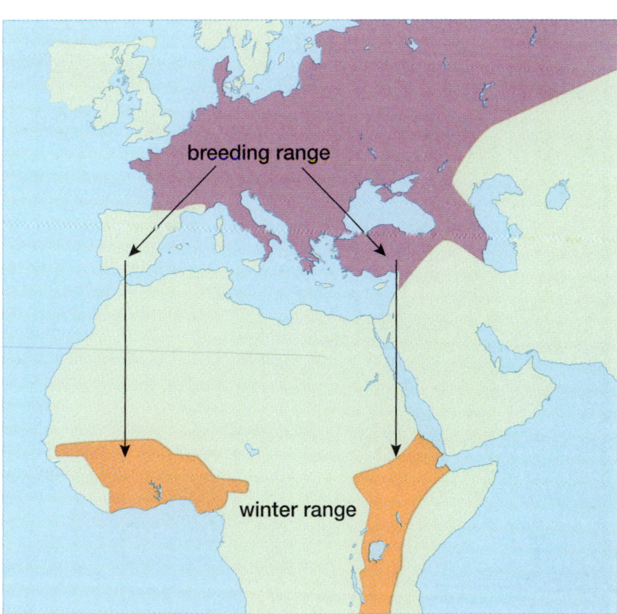

Figure 25-7 Migration map

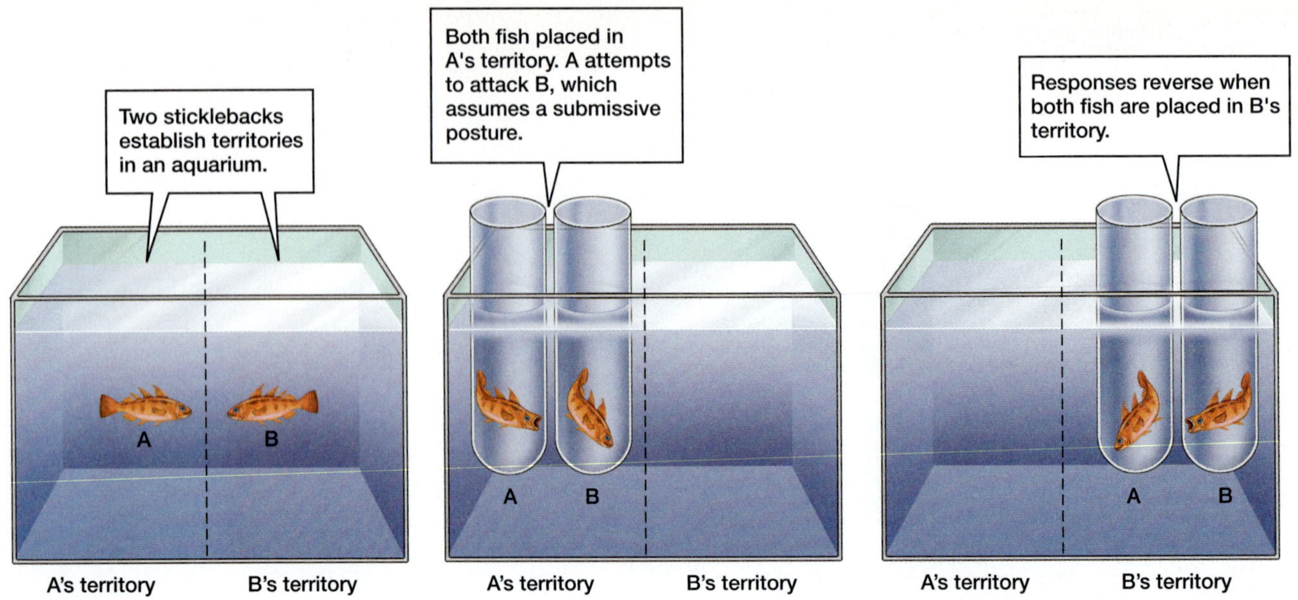

Figure 25-19 Stickleback territory experiment

(a) A male, inconspicuously colored, leaves the school of males and females to establish a breeding territory.

(b) As his belly takes on the red color of the breeding male, he displays aggressively at other red-bellied males, exposing his red underside.

(c) Having established a territory, the male begins nest construction by digging a shallow pit that he will fill with bits of algae cemented together by a sticky secretion from his kidneys.

(d) After he tunnels through the nest to make a hole, his back begins to take on the blue courting color that makes him attractive to females.

(e) An egg-carrying female displays her enlarged belly to him by assuming a head-up posture. Her swollen belly and his courting colors are passive visual displays.

(f) Using a zigzag dance, he leads her to the nest.

(g) After she enters, he stimulates her to release eggs by prodding at the base of her tail.

(h) He enters the nest as she leaves and deposits sperm, which fertilize the eggs.

Figure 25-23 Stickleback courtship

Figure 25-26 Honeybee waggle dance

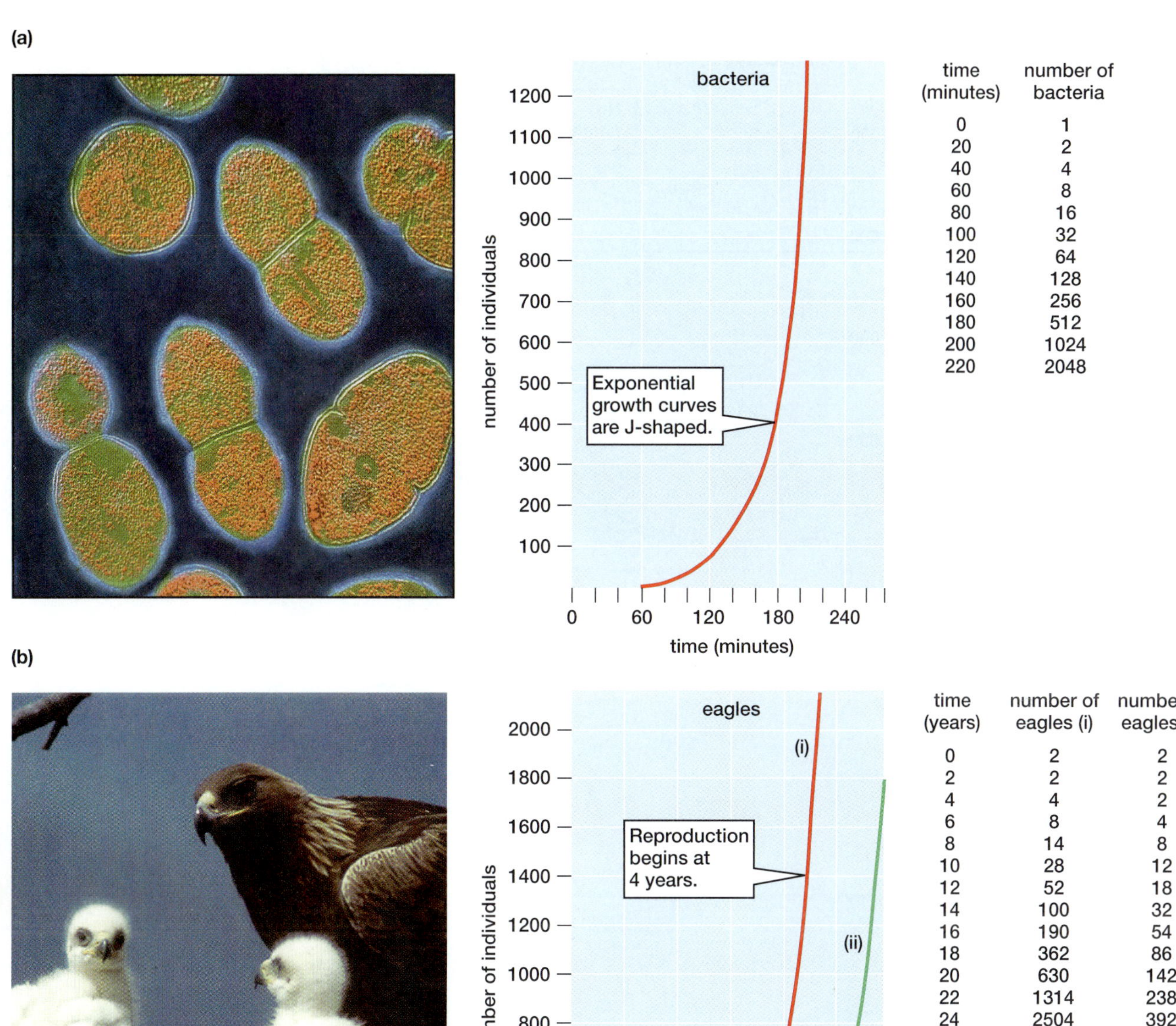

time (minutes)	number of bacteria
0	1
20	2
40	4
60	8
80	16
100	32
120	64
140	128
160	256
180	512
200	1024
220	2048

time (years)	number of eagles (i)	number of eagles (ii)
0	2	2
2	2	2
4	4	2
6	8	4
8	14	8
10	28	12
12	52	18
14	100	32
16	190	54
18	362	86
20	630	142
22	1314	238
24	2504	392
26	4770	644
28	9088	1066
30	17314	1764

Figure 26-1 Population growth

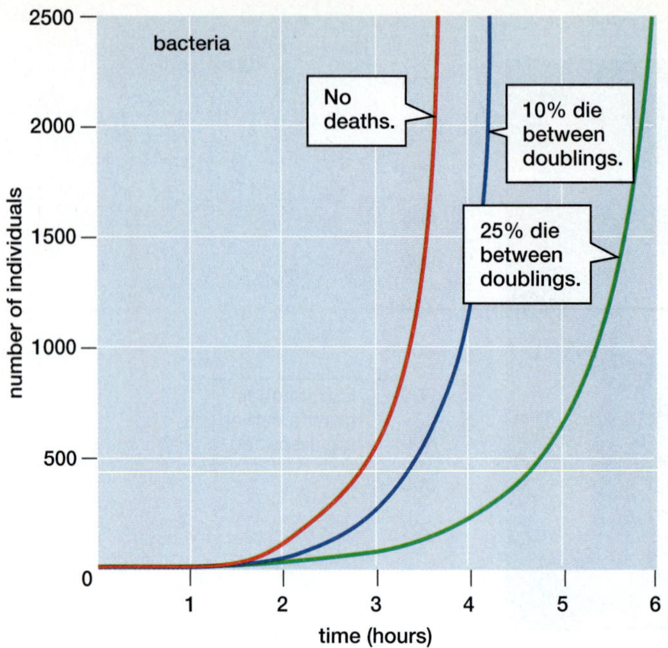

Figure 26-2 Survival rate curves

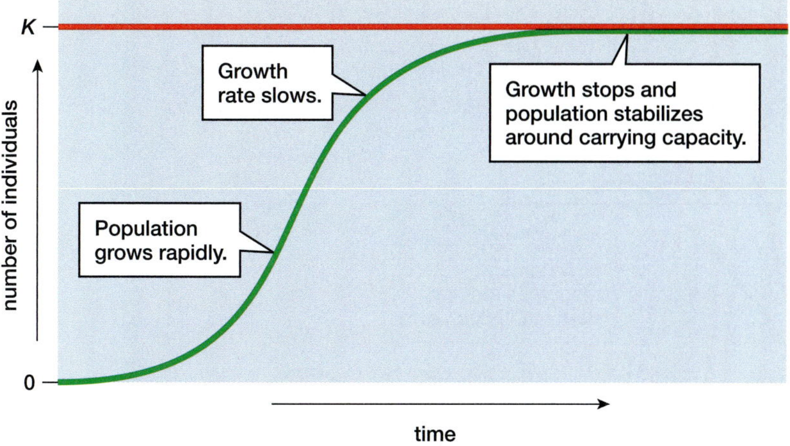

Figure 26-6a Sigmoid growth curve

Consequences of exceeding *K*.

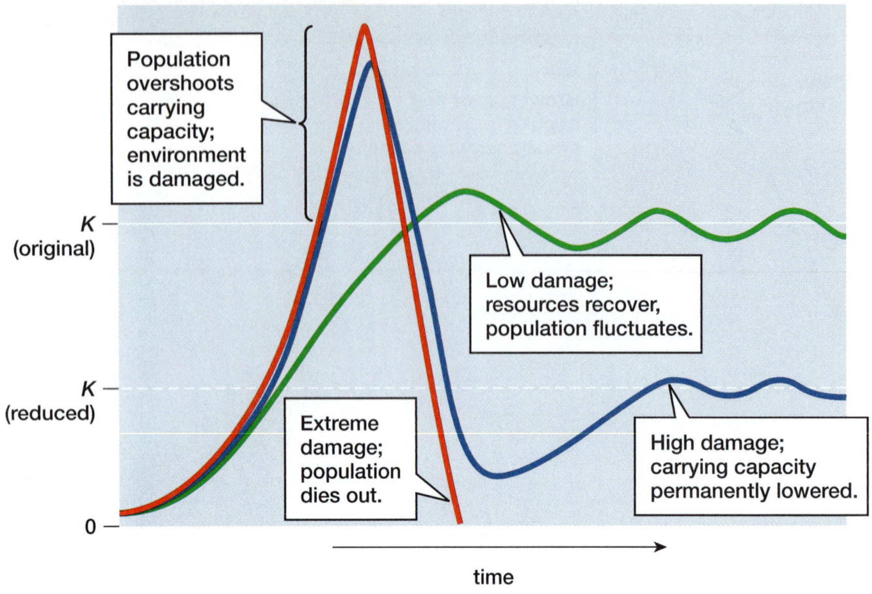

Population overshoots carrying capacity; environment is damaged.
K (original)
K (reduced)
0
Low damage; resources recover, population fluctuates.
High damage; carrying capacity permanently lowered.
Extreme damage; population dies out.
time

Figure 26-6b Exceeding *K* curves

1000
100
10
0
number of survivors
late loss (human)
constant loss (*American robin*)
early loss (dandelion)
age (in percentage of maximum life span)

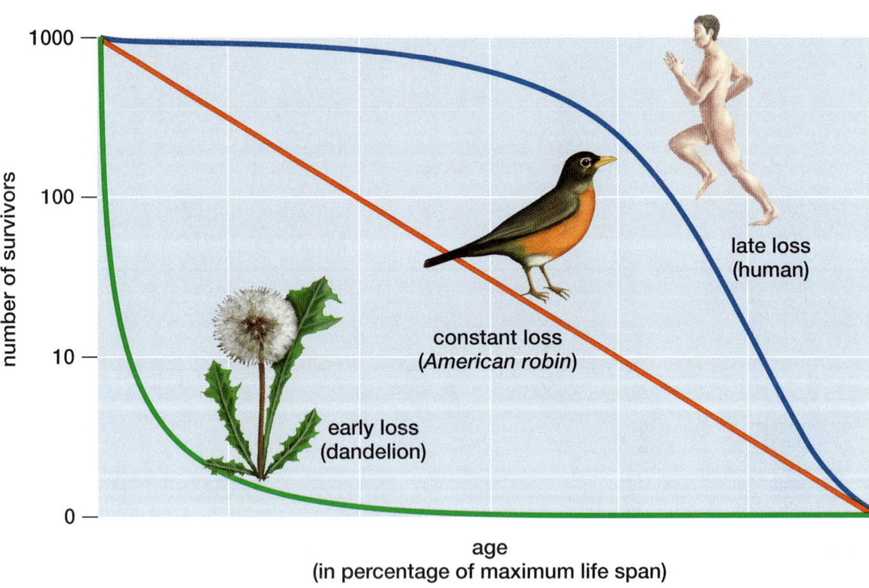

Figure 26-14b Survivorship curves

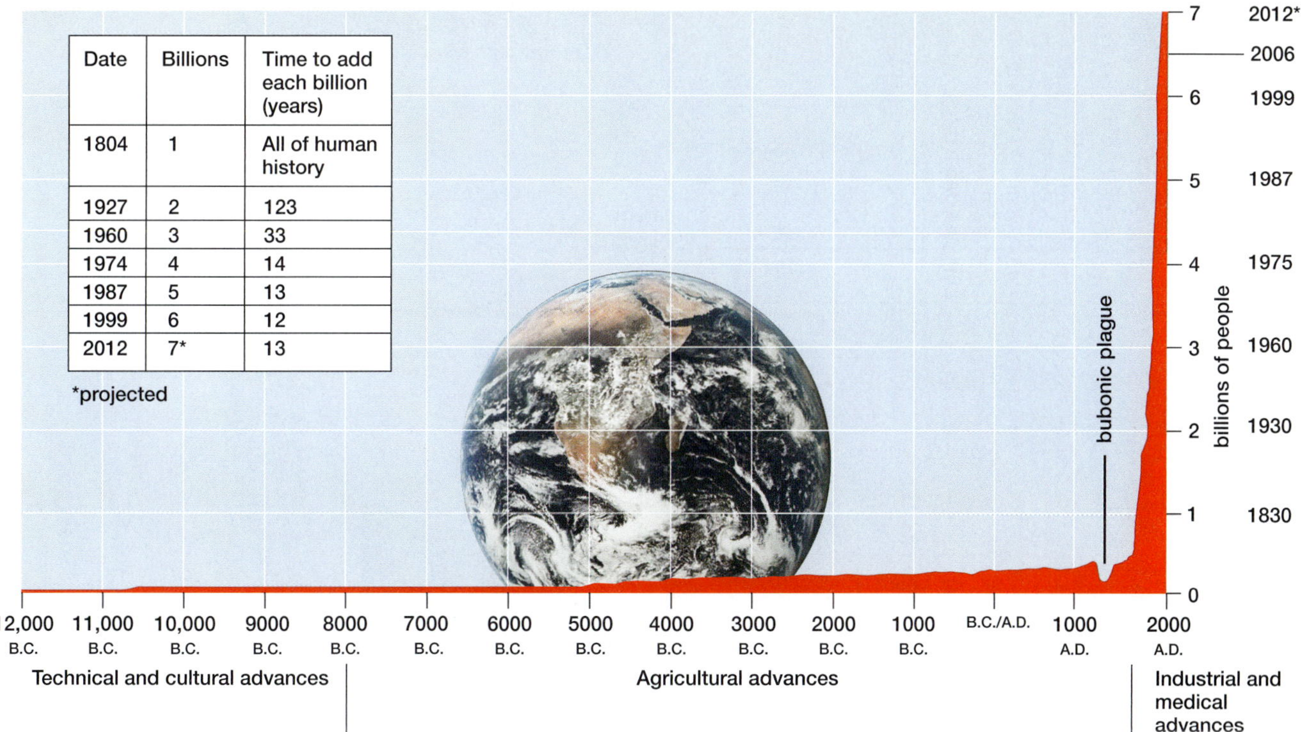

Date	Billions	Time to add each billion (years)
1804	1	All of human history
1927	2	123
1960	3	33
1974	4	14
1987	5	13
1999	6	12
2012	7*	13

*projected

Figure 26-15 World population growth graph

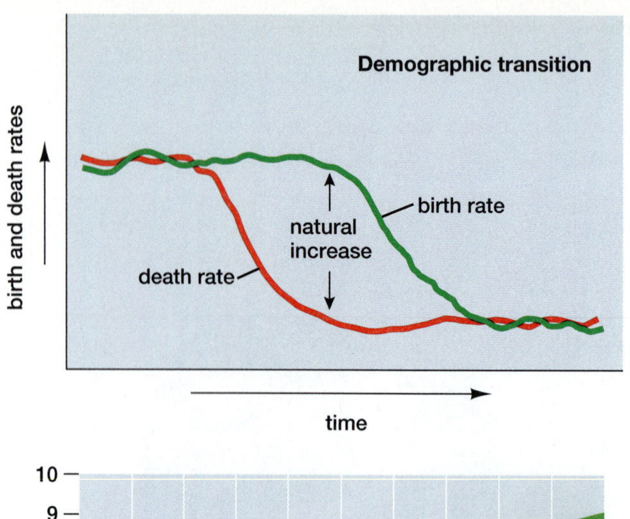

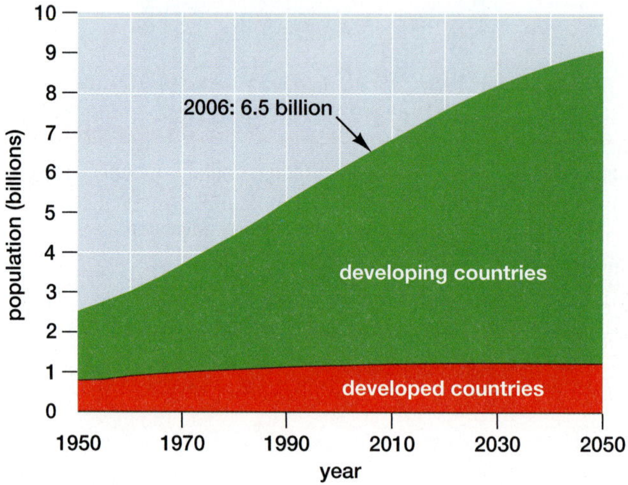

Figures 26-16, 26-17 Demographic transition, Population projections to 2050

(a) Population pyramids for Mexico

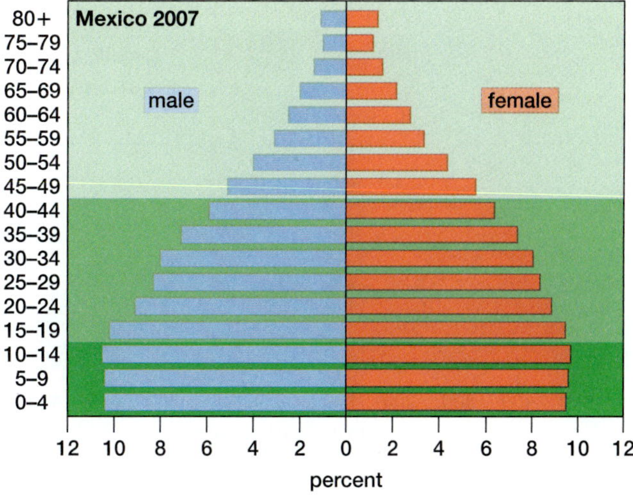

(b) Population pyramids for Sweden

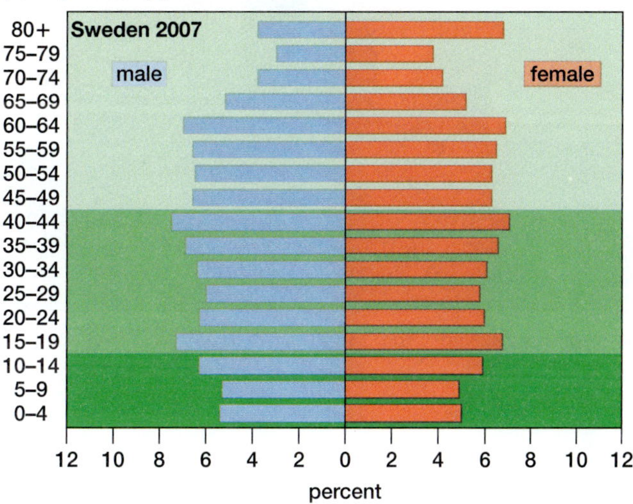

(c) Population pyramids for Italy

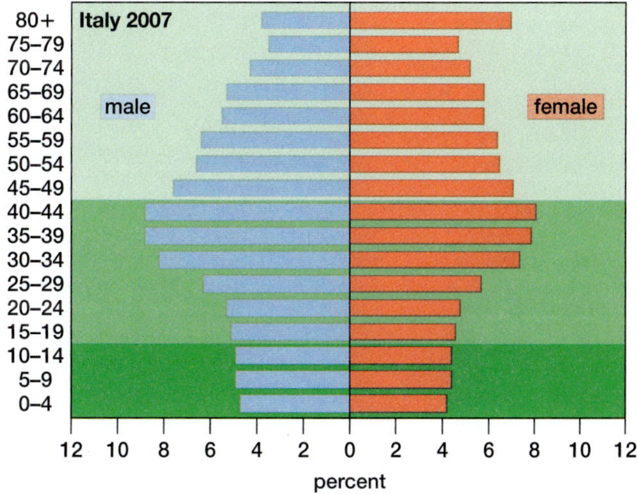

Figure 26-18 Age structures: Mexico, Sweden, Italy

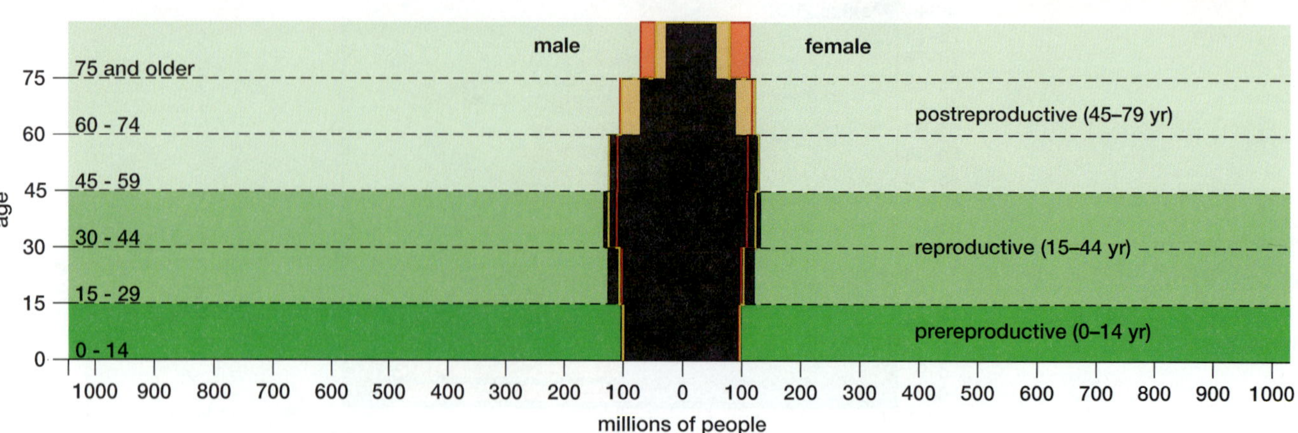

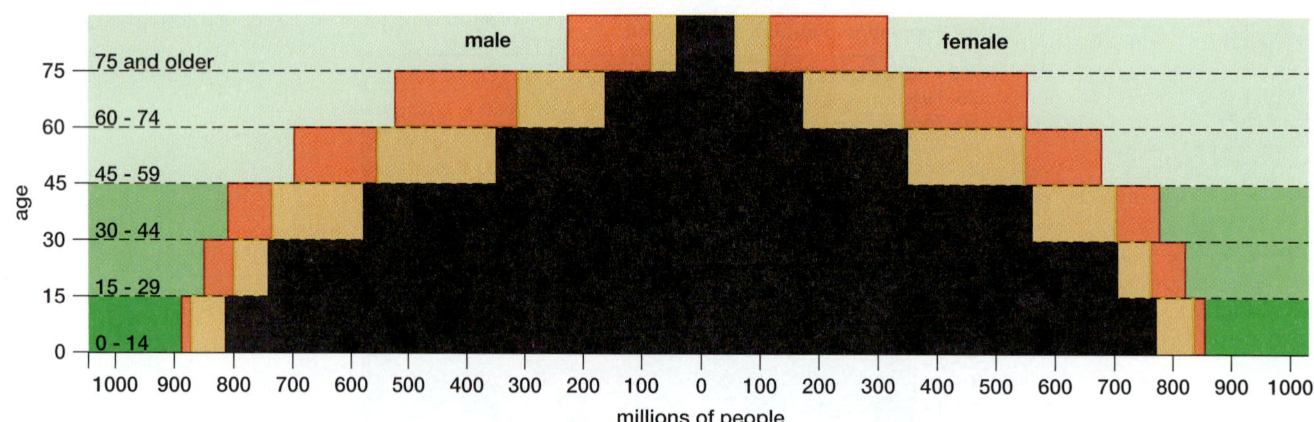

Figure 26-19 Age structures

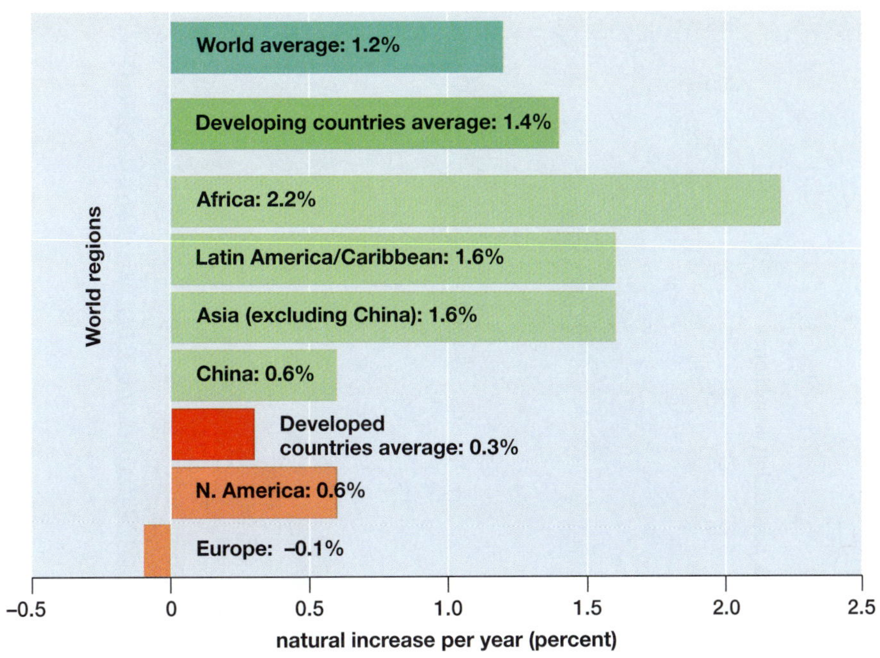

Figure 26-20 Natural increase graph

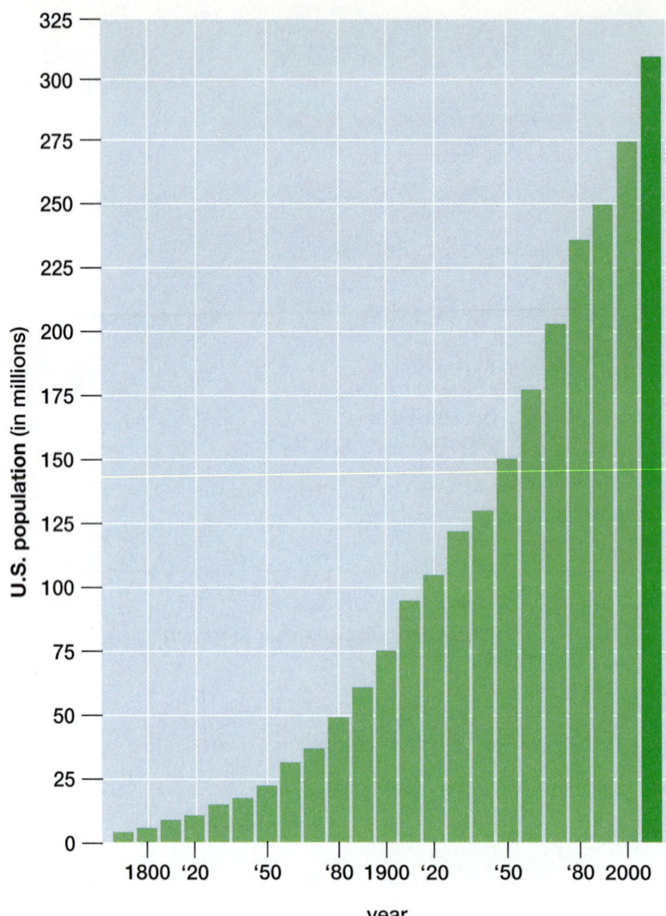

Figure 26-21 U.S population graph

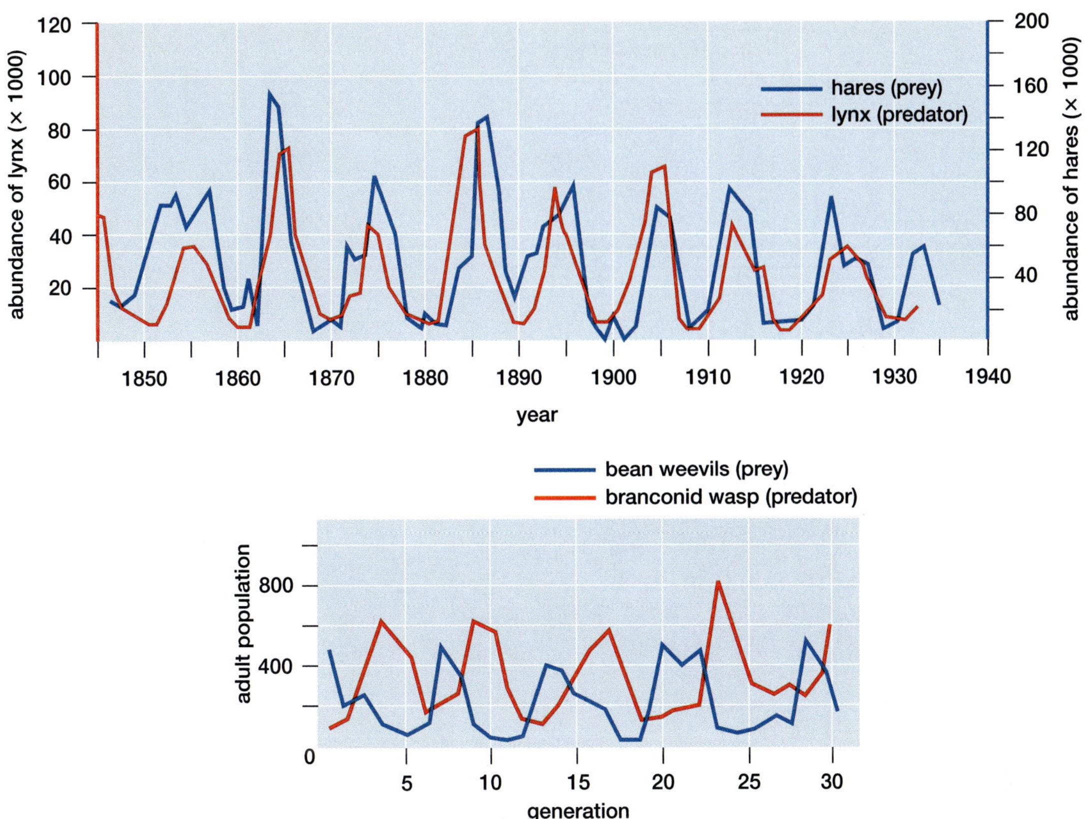

Figures E26-1, E26-2 Hare/lynx graph, Braconid wasp/bean weevil graph

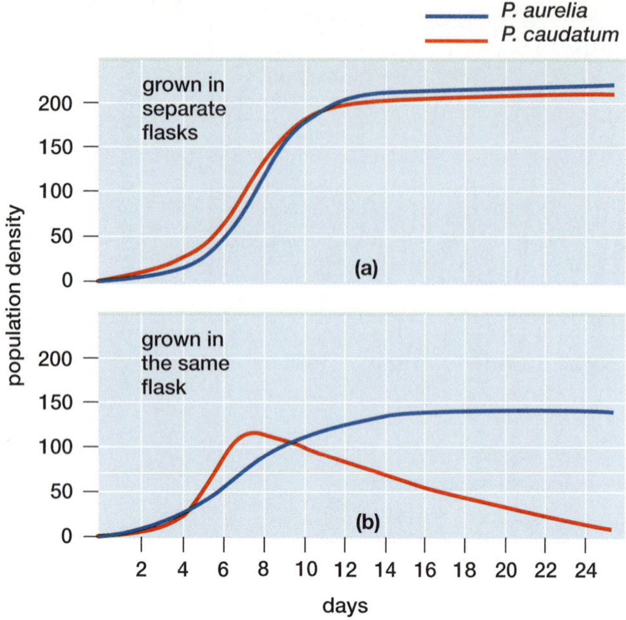

Figure 27-1 Competitive exclusion

Figure 27-2 Resource partitioning

(a) sand dab (fish)

(b) nightjar (bird)

Figure 27-4 Camouflage - blending

(a) moth

(b) leafy sea dragon

(c) treehoppers

(d) cactus

Figure 27-5 Camouflage - resembling

(a) cheetah

(b) frogfish

Figure 27-6 Camouflage assists predators

(a) bombardier beetle

(b) monarch caterpillar

Figure 27-12 Chemical warfare: beetle and caterpillar

(a) lichen

(b) clownfish

Figure 27-13 Mutualism

Mt. Kilauea, Hawaii

Figure 27-15a Succession in progress

Yellowstone, Wyoming

Figure 27-15c Yellowstone fire succession

Figure 27-16 Primary succession

Figure 27-17 Secondary succession

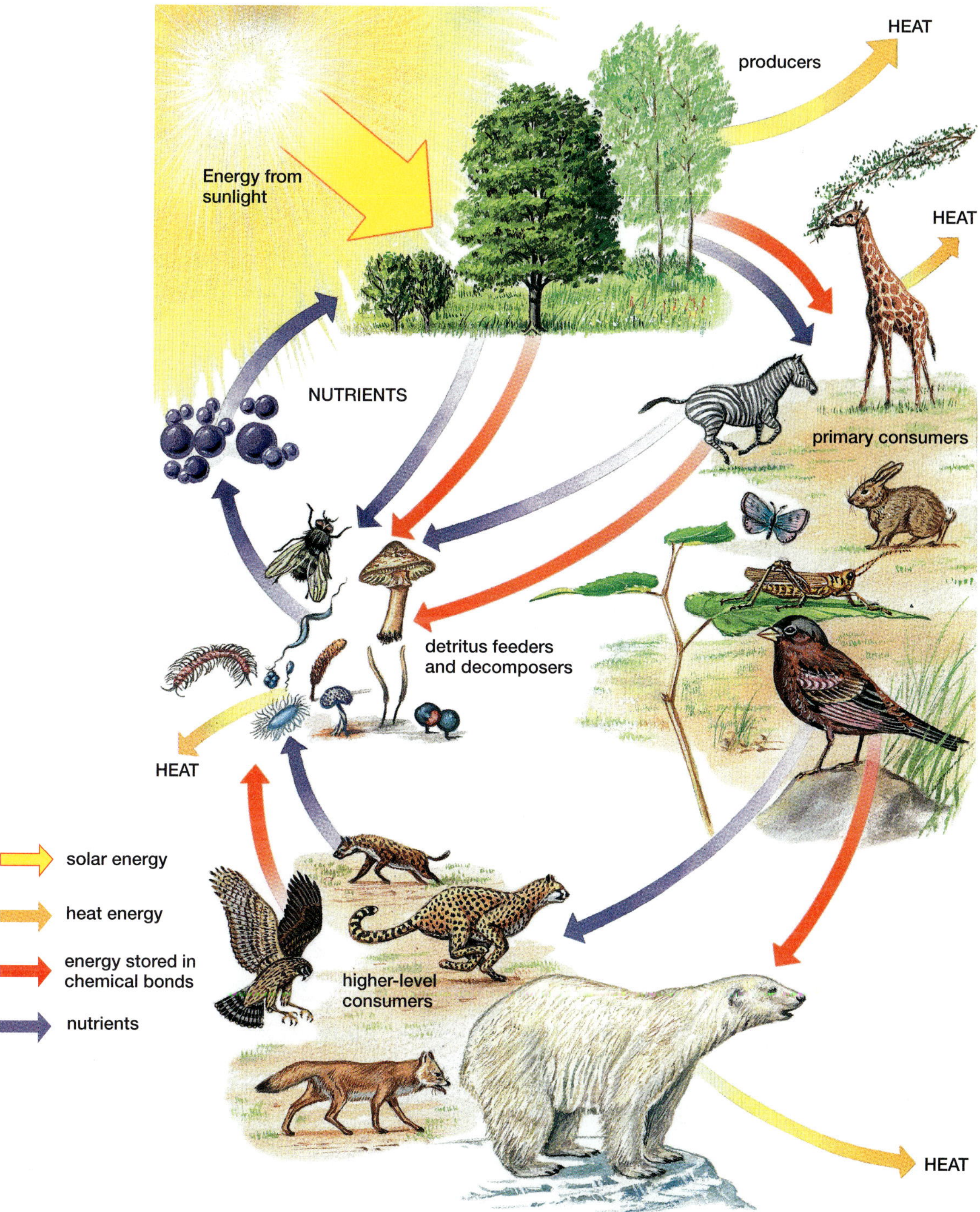

Figure 28-1 Energy flow and nutrient cycling

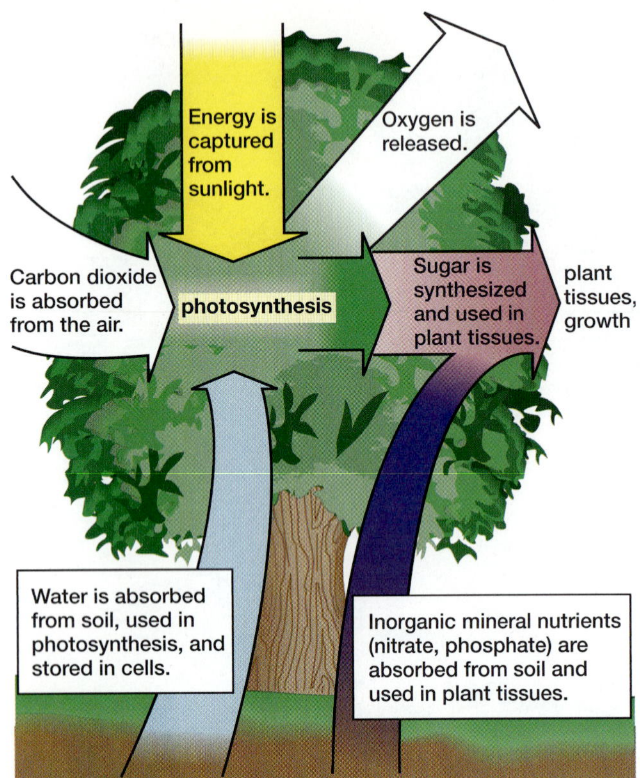

Figure 28-2 Primary productivity

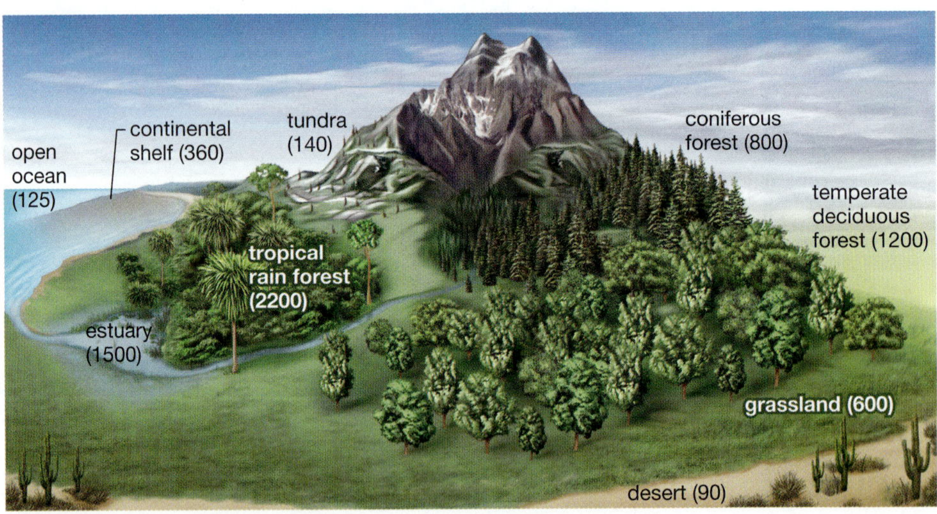

Figure 28-3 Ecosystem productivity

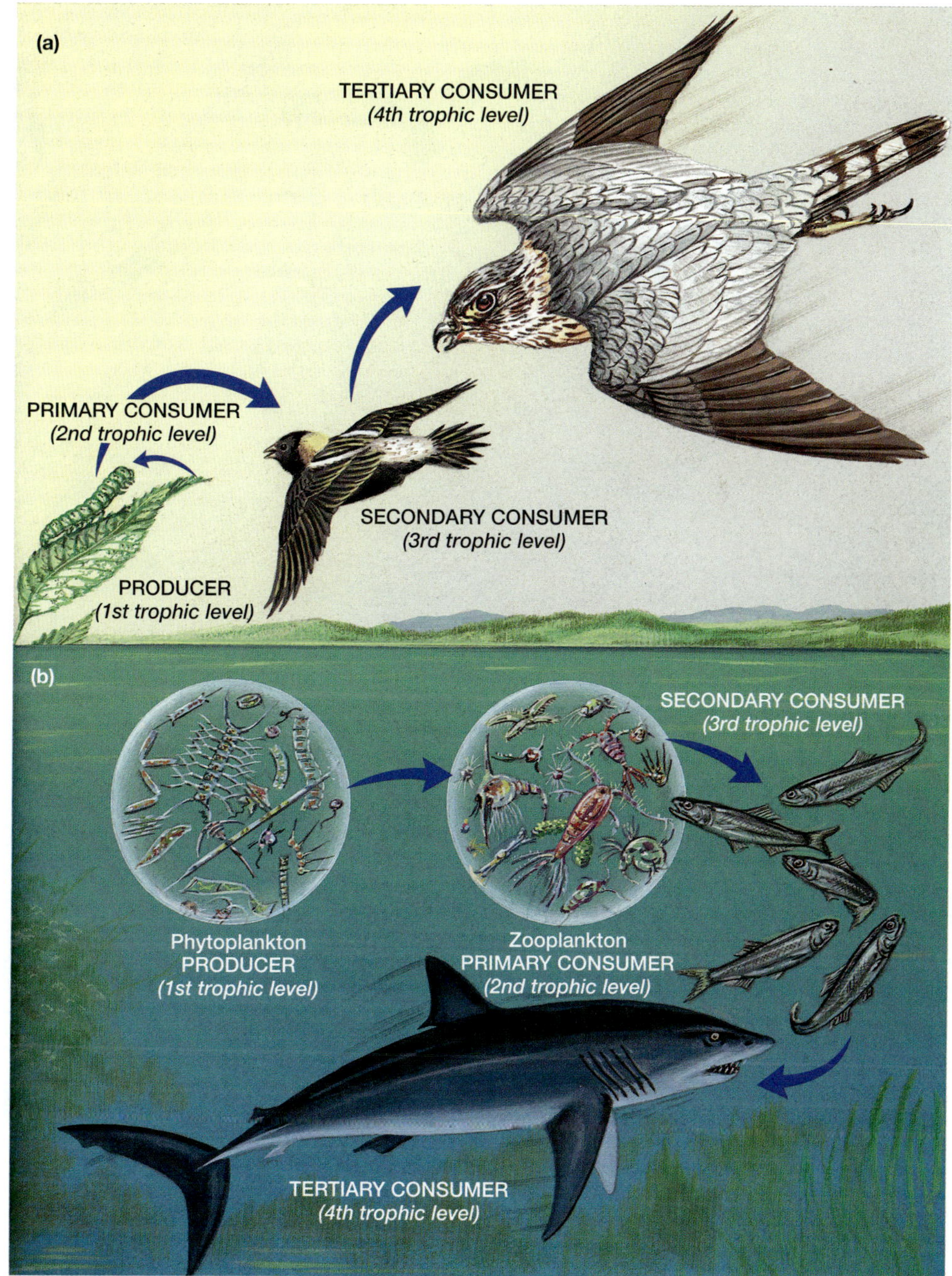

Figure 28-4 Food chains

Figure 28-5 Grassland food web

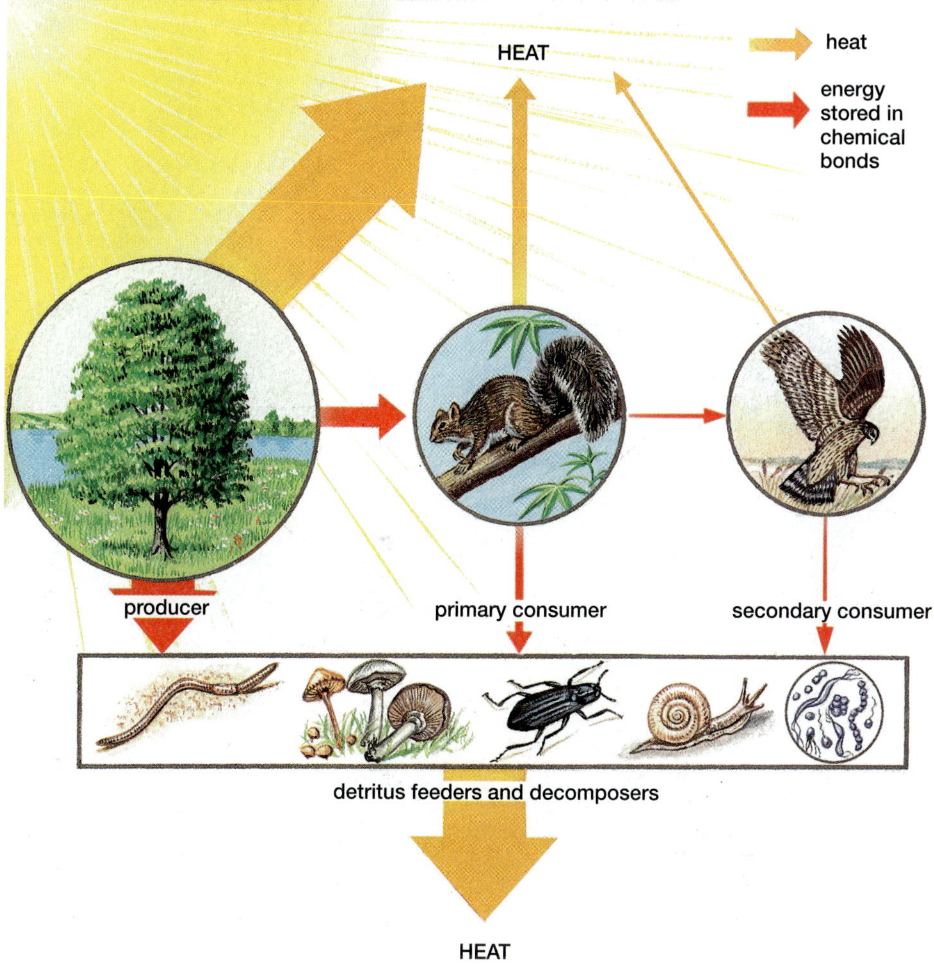

Figure 28-6 Energy transfer

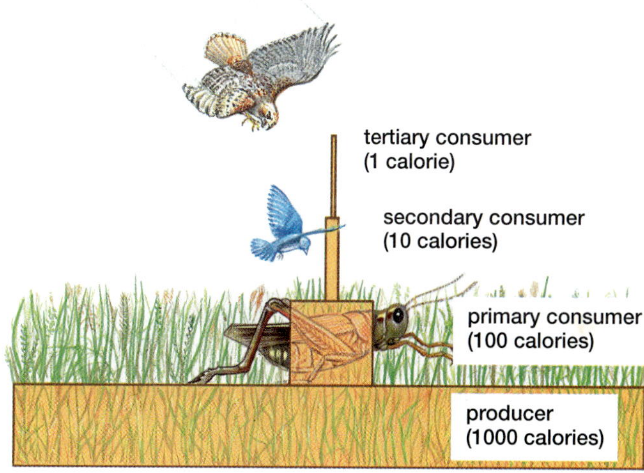

Figure 28-7 Energy pyramid

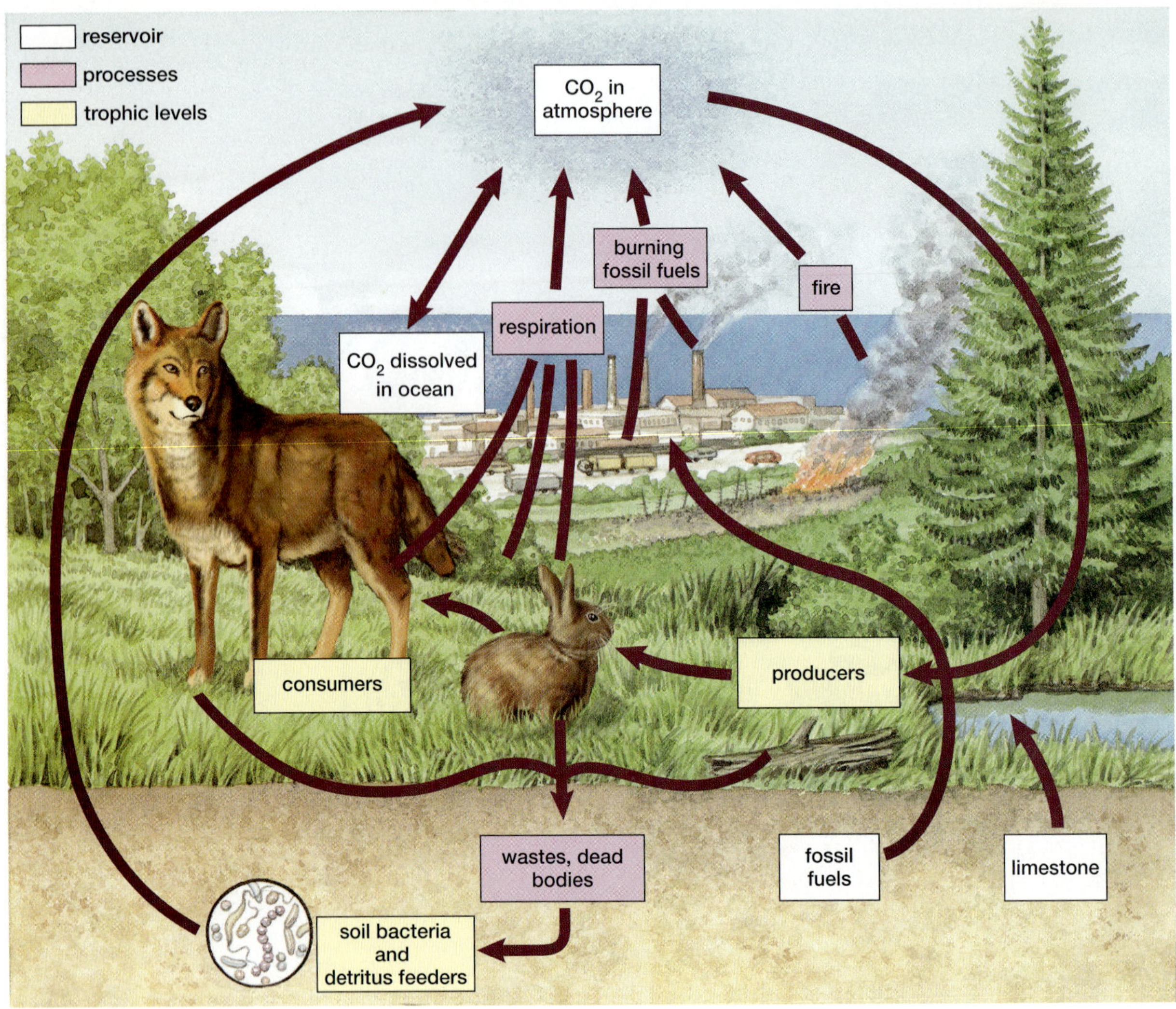

Figure 28-8 Carbon cycle

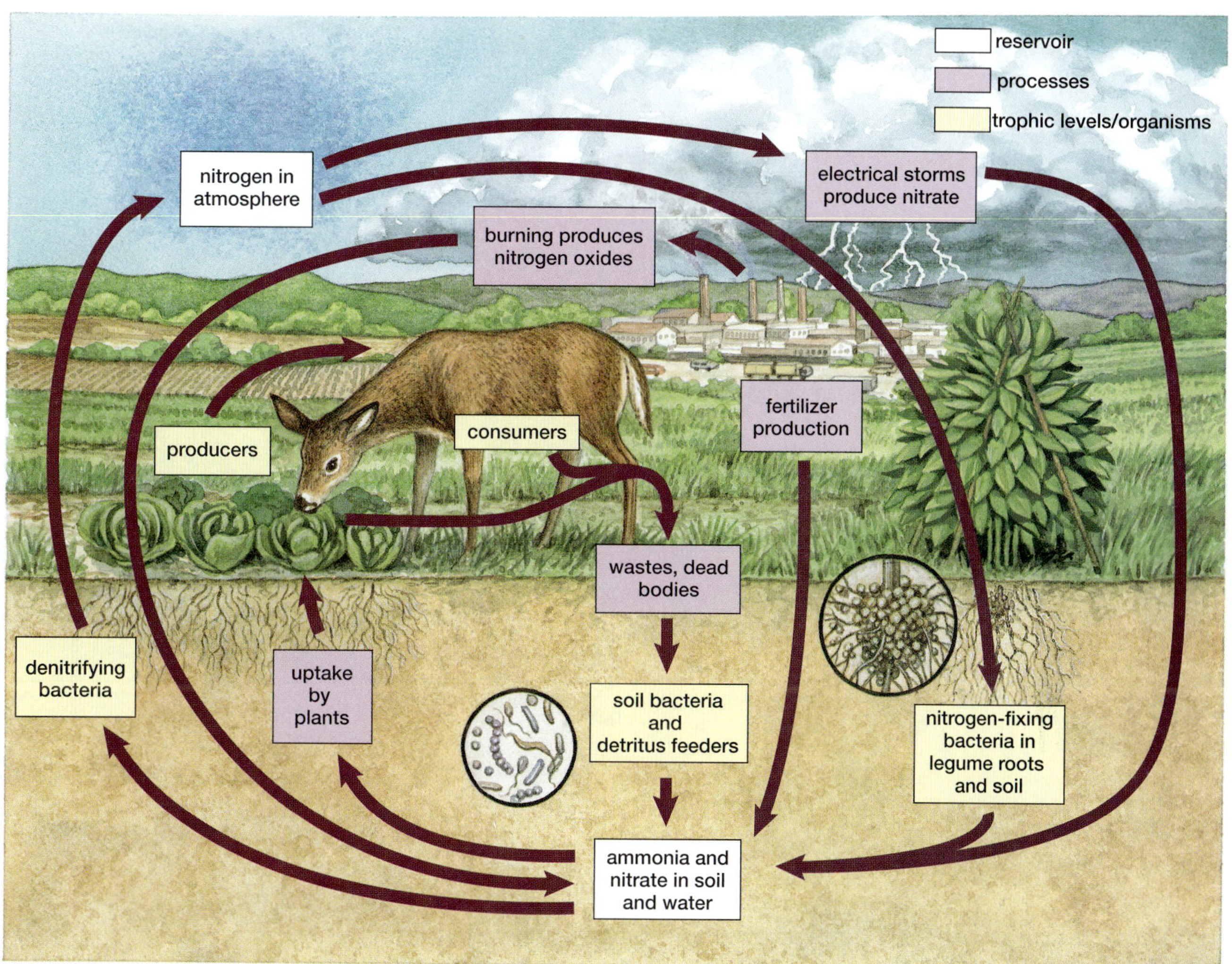

Figure 28-9 Nitrogen cycle

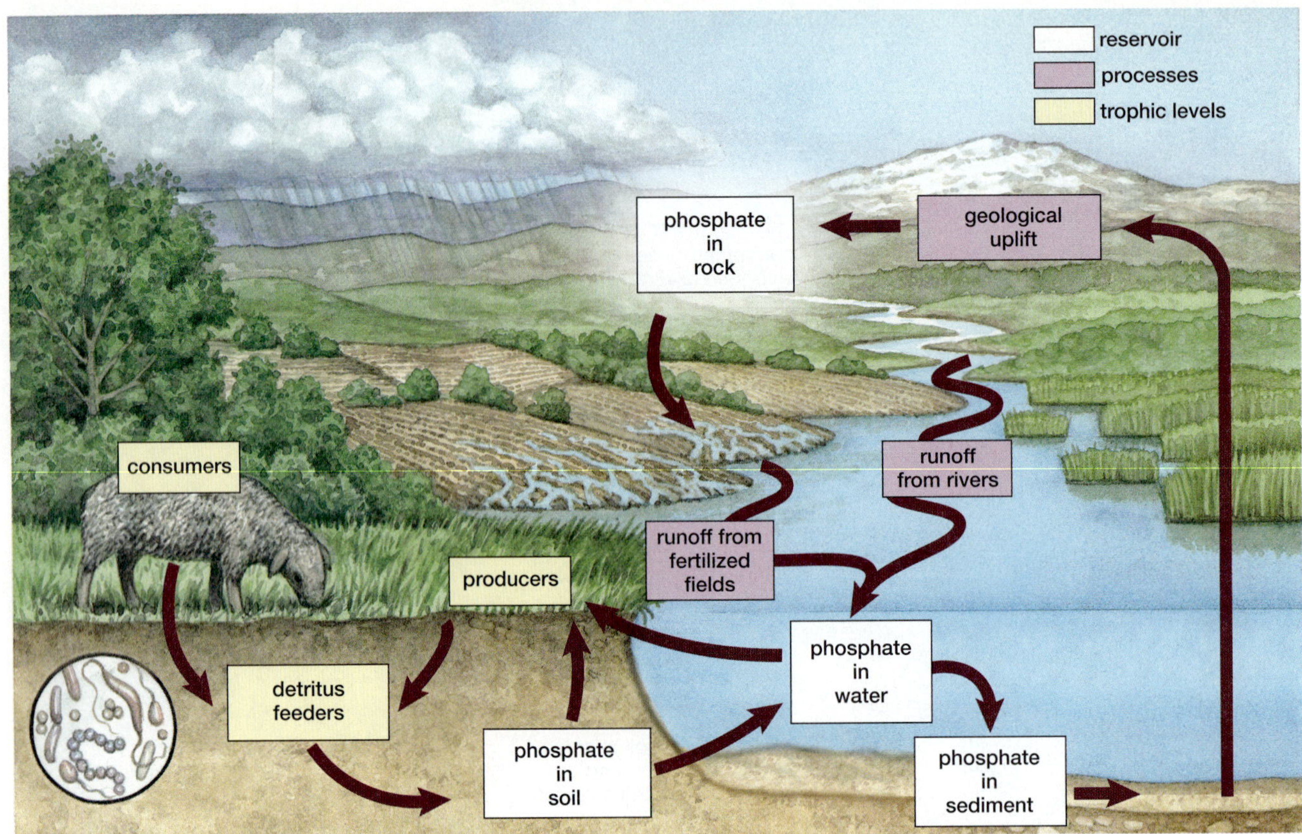

Figure 28-10 Phosphorus cycle

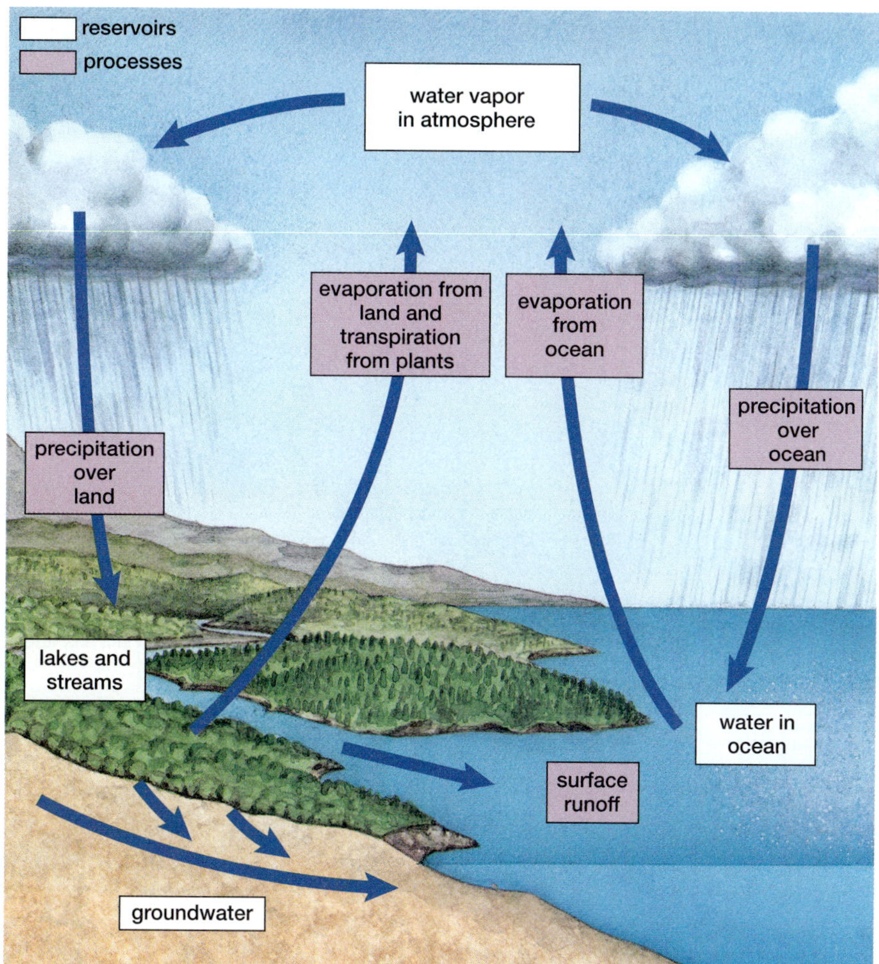

Figure 28-11 Hydrologic cycle

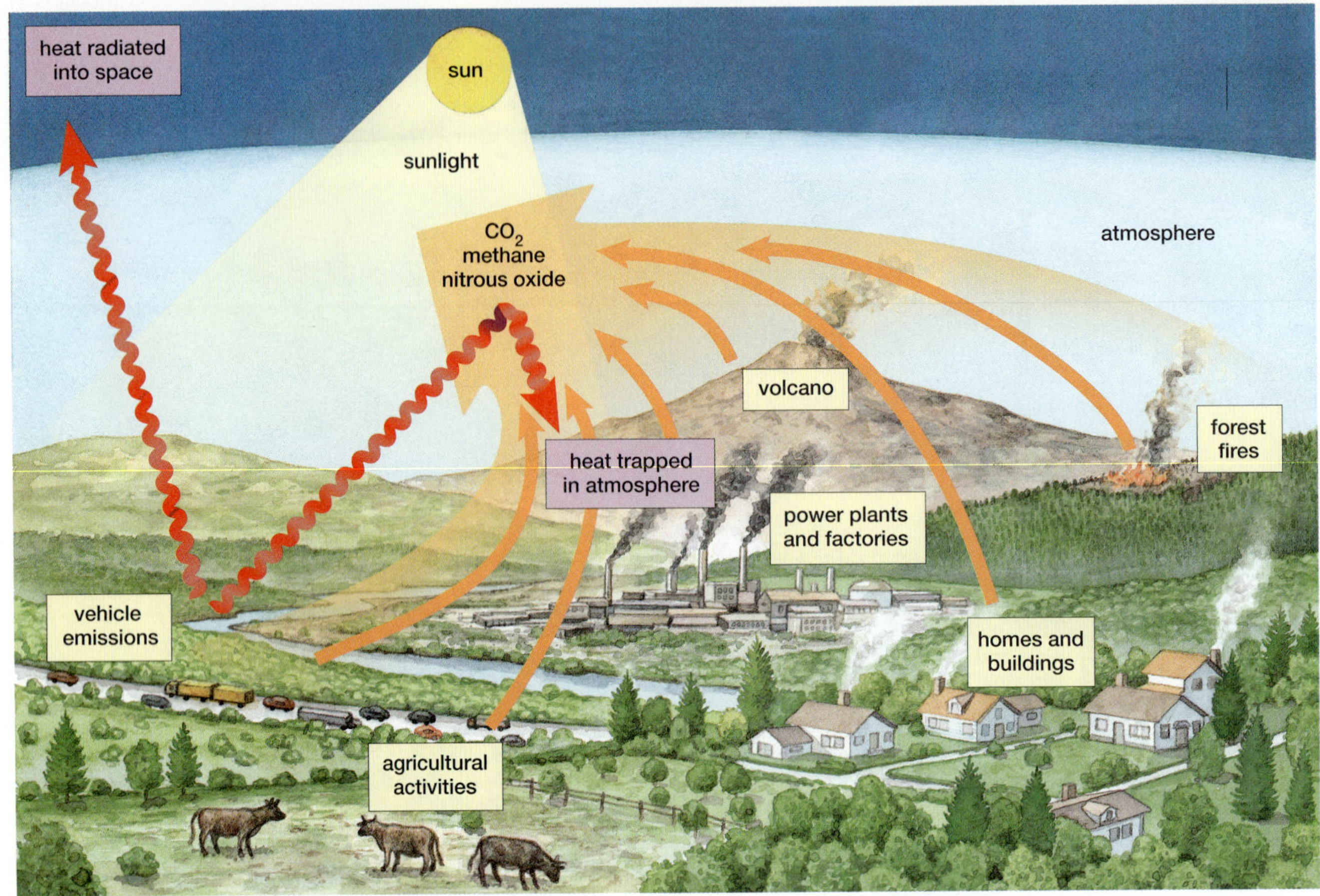

Figure 28-15 Greenhouse gases and global warming

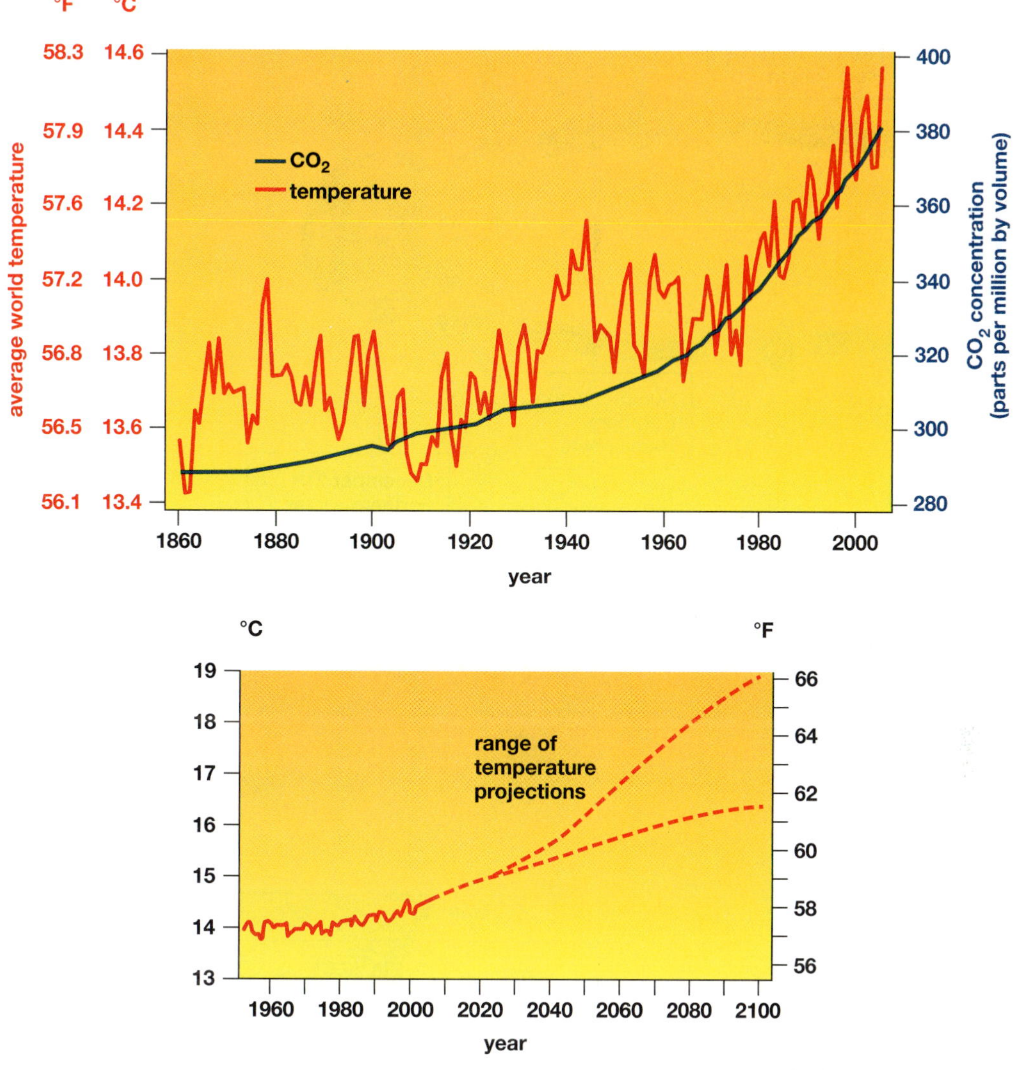

Figures 28-16, 28-17 Global warming and CO_2 increases, Projected temperature increases

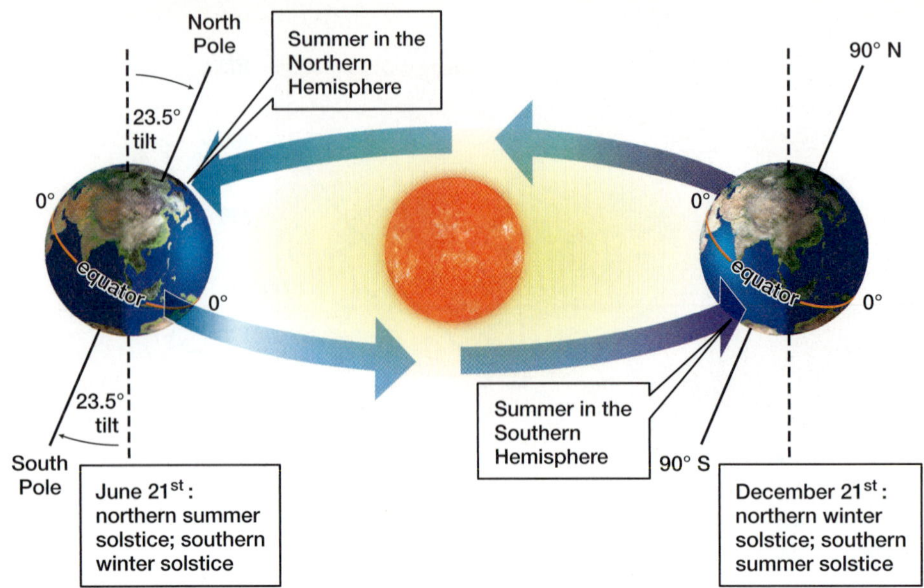

Figure 29-1 Sun hitting earth

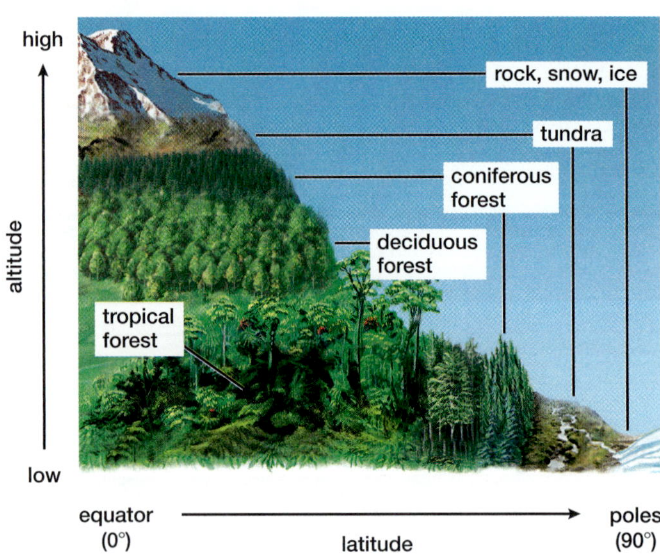

Figure 29-4 Altitude/latitude

Figure 29-5 Rain shadow

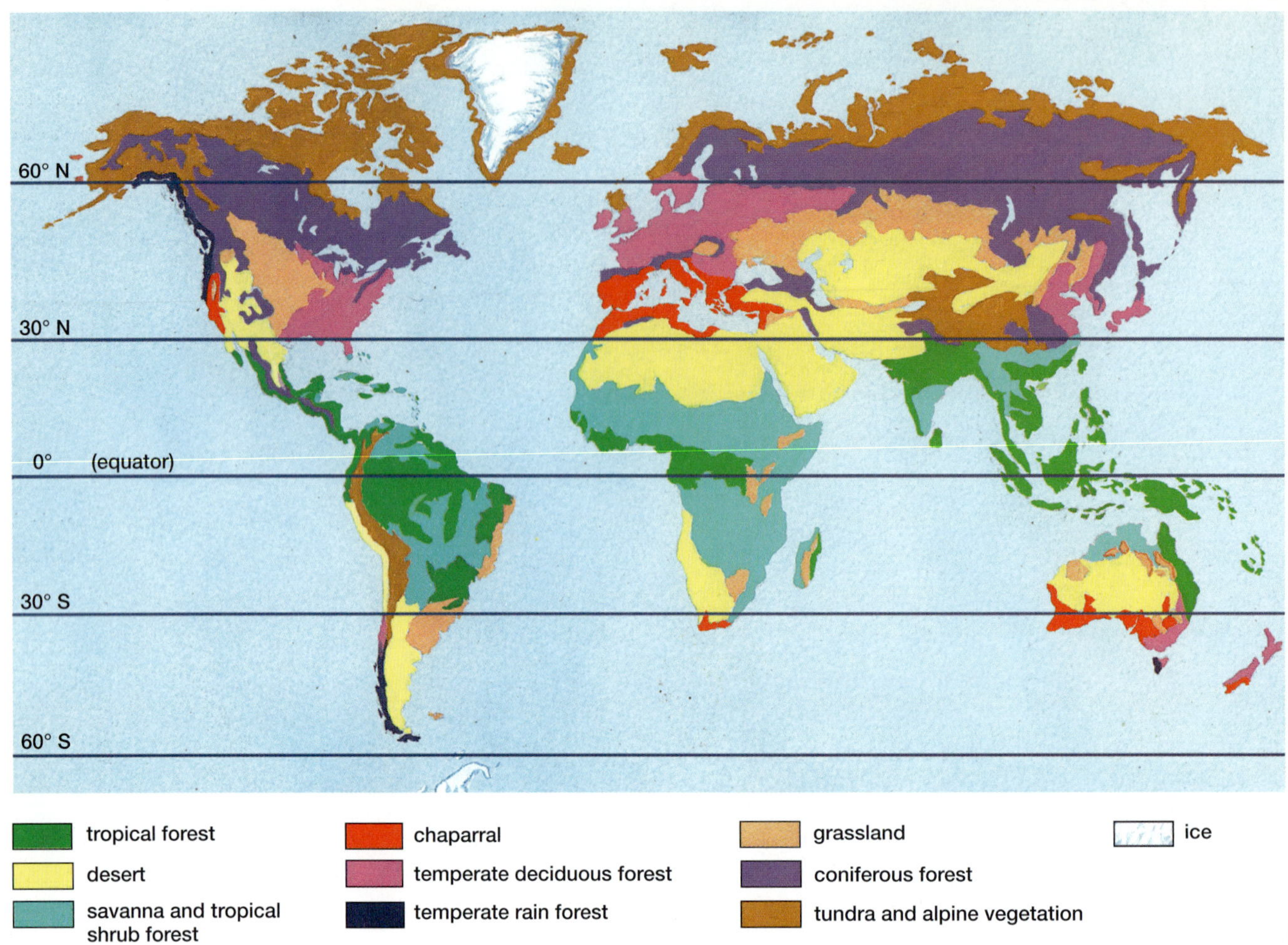

Figure 29-7 Biomes

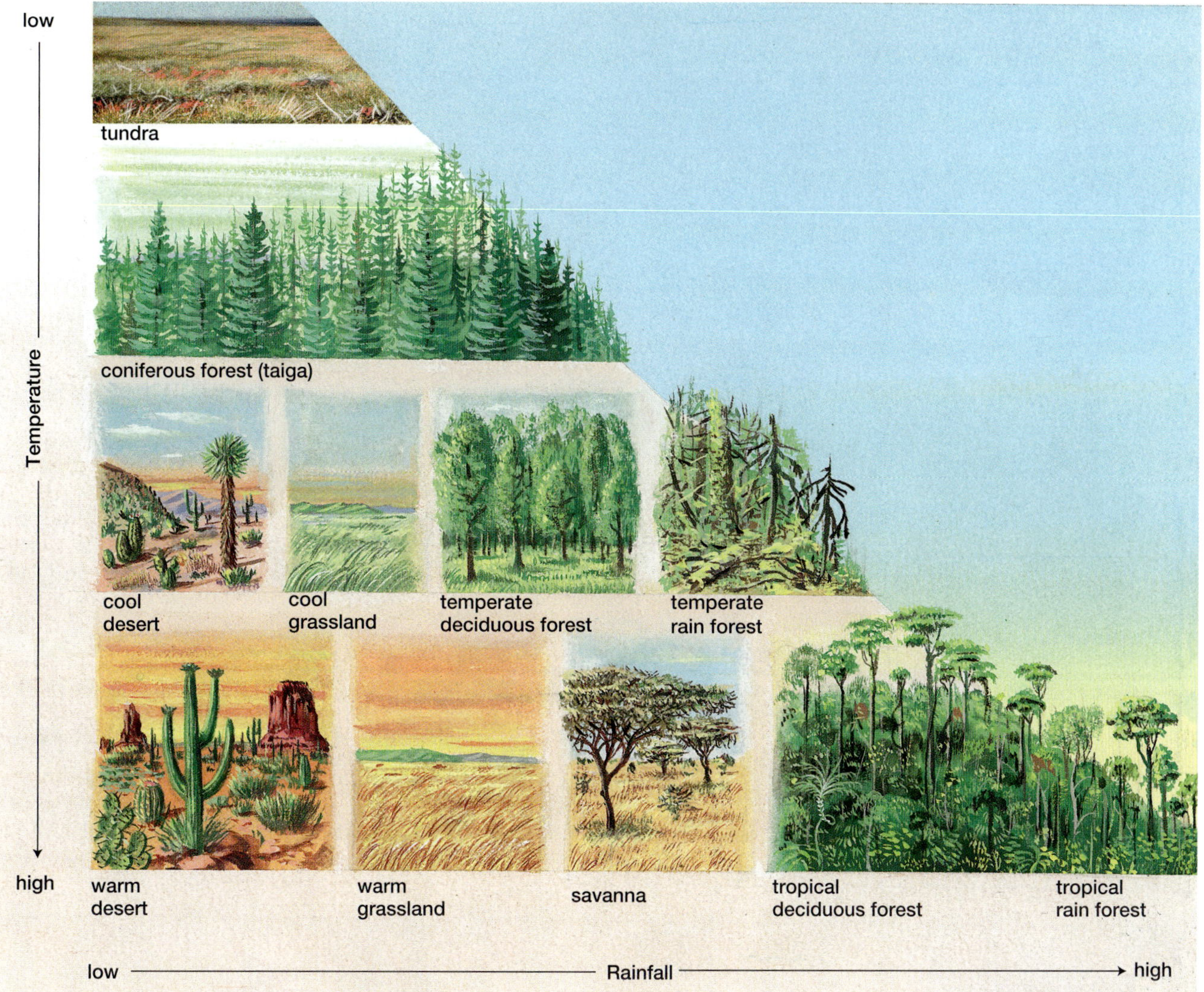

Figure 29-8 Temperature/rainfall

Figure 29-11 African savanna

Figure 29-14 Desert in bloom

Figure 29-18 Shortgrass prairie

Figure 29-20 Deciduous forest

Figure 29-21 Temperate rainforest

Figure 29-22 Taiga

Figure 29-24 Tundra

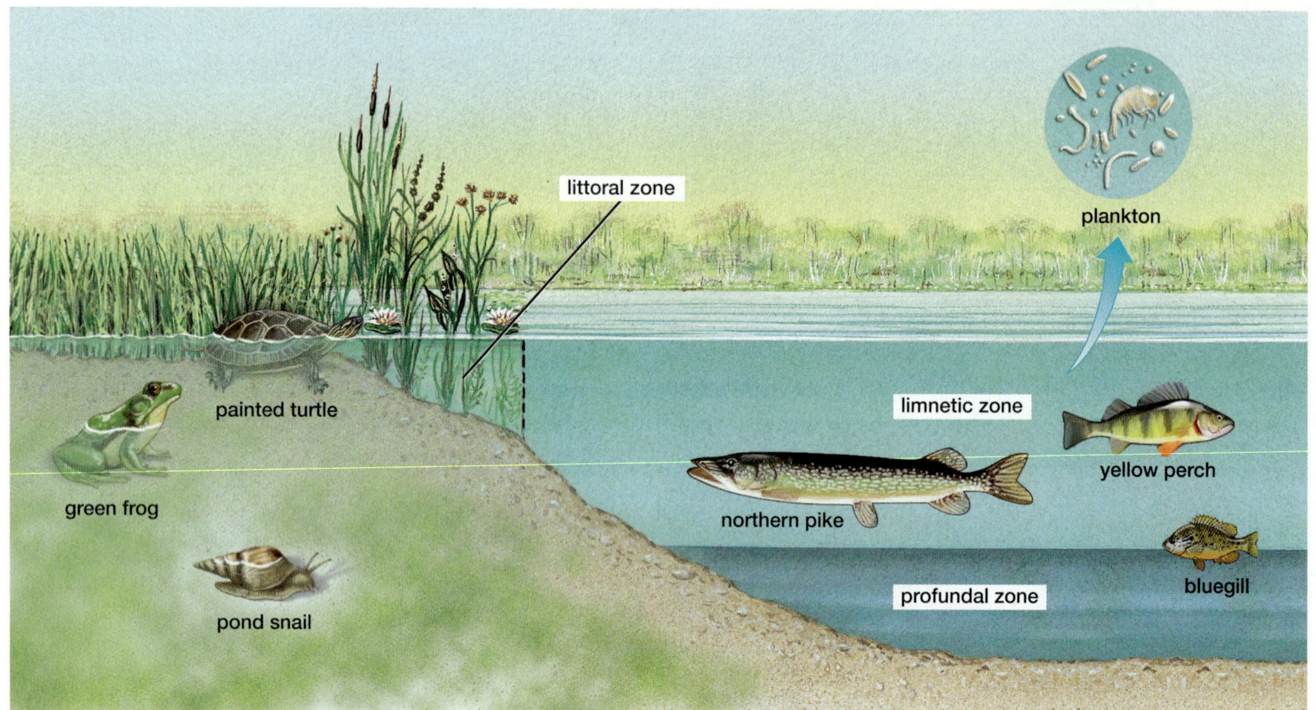

Figure 29-25 Freshwater ecosystems

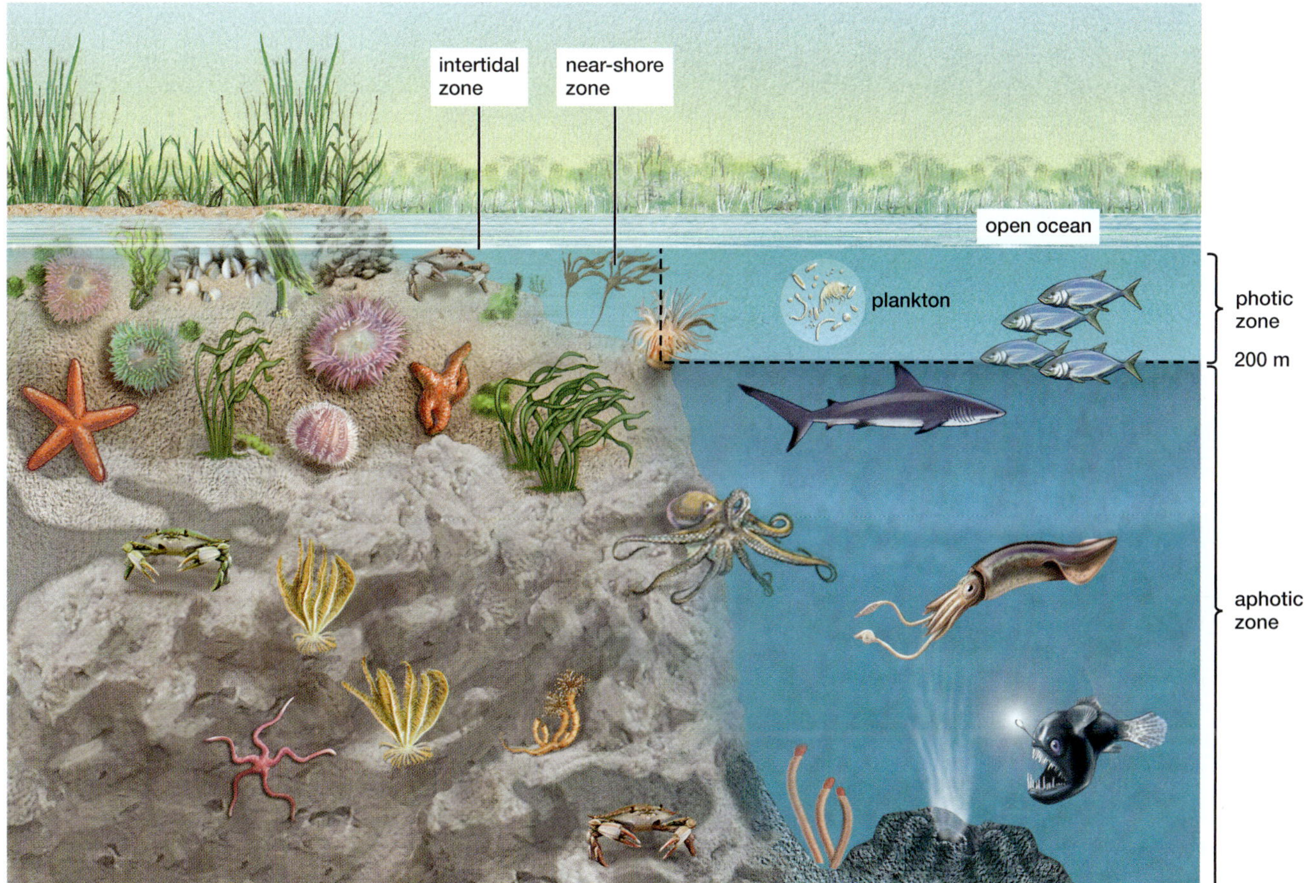

Figure 29-28 Ocean subdivisions

Figure 29-29 Nearshore ecosystems

Figure 29-30 Coral reef

Figure 29-31 Open ocean

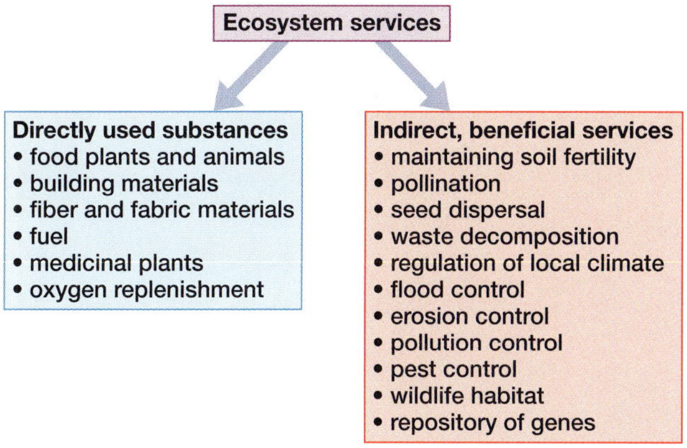

Ecosystem services
Directly used substances
• food plants and animals
• building materials
• fiber and fabric materials
• fuel
• medicinal plants
• oxygen replenishment
Indirect, beneficial services
• maintaining soil fertility
• pollination
• seed dispersal
• waste decomposition
• regulation of local climate
• flood control
• erosion control
• pollution control
• pest control
• wildlife habitat
• repository of genes

Figure 30-1 Ecosystem services

least concern: 55%
not enough data
critically endangered: 3%
endangered: 7%
vulnerable: 12%
near threatened: 12%
depends on conservation projects: 1%

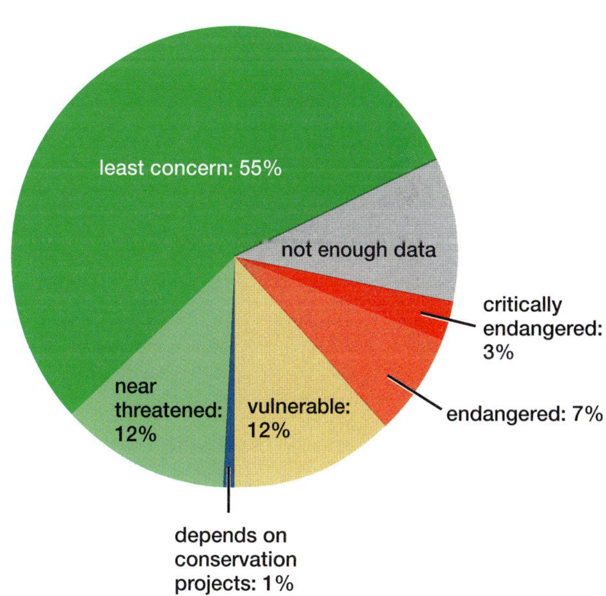

Figure 30-6 Pie chart

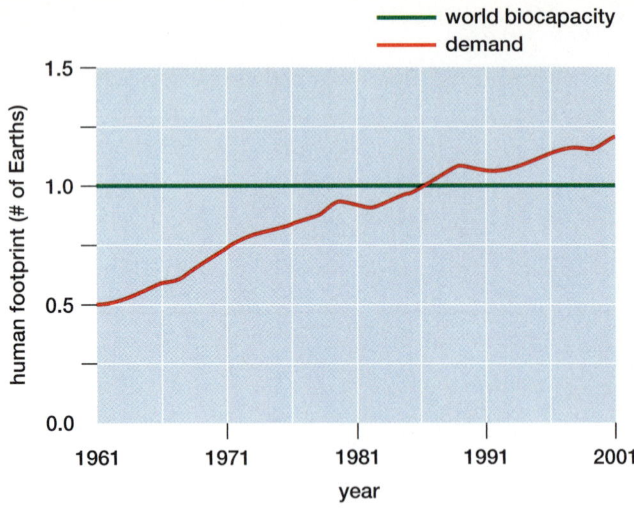

Figure 30-7 Earth's estimated biocapacity

Figure 30-8b Crop patterns in rainforest

Figure 30-9 Patches of forest

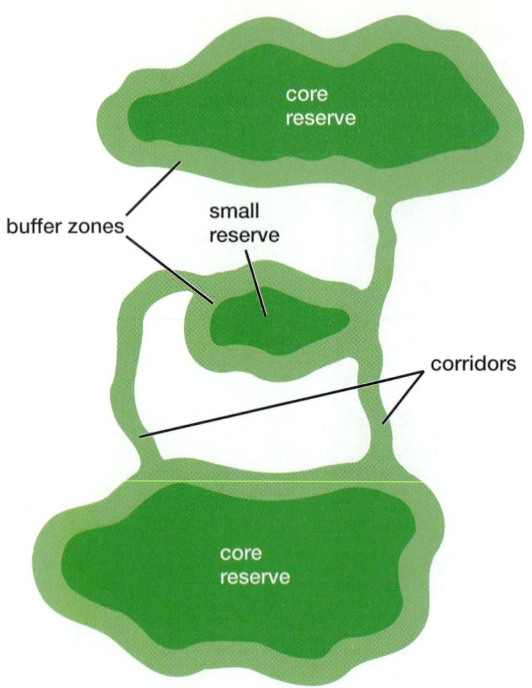

Figure 30-12 Core and corridors

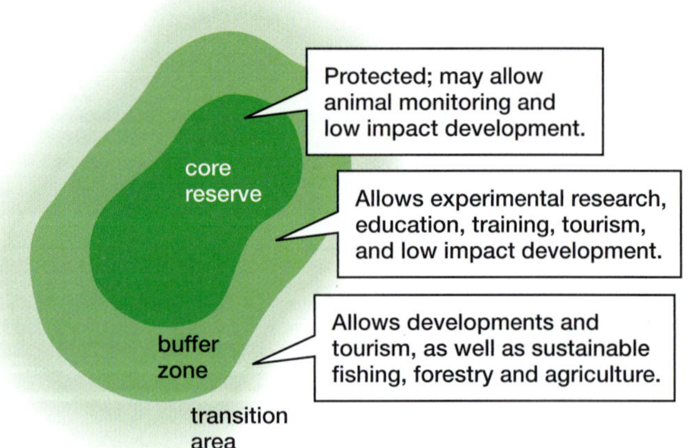

Figure 30-14 Core reserve

Figure E30-2b Florida Everglades restoration map

Figure E30-4a Turtle digging on beach

Figure 30-4b Baby turtles heading to sea

Figure 30-4c Map of Brazil showing Tamar location

Figure E30-5a Map of Yellowstone National Park

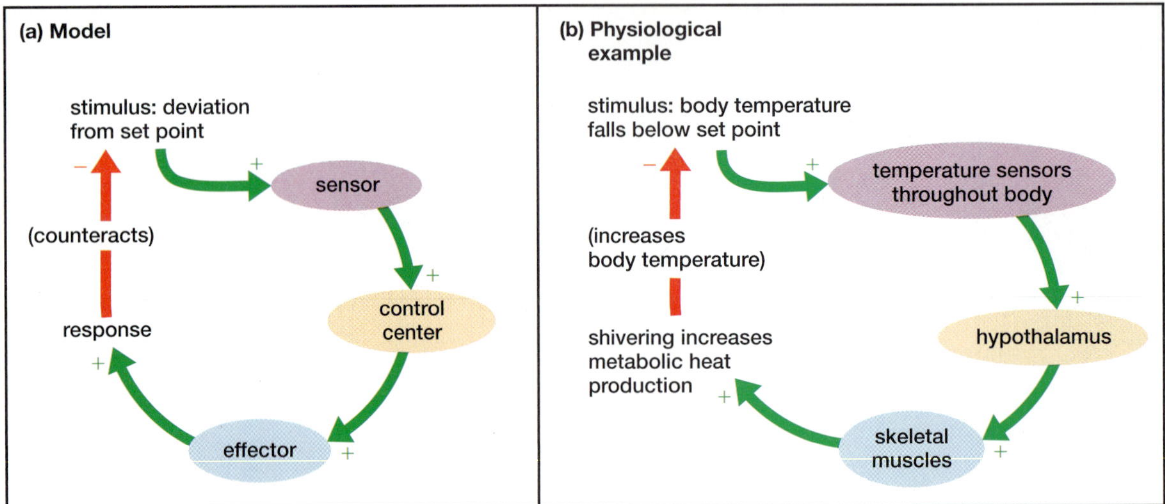

Figure 31-2 Positive and negative feedback

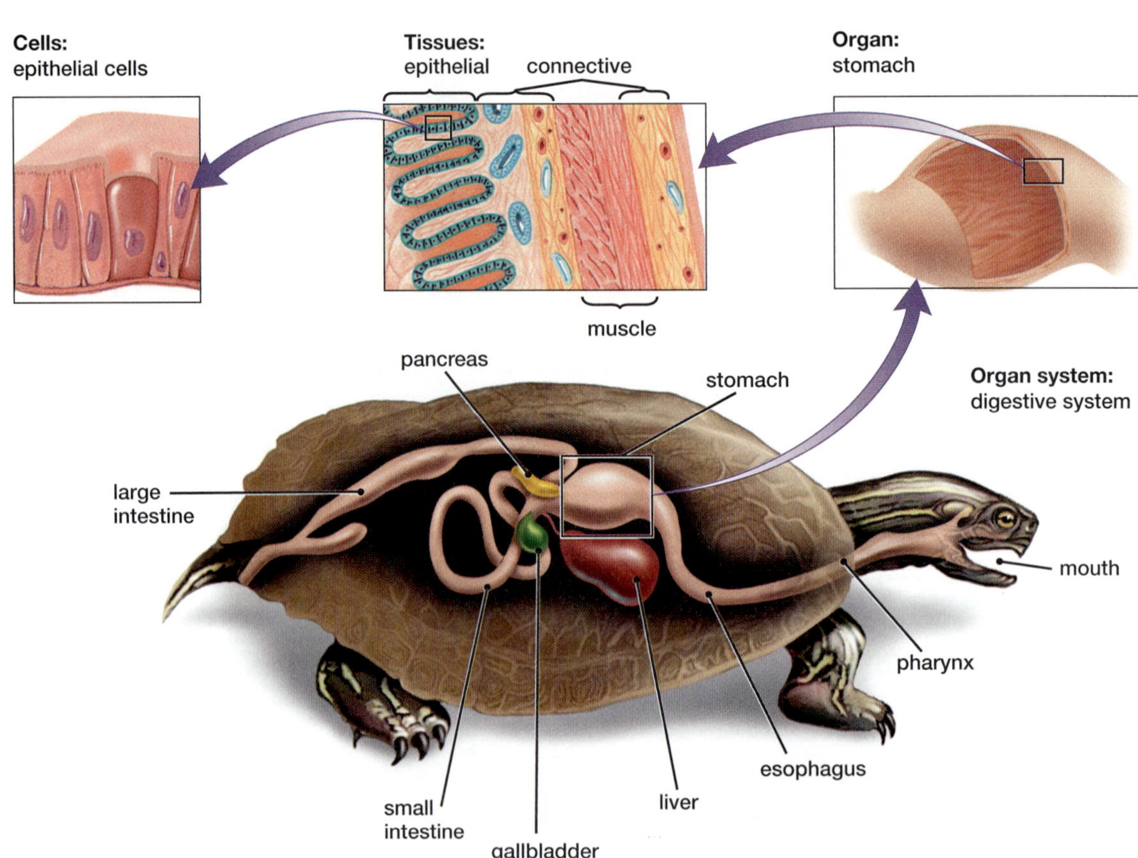

Figure 31-3 Cell/tissue/organ/organ system

(a) **Lining of lungs (simple)**

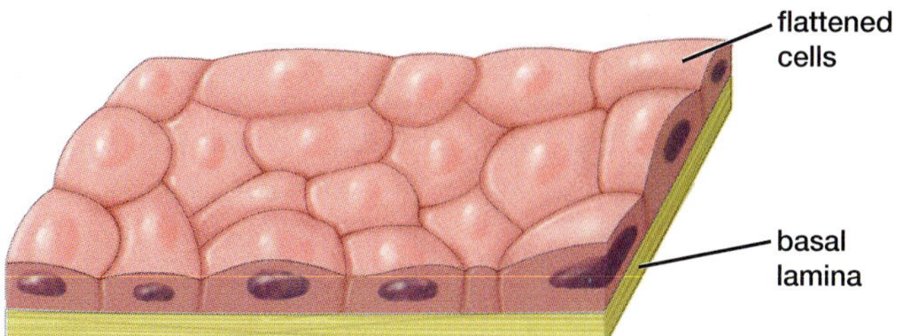

(b) **Lining of trachea (simple)**

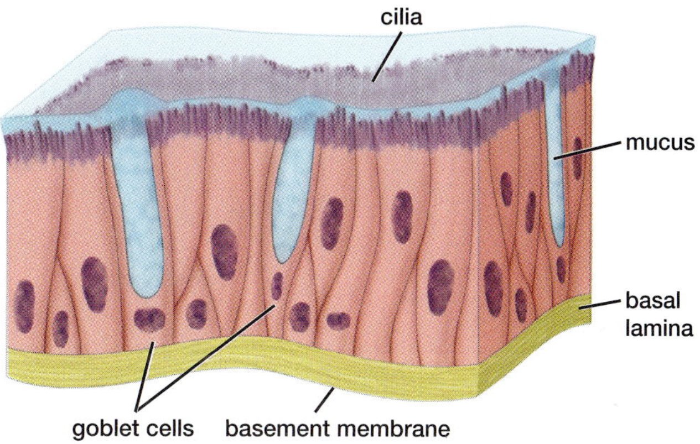

(c) **Skin epidermis (stratified)**

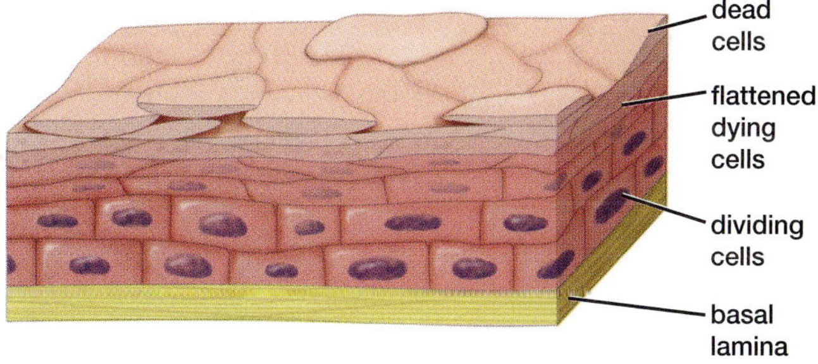

Figure 31-4 Epithelial tissue

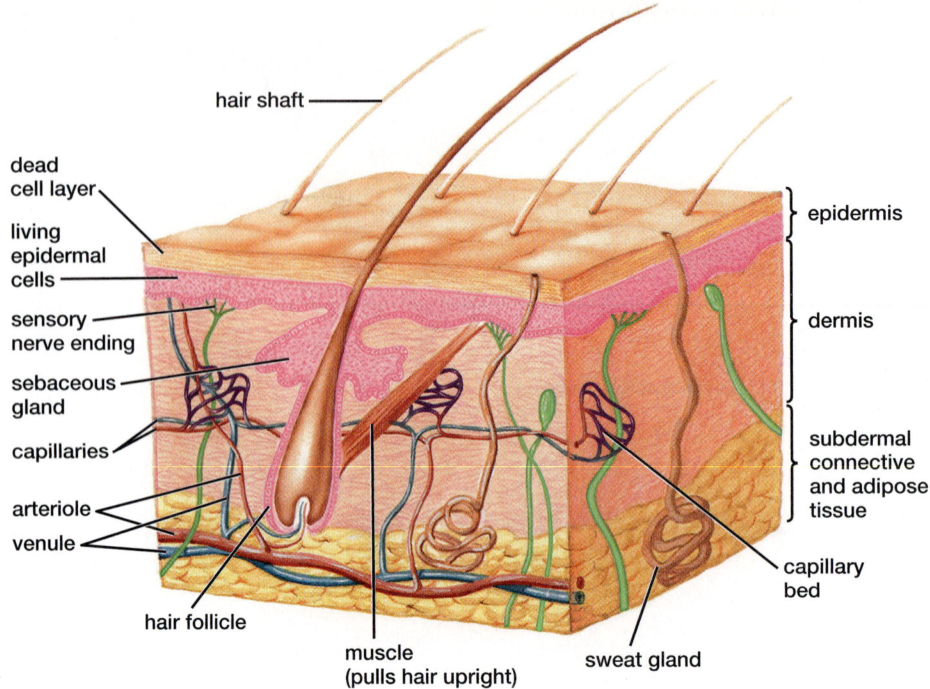

Figure 31-11 Skin is an organ

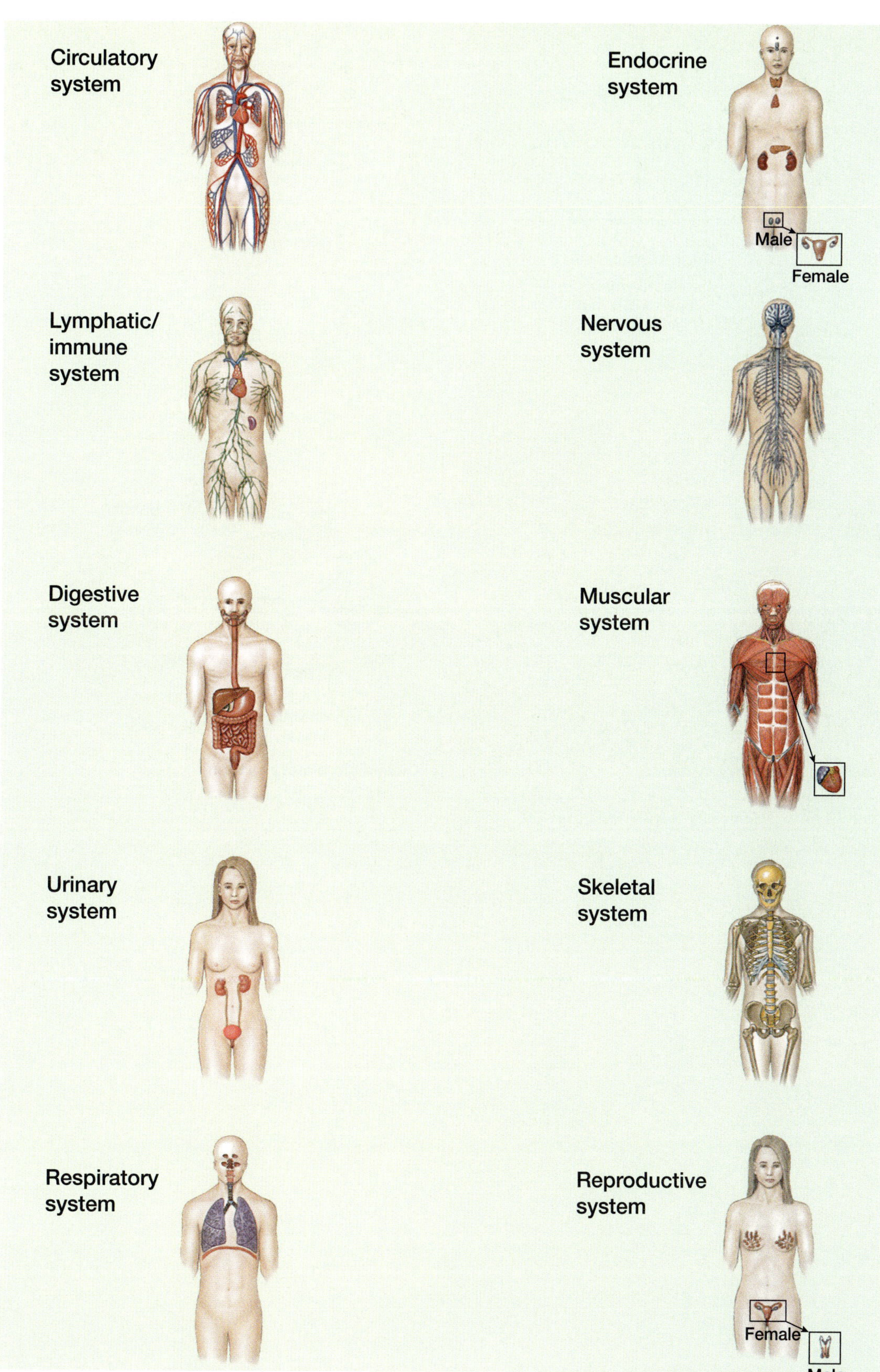

Table 31-1 Organ systems

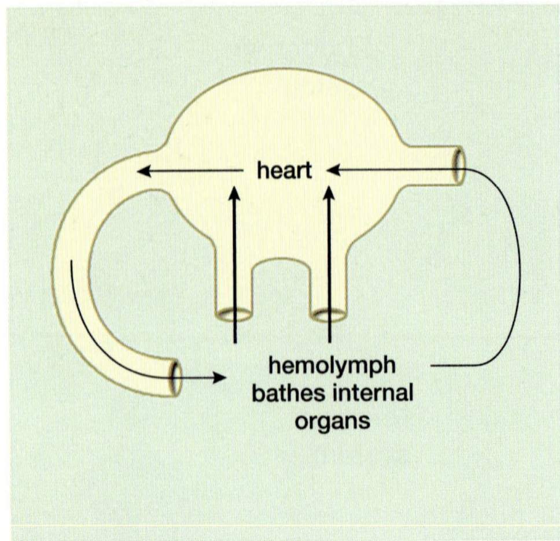

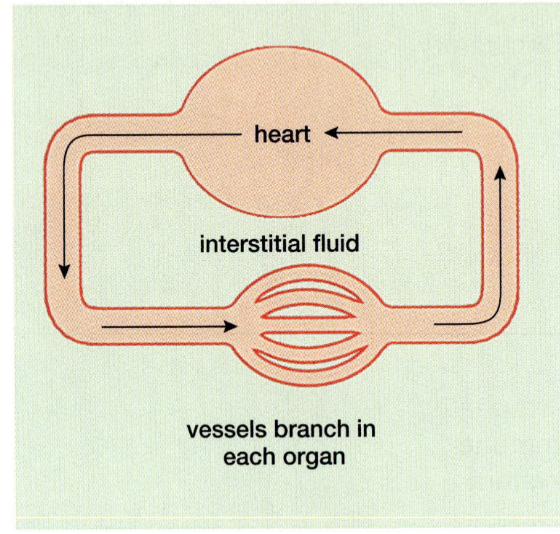

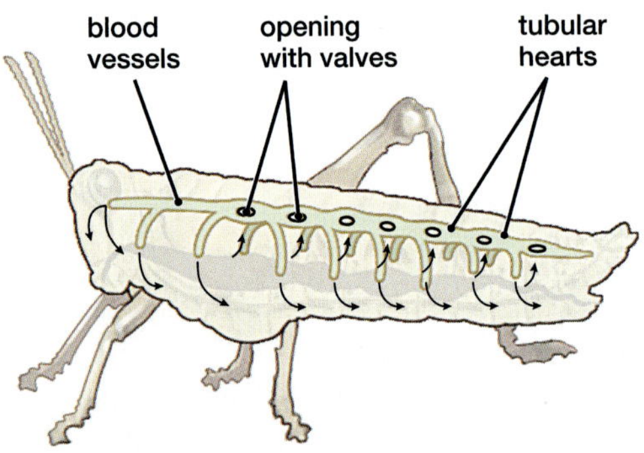

(a) Open circulatory system

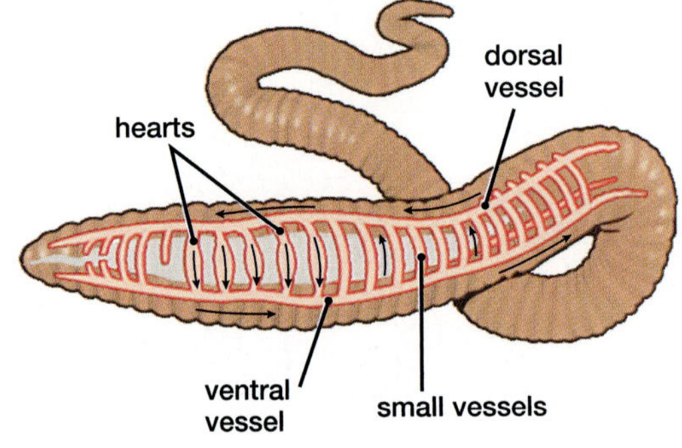

(b) Closed circulatory system

Figure 32-1 Open, closed circulatory systems

(a) Fish
gill capillaries
ventricle
atrium
body capillaries
(b) Amphibians, most reptiles
lung capillaries
atria
ventricle
body capillaries
(c) Mammals, birds
lung capillaries
atria
ventricle
body capillaries

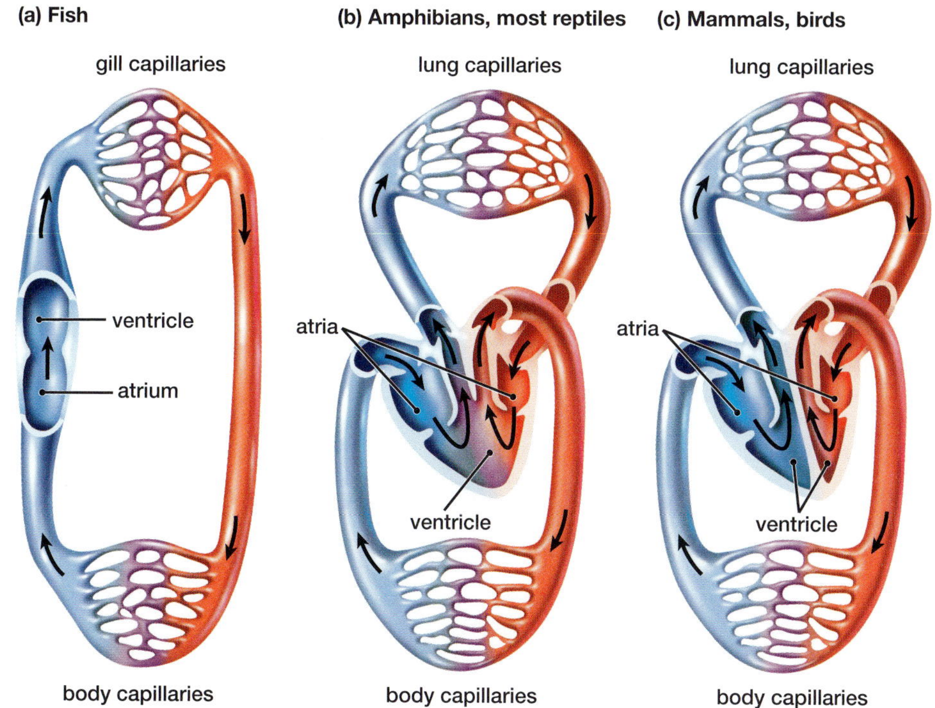

Figure 32-2 Evolution of vertebrate heart

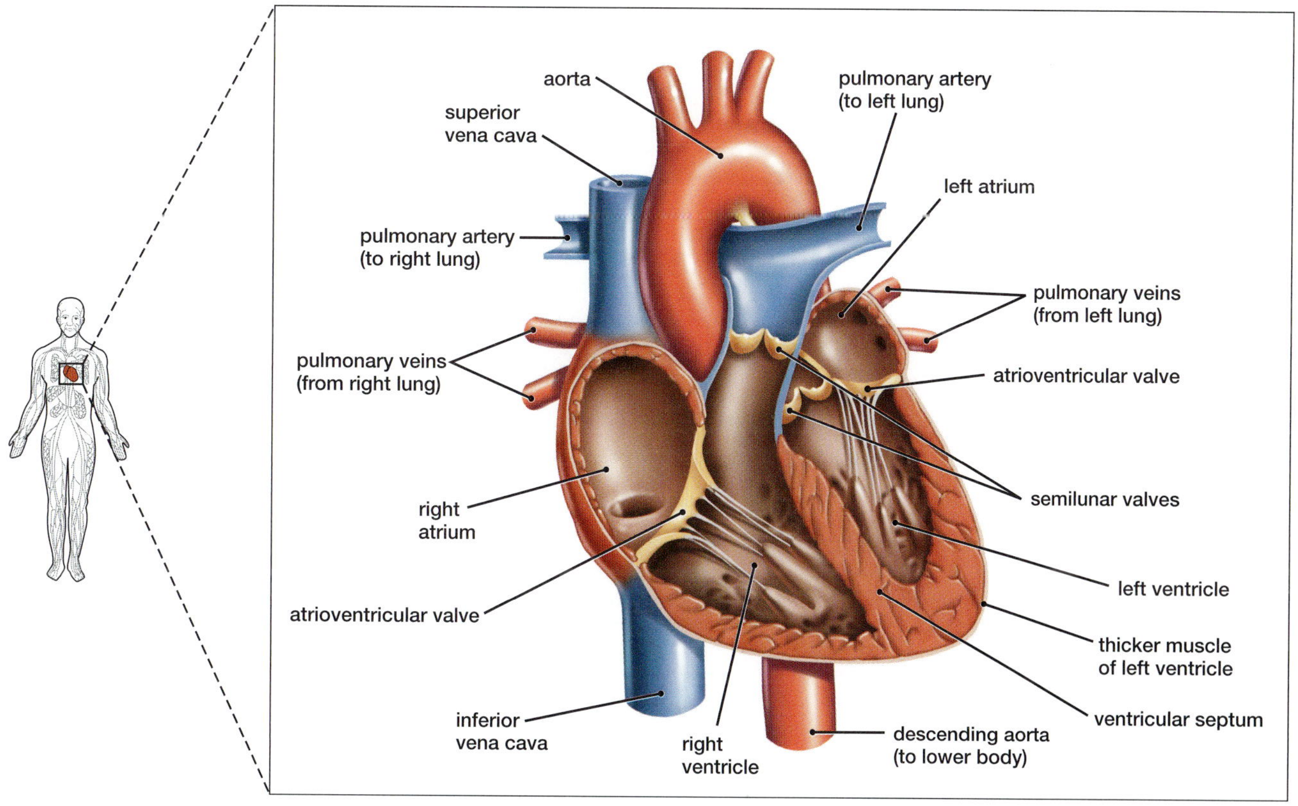

aorta
superior vena cava
pulmonary artery (to right lung)
pulmonary veins (from right lung)
right atrium
atrioventricular valve
inferior vena cava
right ventricle
pulmonary artery (to left lung)
left atrium
pulmonary veins (from left lung)
atrioventricular valve
semilunar valves
left ventricle
thicker muscle of left ventricle
ventricular septum
descending aorta (to lower body)

Figure 32-3 Human heart

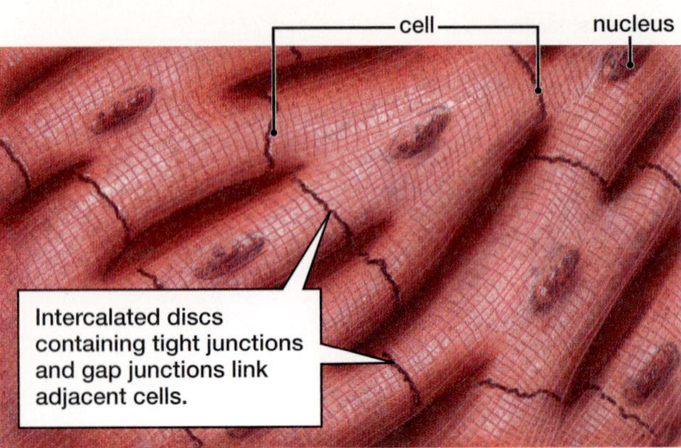

Figure 32-4 Cardiac muscle

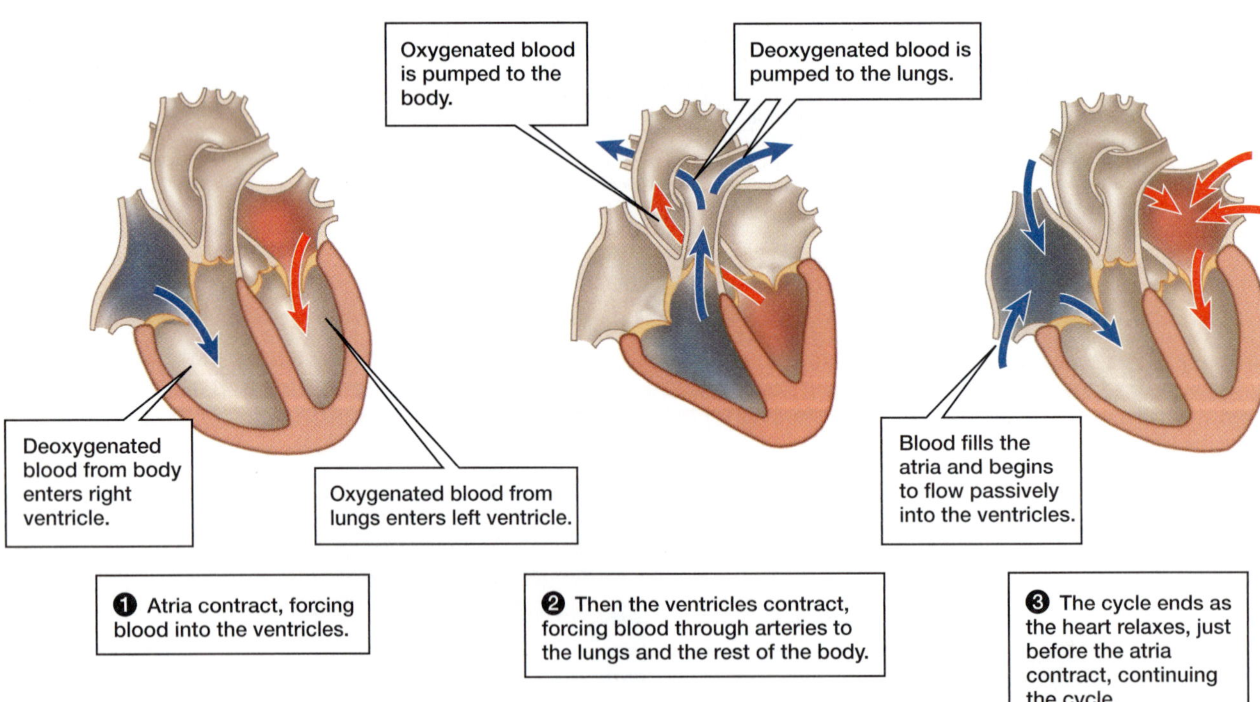

Figure 32-5 Cardiac cycle

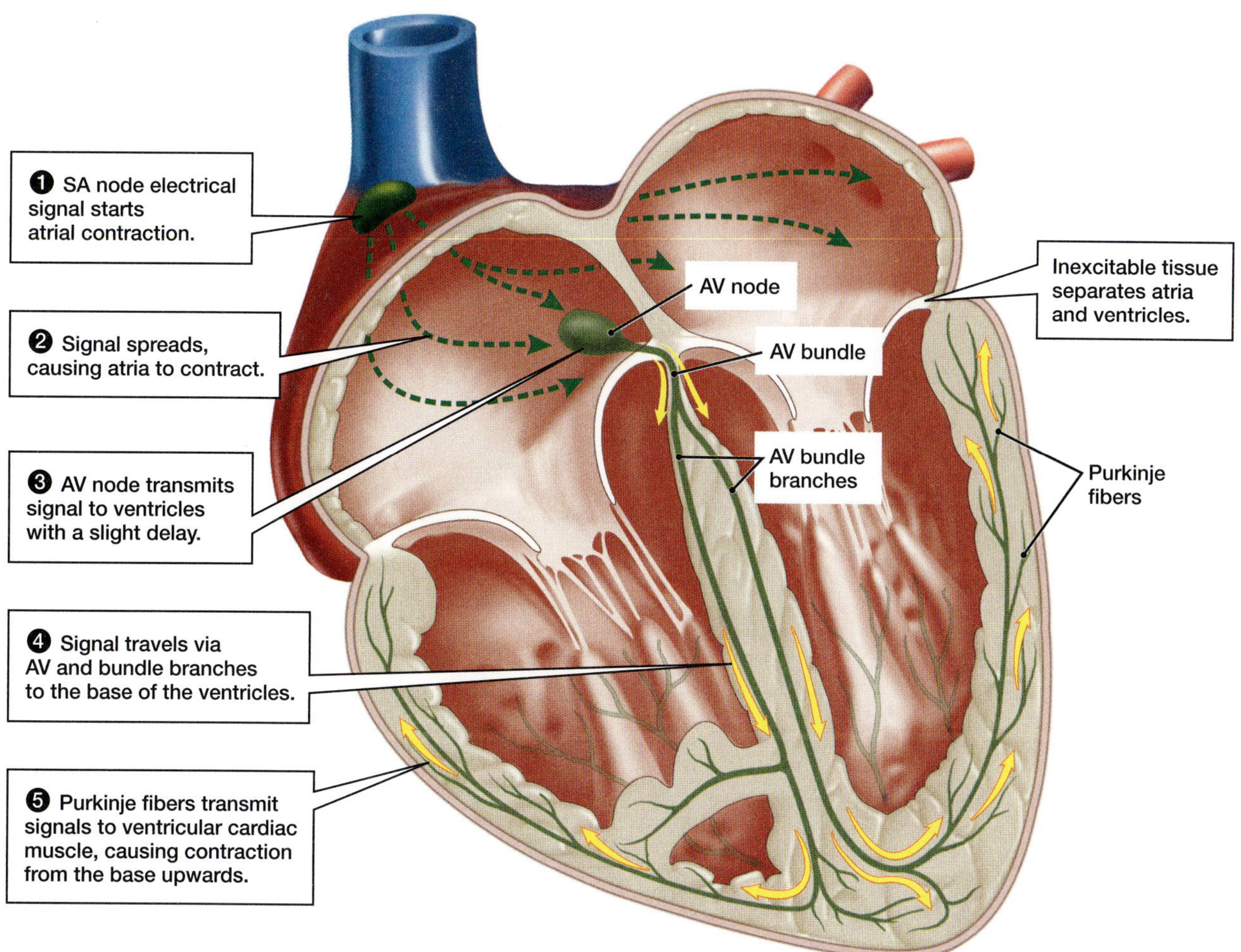

Figure 32-7 Heart's Pacemaker

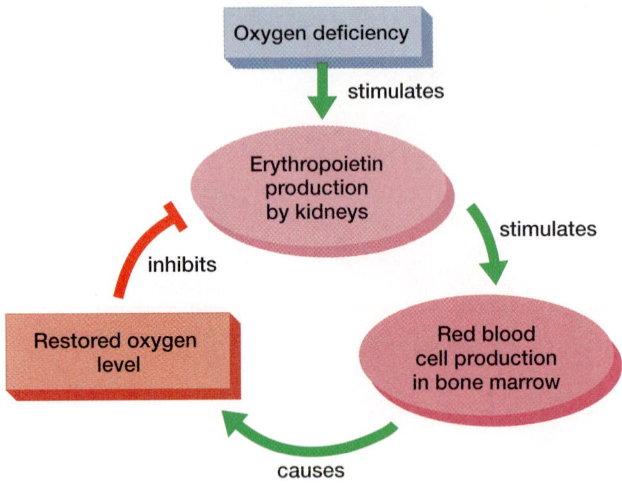

Figure 32-10 Red blood cell regulation

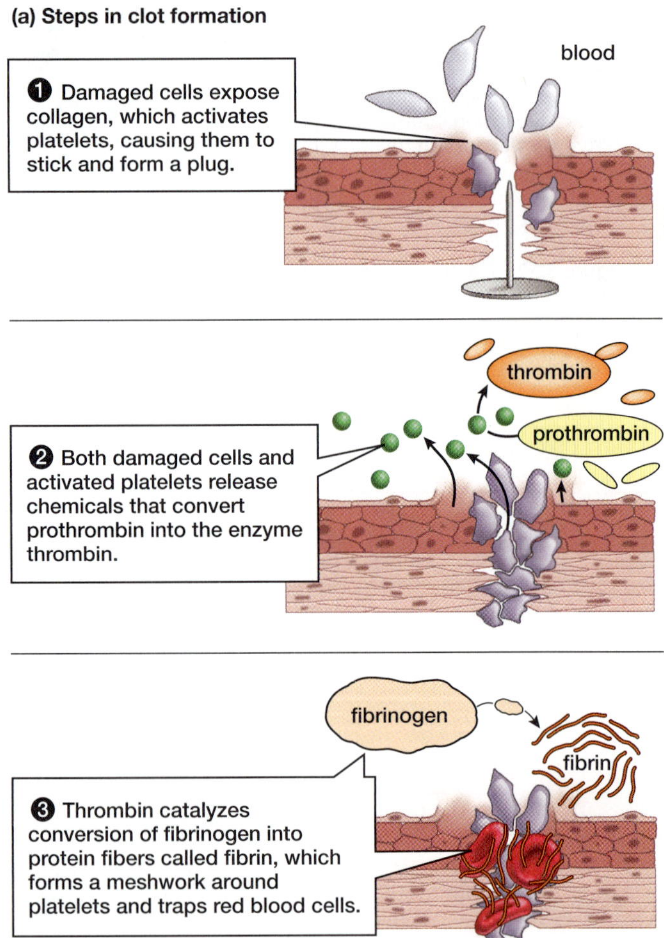

Figure 32-12a Steps in clot formation

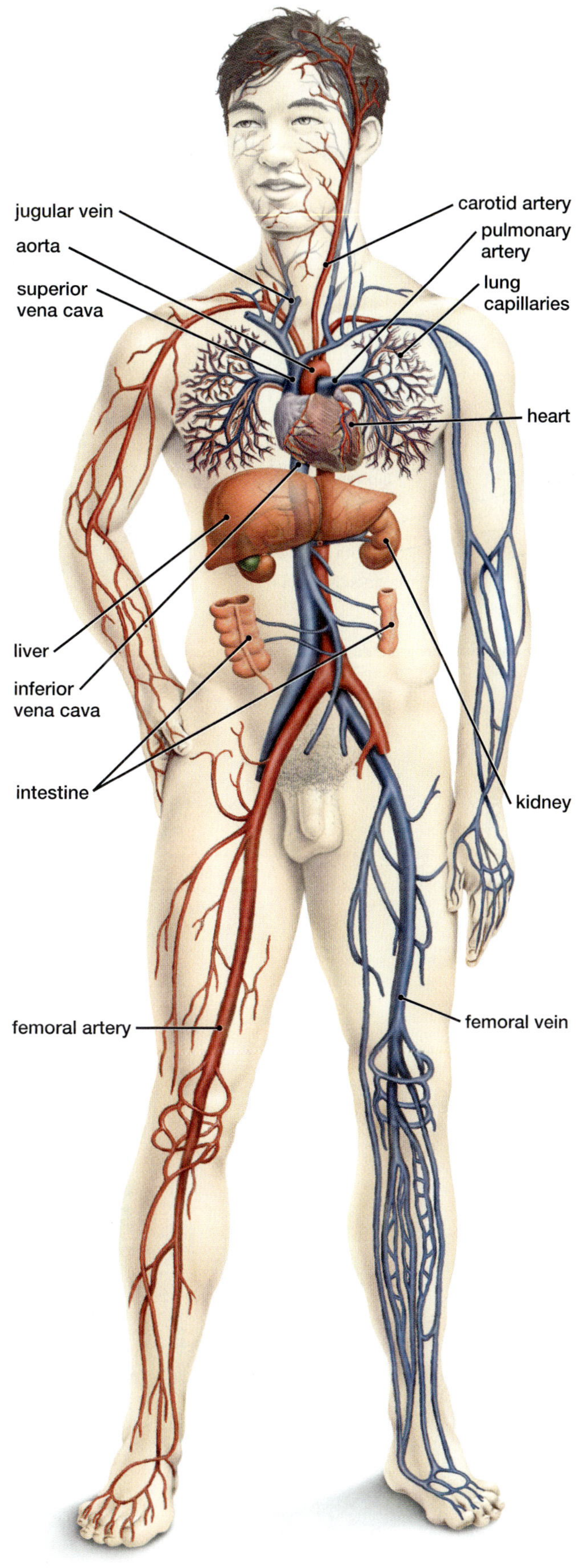

Figure 32-13 Human Circulatory System

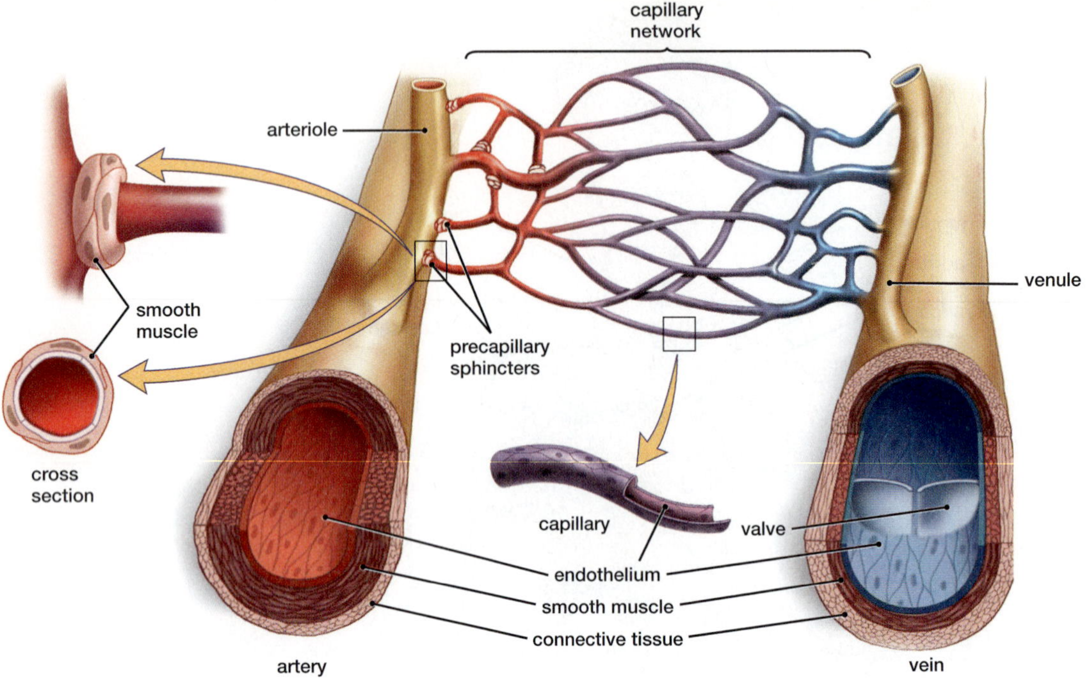

Figure 32-14 Structures and interconnections of blood vessels

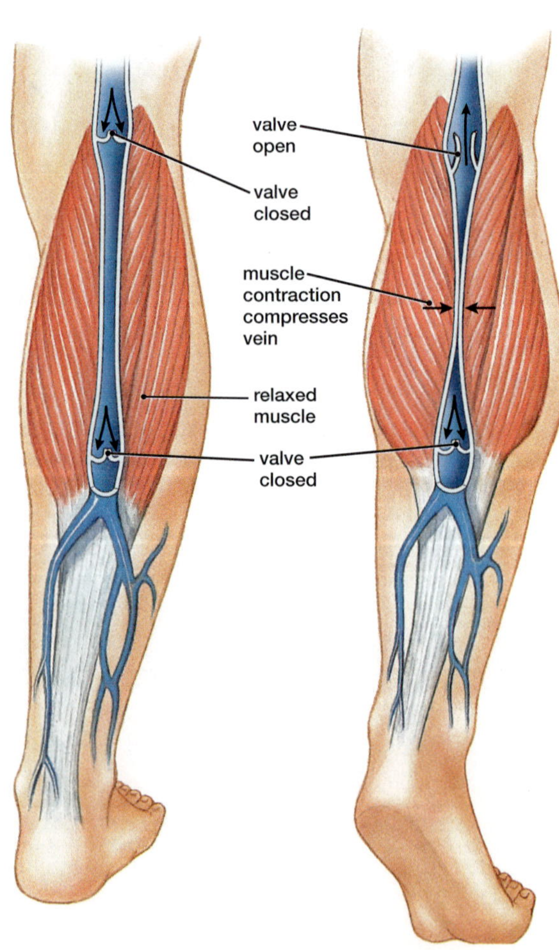

Figure 32-16 Valves direct blood flow in veins

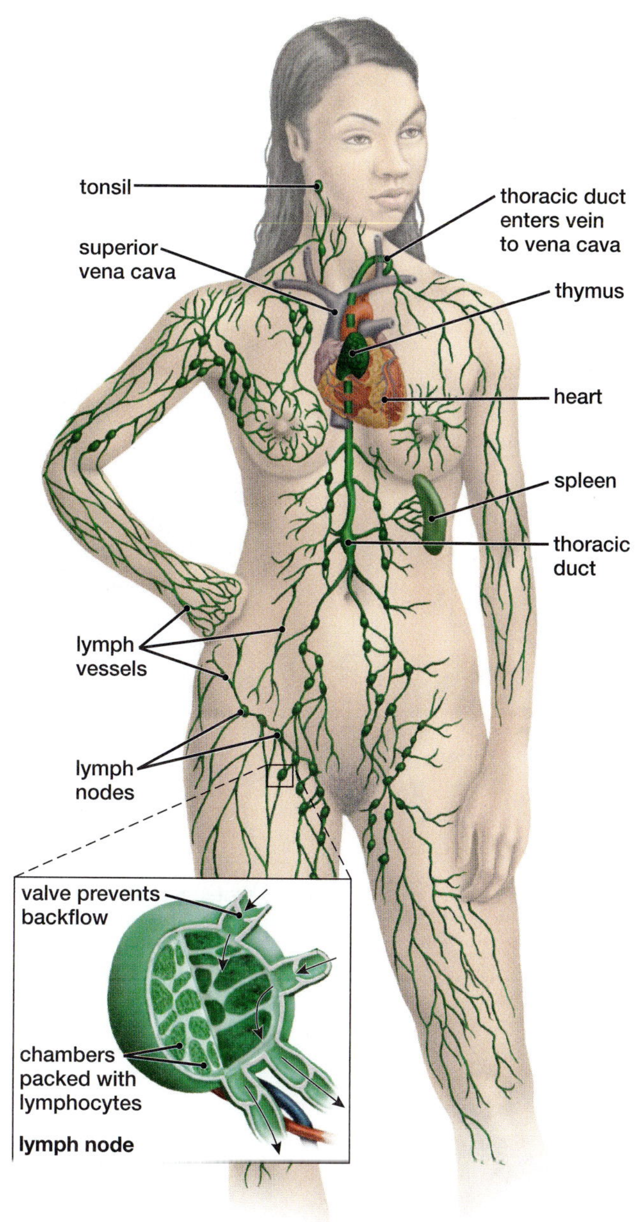

Figure 32-17 Human lymphatic system

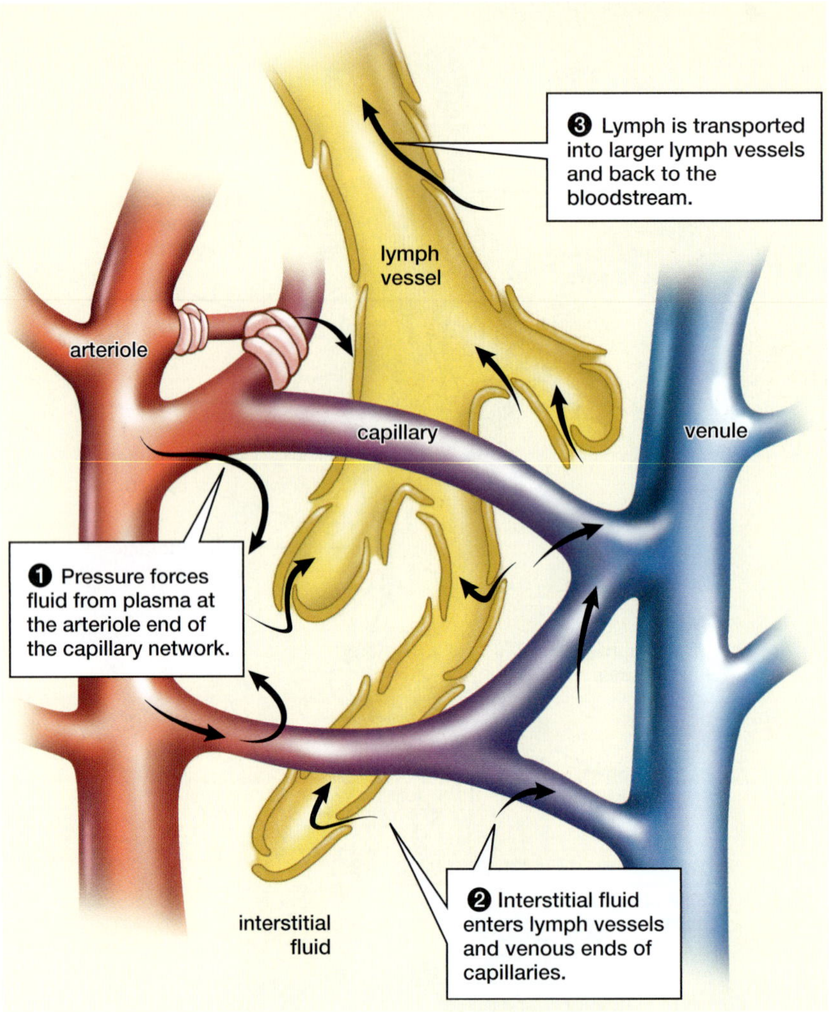

Figure 32-18 Lymph capillary structure

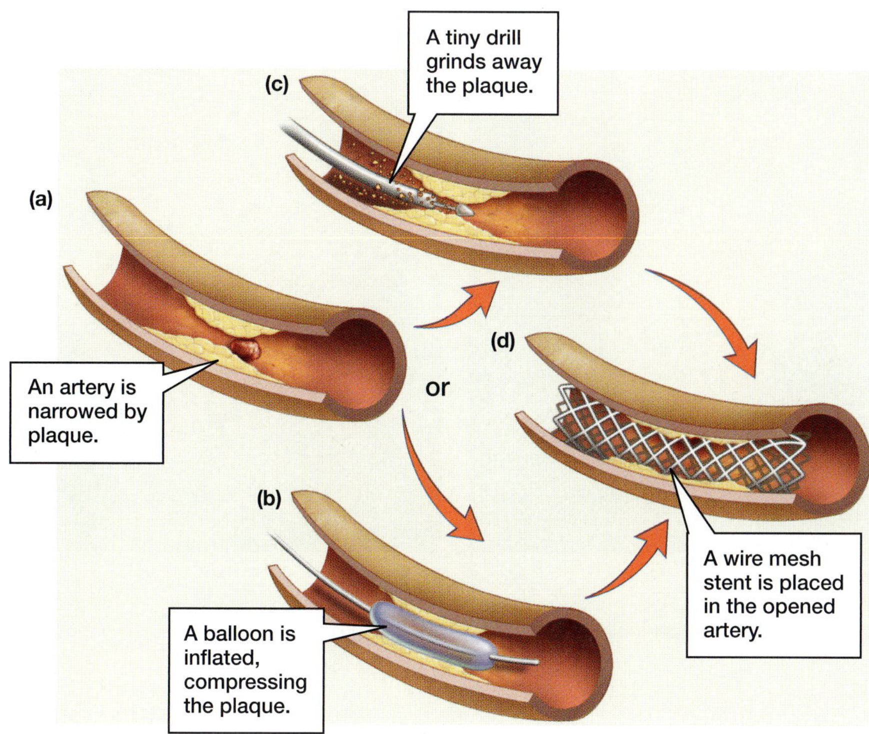

Figure E32-2 Surgery for plaque removal

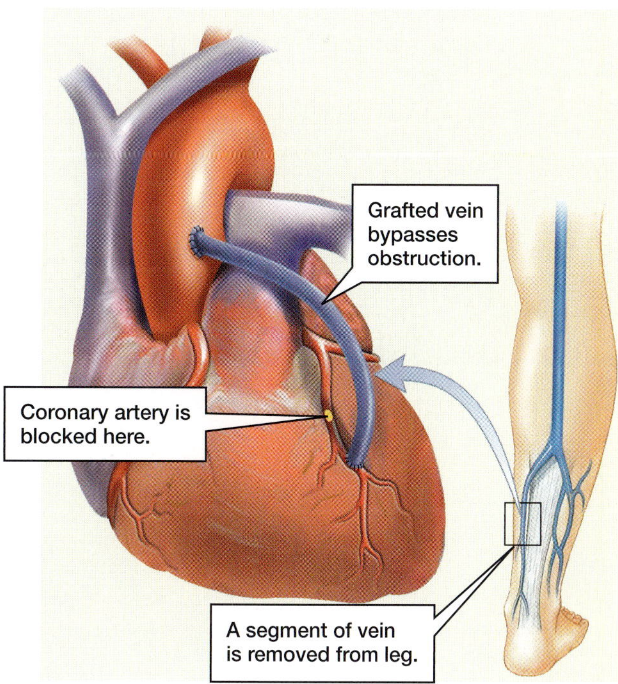

Figure E32-3 Bypass surgery

Figure 33-1 Some animals lack specialized respiratory structures: a) flatworm b) sea jelly c) sponge

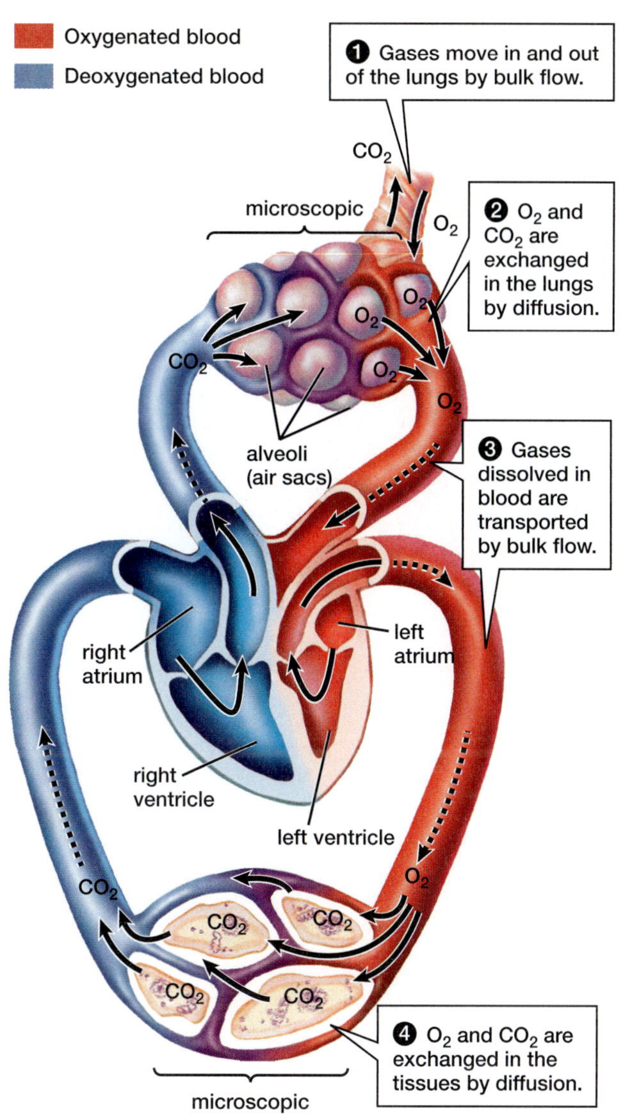

Figure 33-2 Overview of gas exchange

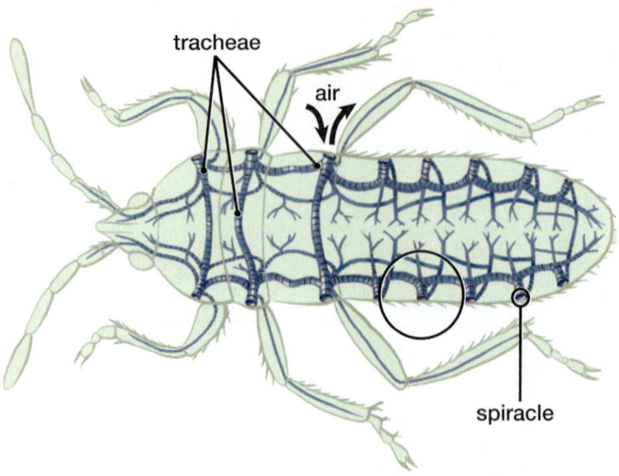

Figure 33-4a Insects breathe with tracheae

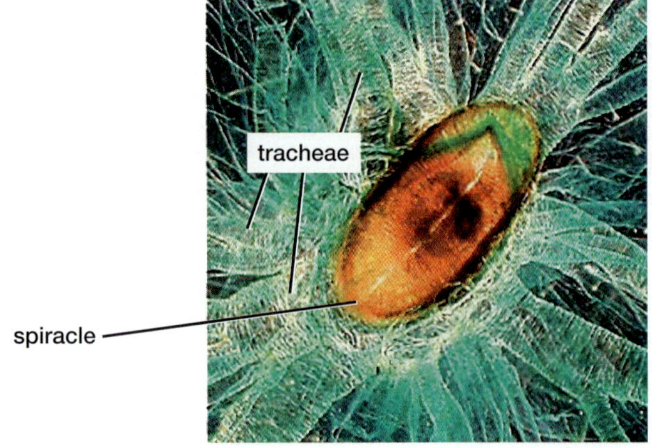

Figure 33-4b Tracheae

Figure 33-5c Snake

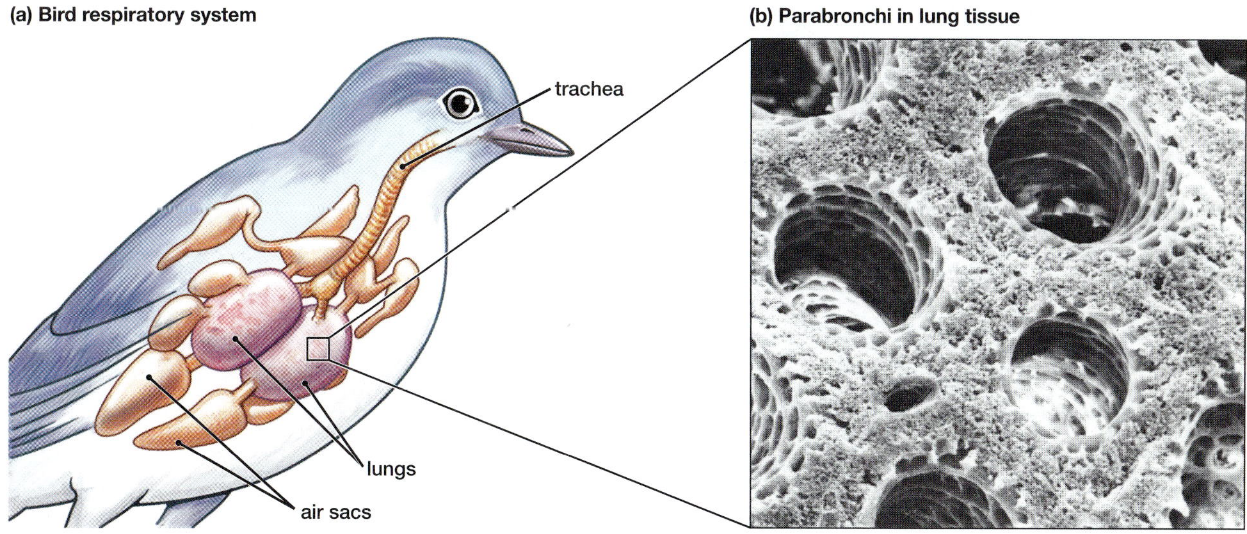

Figure 33-6 Bird respiratory system, parabronchi

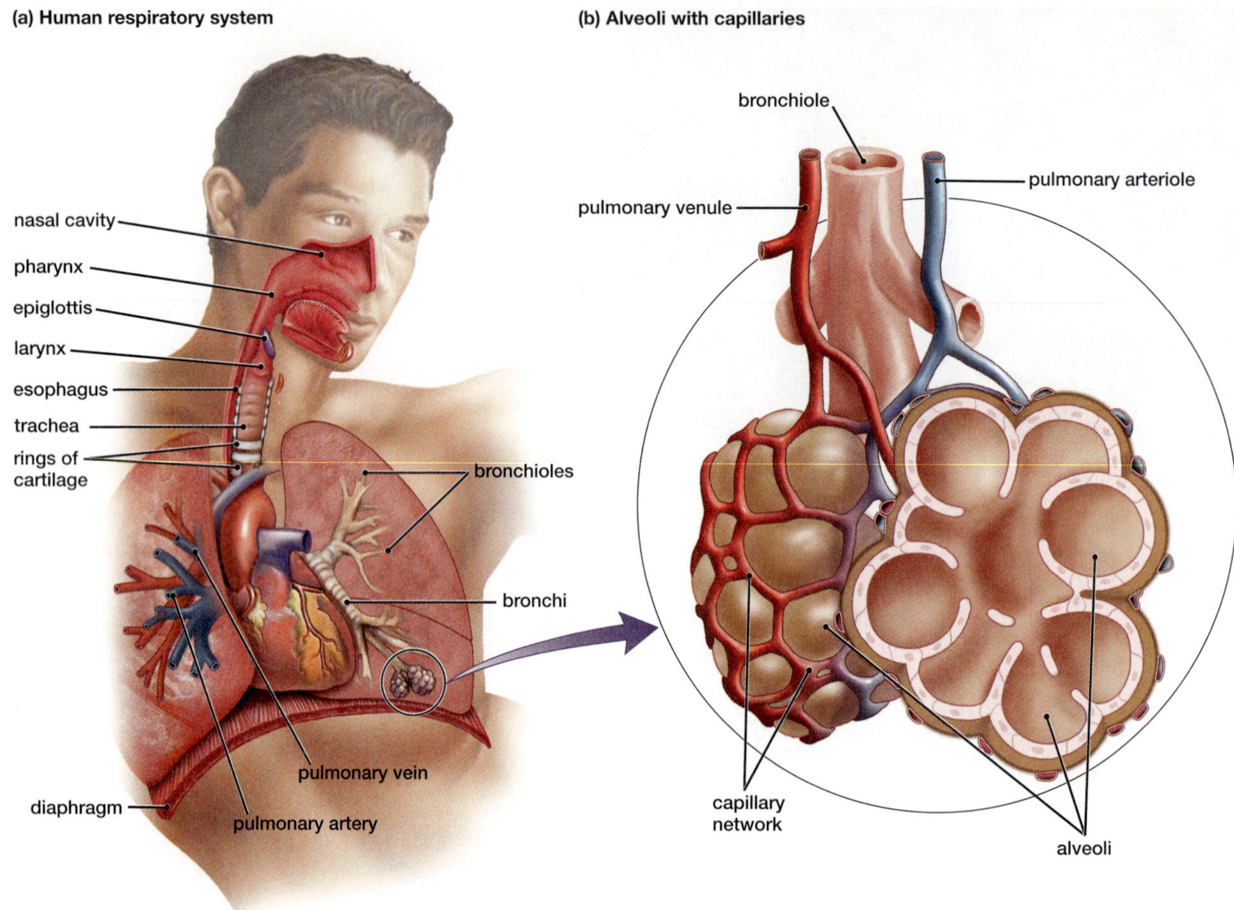

Figure 33-7 Human respiratory system

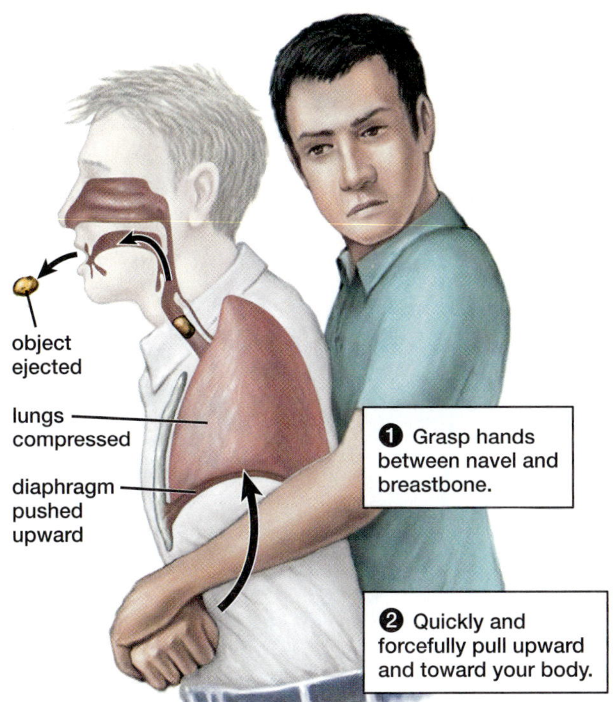

Figure 33-8 Heimlich maneuver

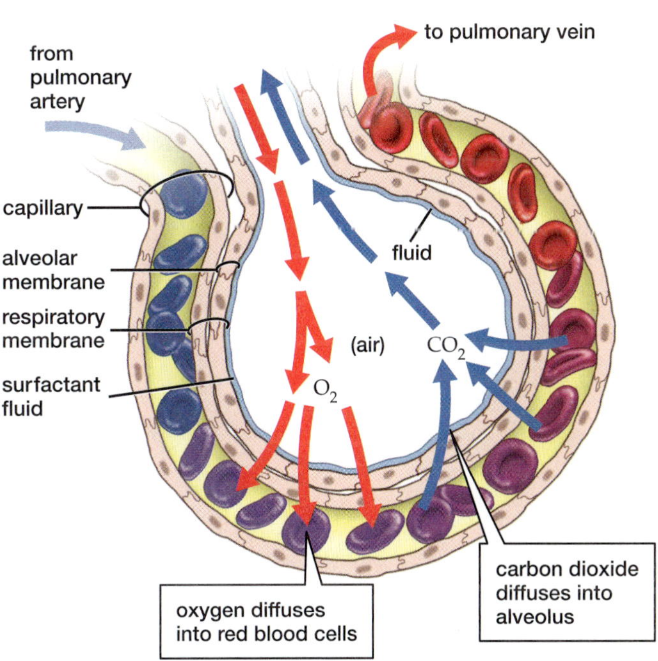

Figure 33-9 Gas exchange

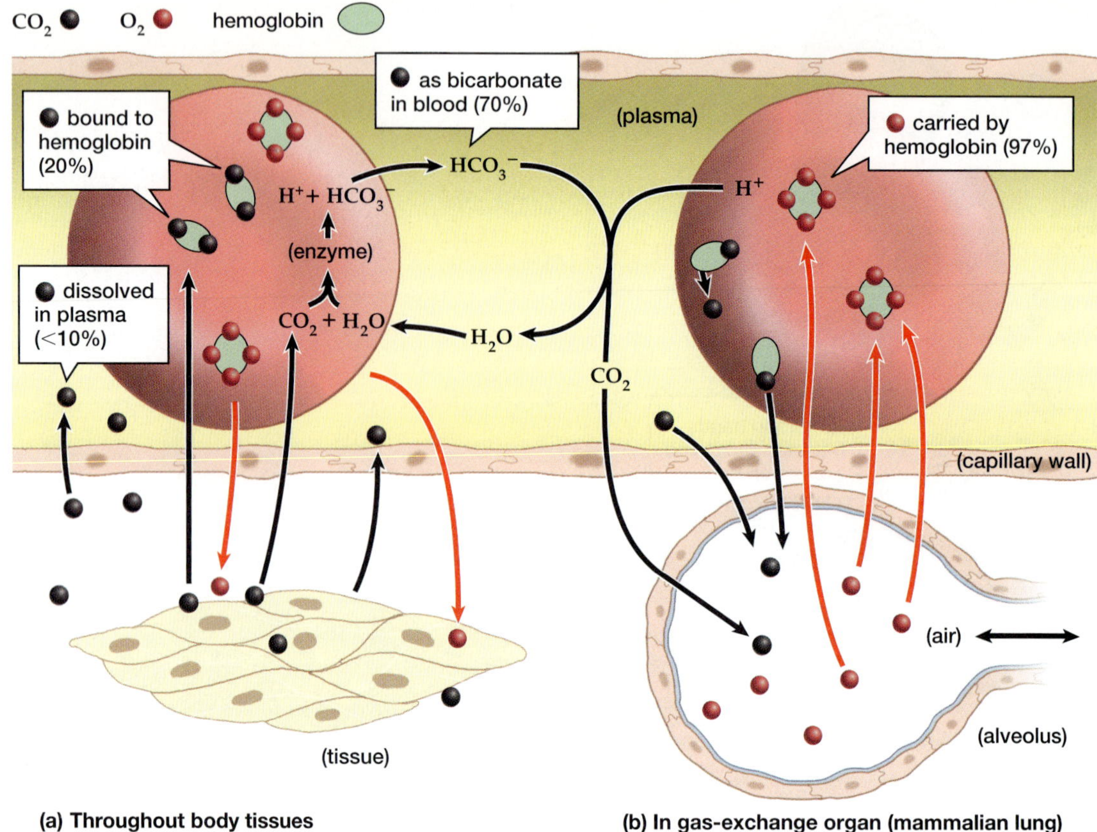

Figure 33-10 Chemistry and mechanism of gas exchange

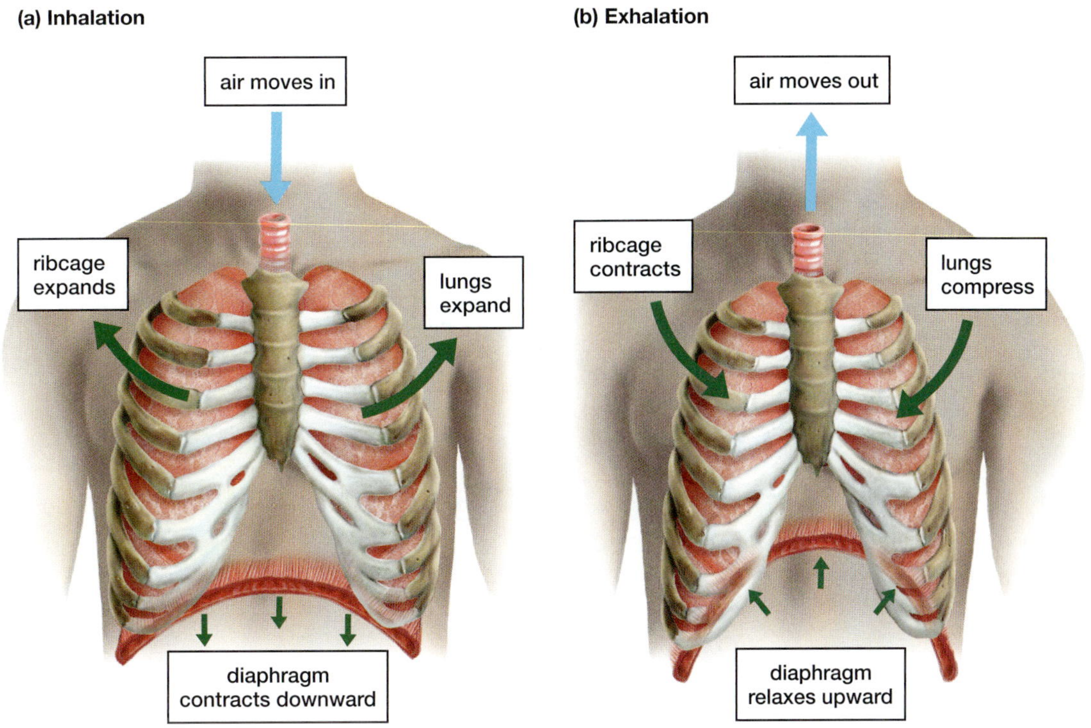

(a) Inhalation
air moves in
ribcage expands
lungs expand
diaphragm contracts downward
(b) Exhalation
air moves out
ribcage contracts
lungs compress
diaphragm relaxes upward

Figure 33-11 Mechanics of breathing

0
25
50
50
100
75
50
50
(a) Concurrent exchange
0
40
90
10
50
100
(b) Countercurrent exchange

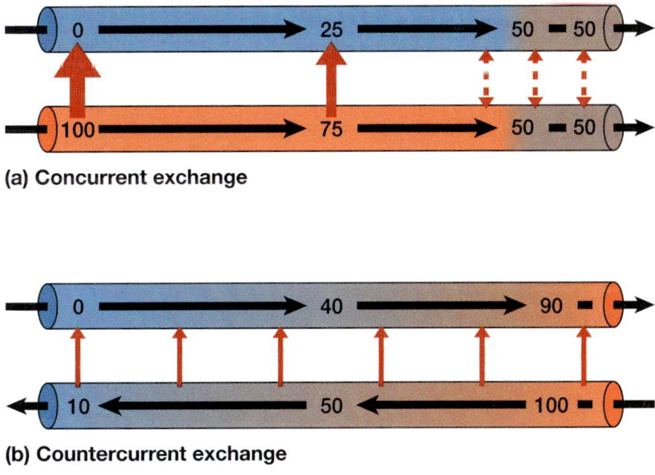

Figure E33-1 Concurrent and countercurrent exchange

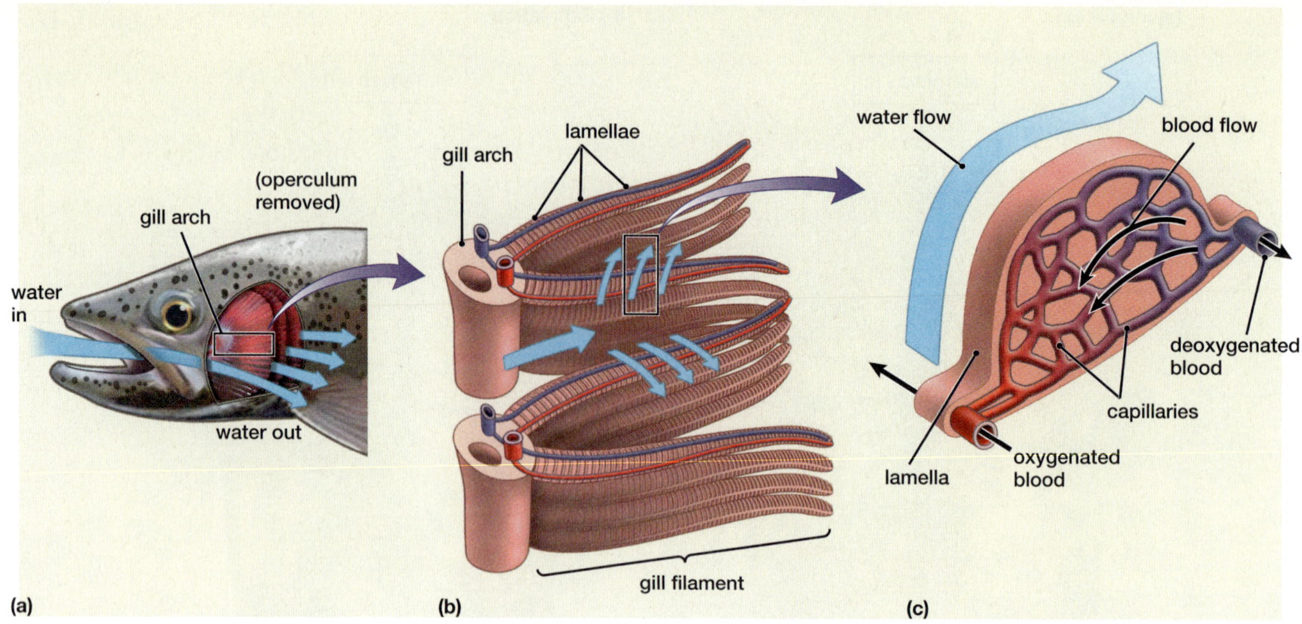

Figure E33-2 How fish gills work

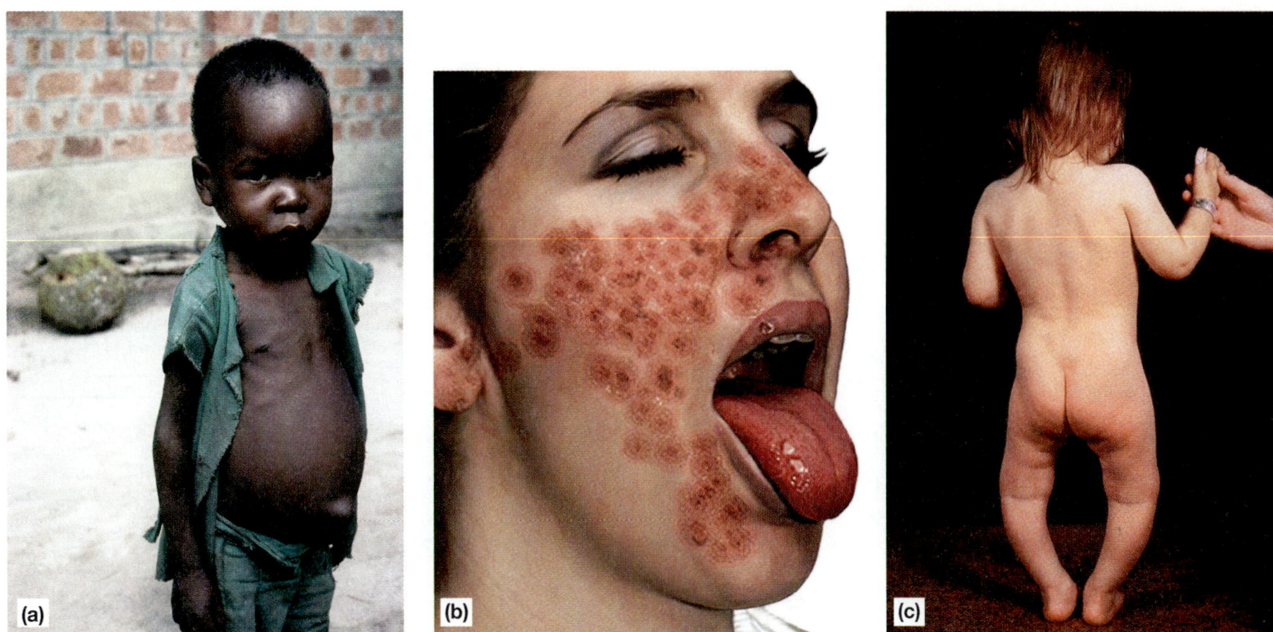

Figure 34-2 Child with kwashiorkor, face of person with pellagra, child with rickets

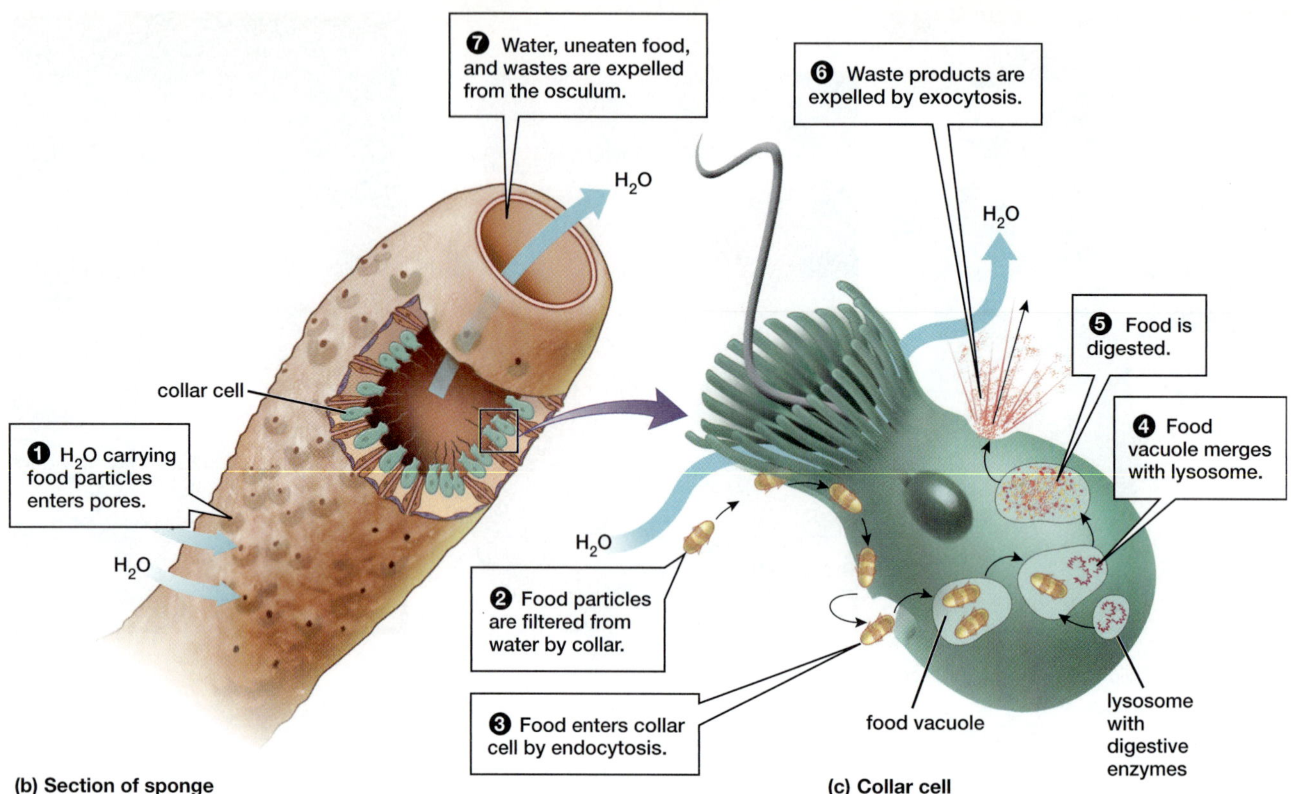

Figure 34-4b,c Intracellular Digestion in sponge

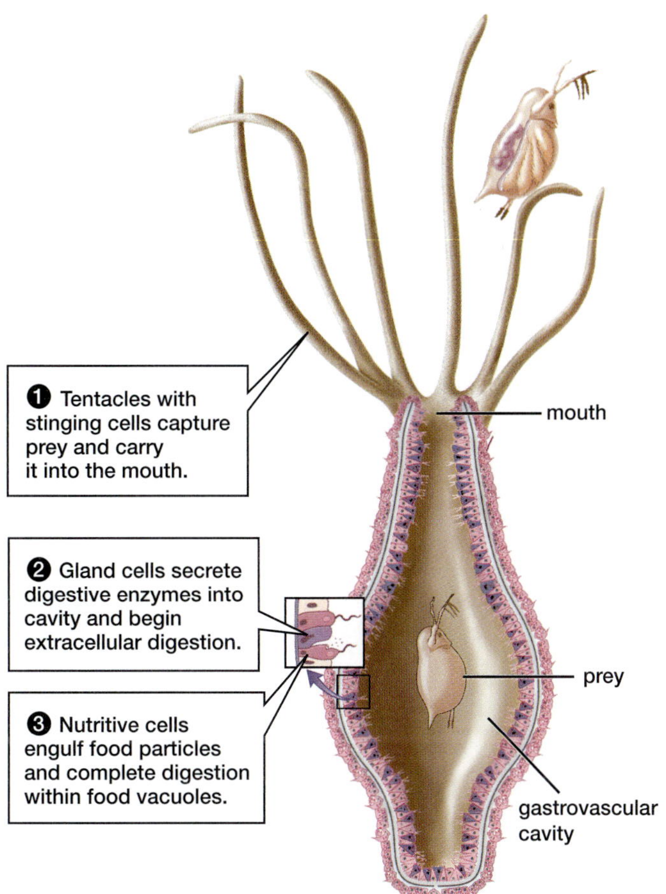

Figure 34-5b Digestion in sac- hydra

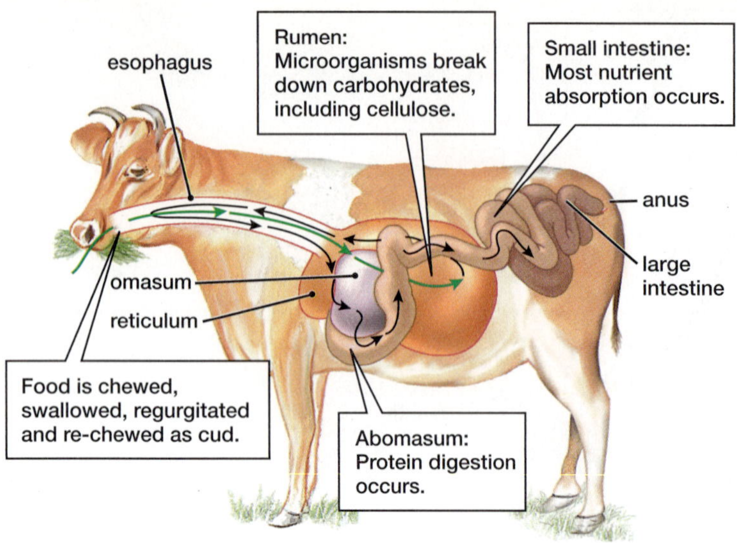

Figure 34-6 Cow digestive tract

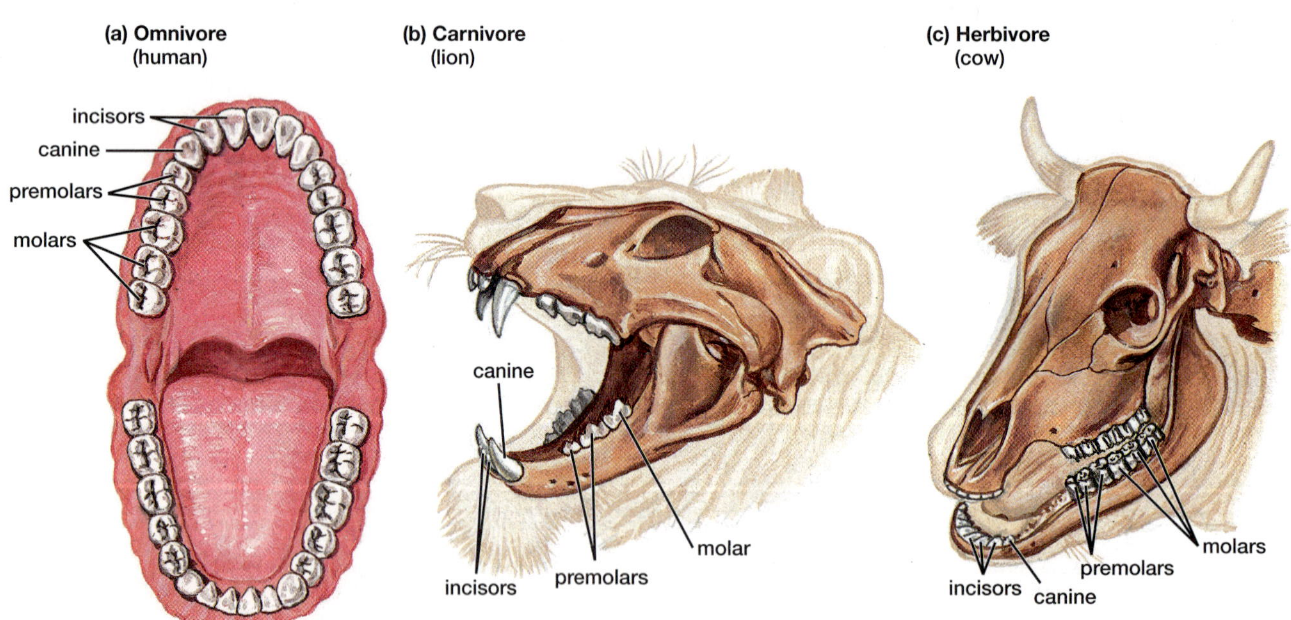

Figure 34-7 Teeth

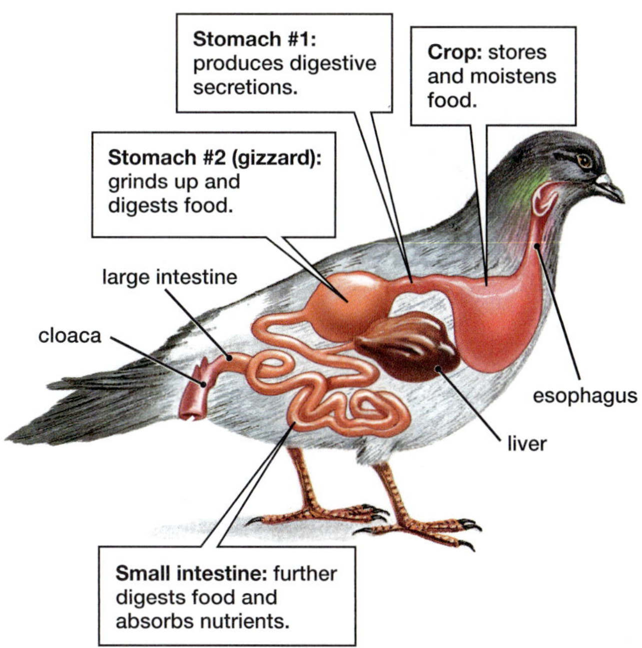

Figure 34-8 Bird digestive tract

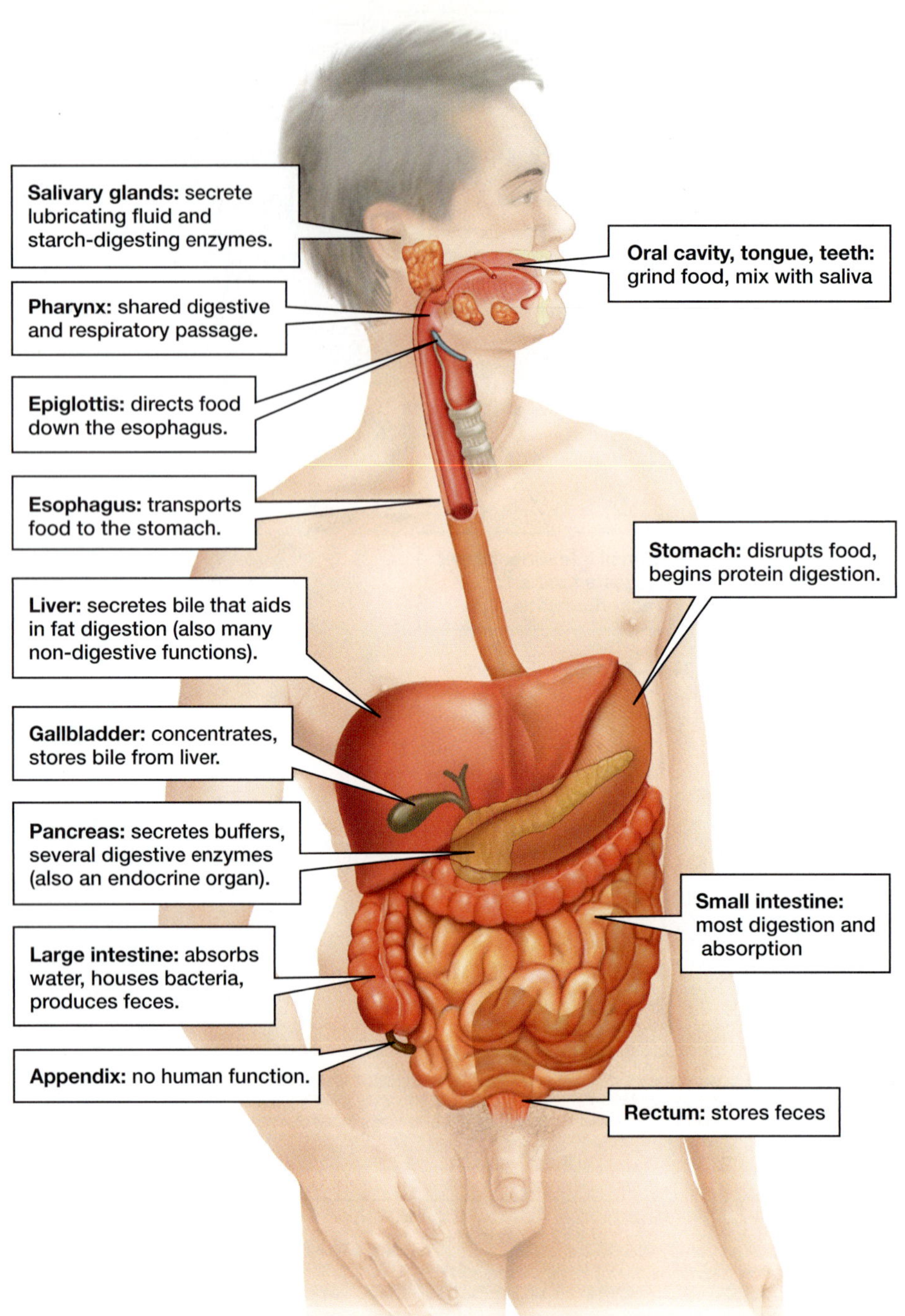

Figure 34-9 Human digestive tract

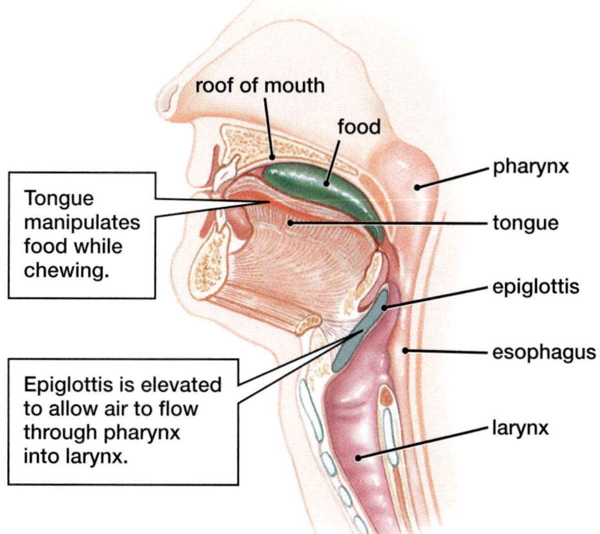

(a) Before swallowing

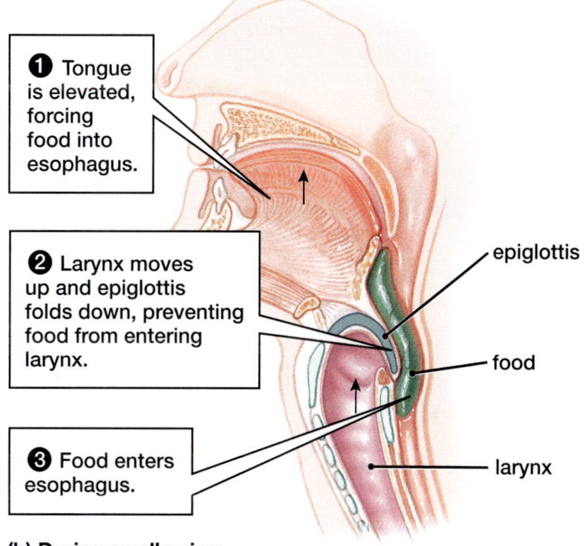

(b) During swallowing

Figure 34-10 Swallowing

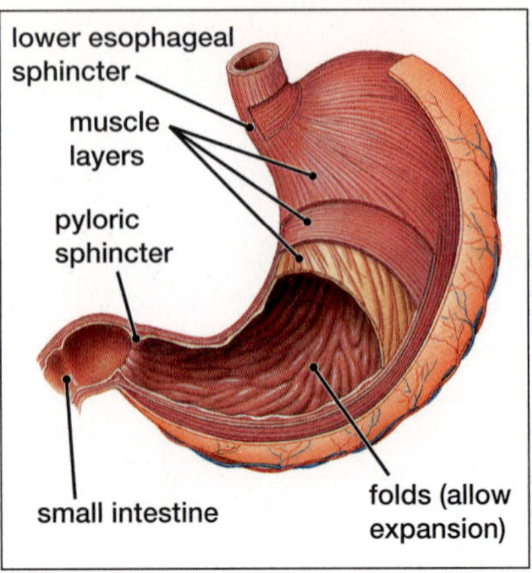

Figure 34-12 Stomach

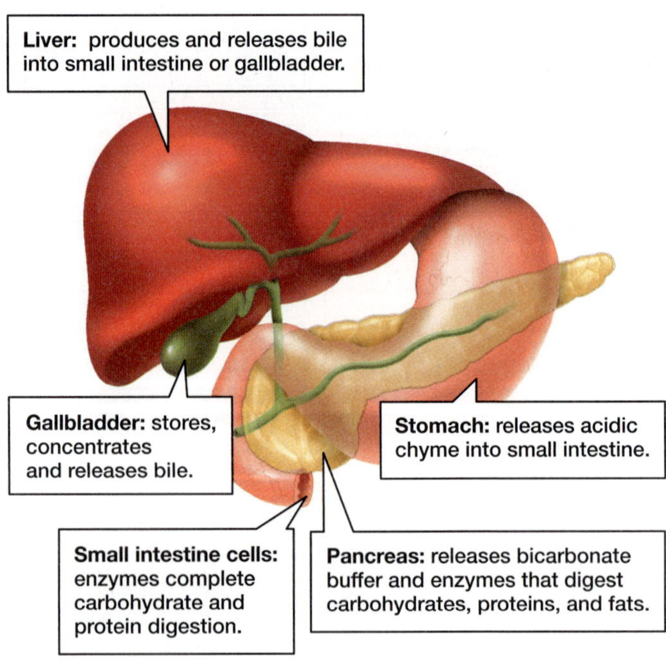

Figure 34-13 Digestion in small intestine

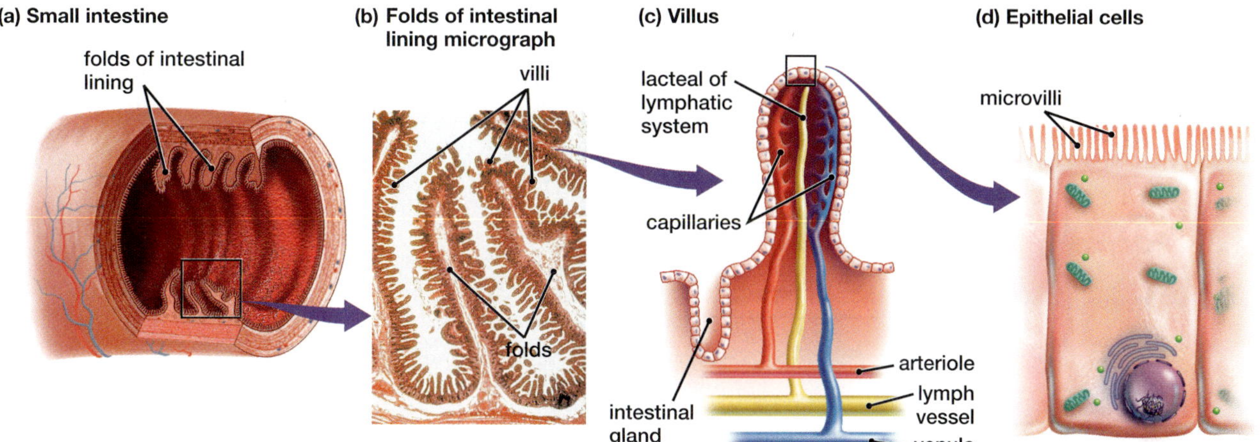

Figure 34-14 Structure of the small intestine

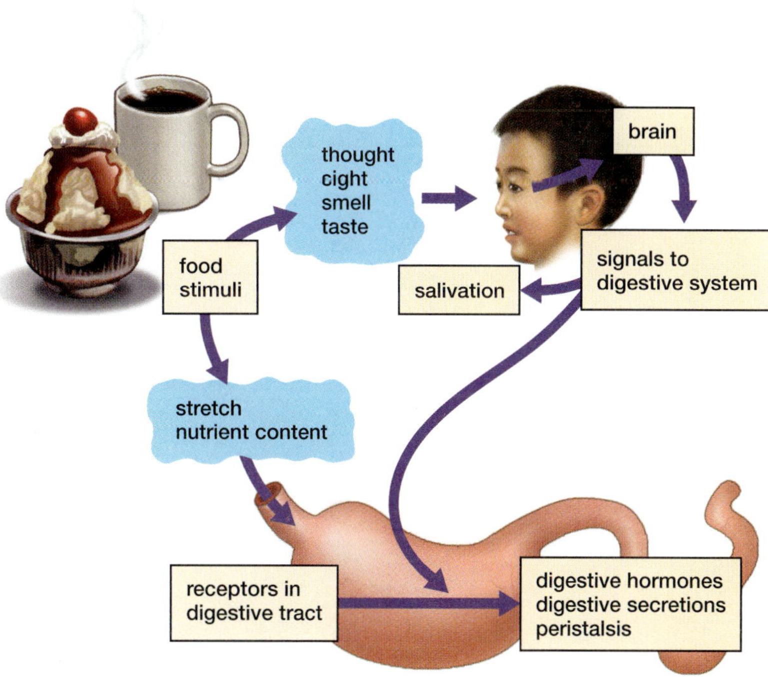

Figure 34-15 Nerves and hormones influence digestion

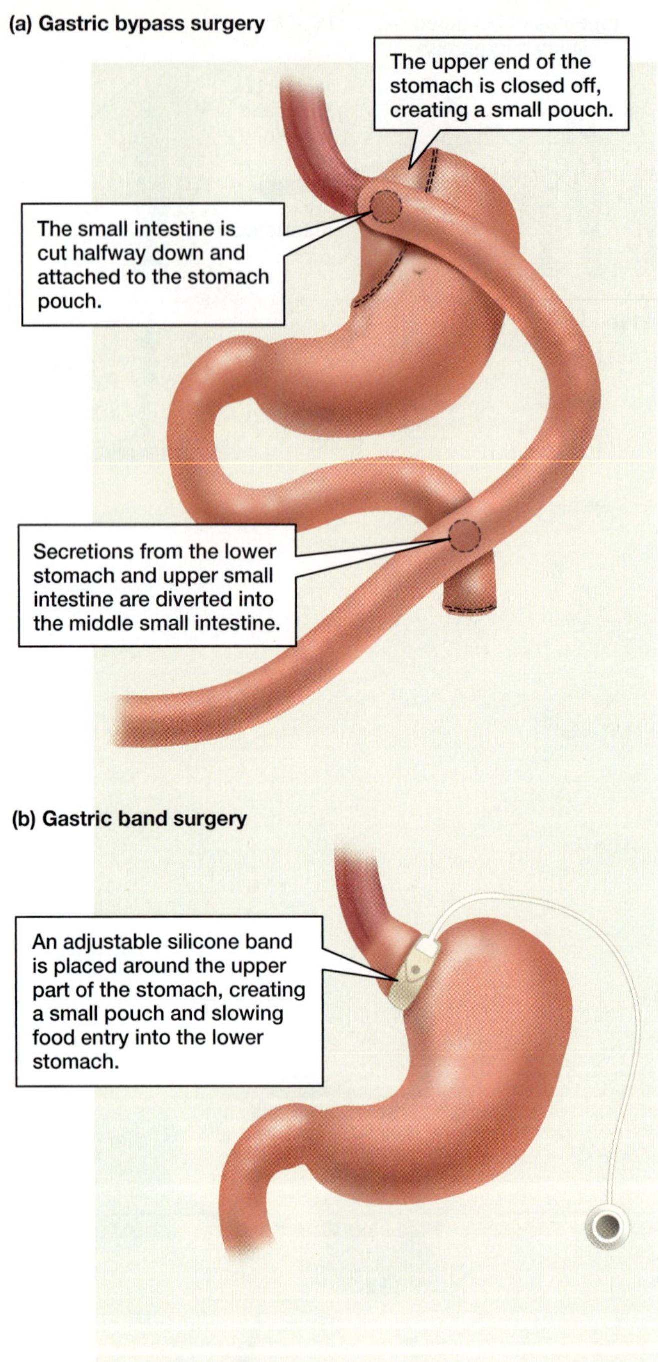

Figure E34-1 Weight-loss surgeries

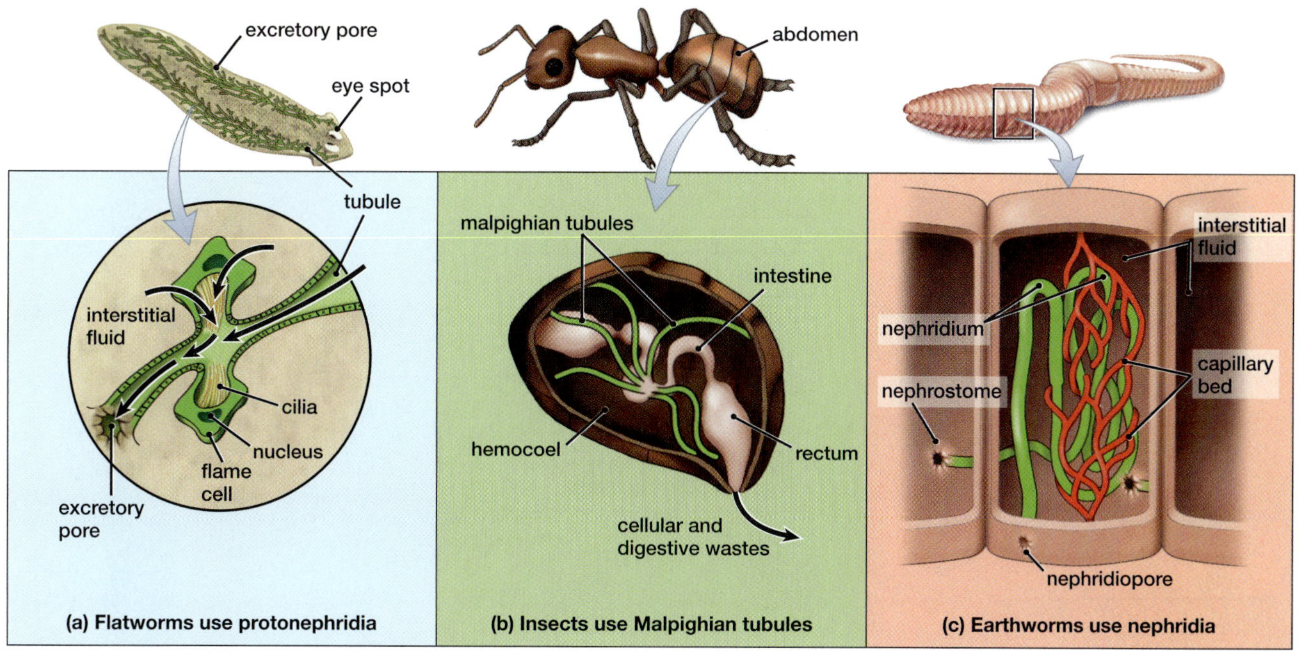

Figure 35-1 Excretory system of flatworm, excretory system of insect, excretory system of earthworm

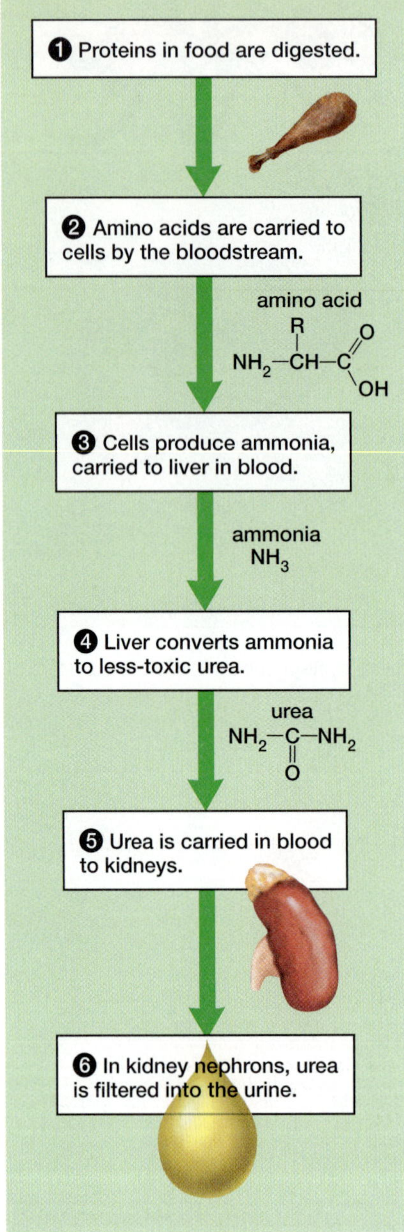

Figure 35-2 Formation/excretion of urea

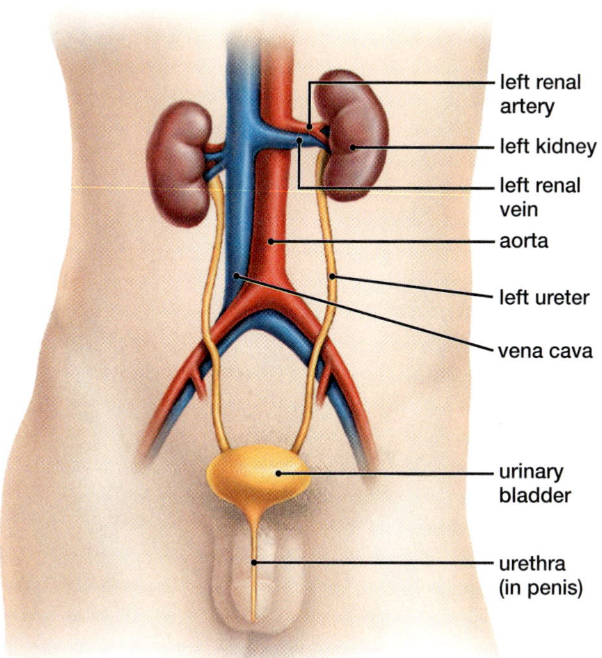

Figure 35-3 Human urinary system

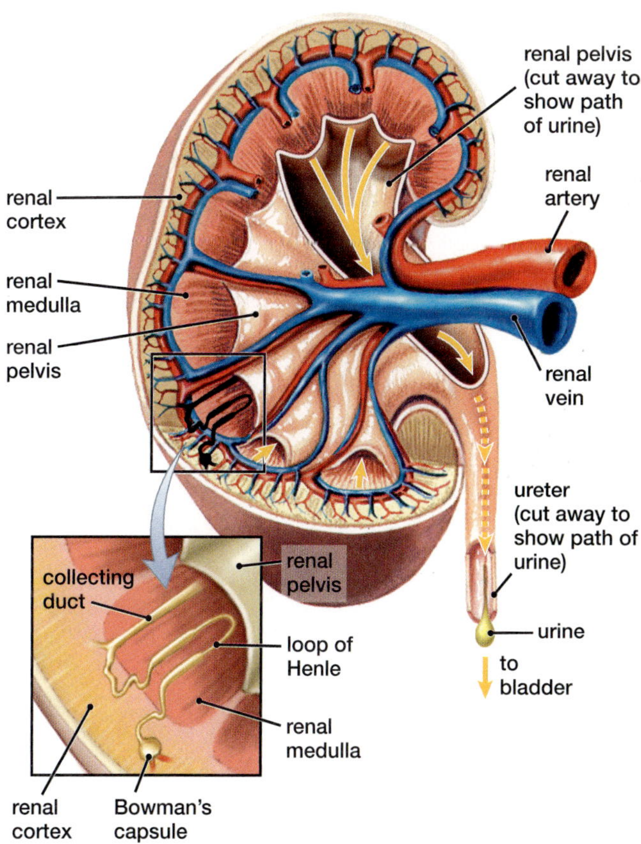

Figure 35-4 Cross section of kidney

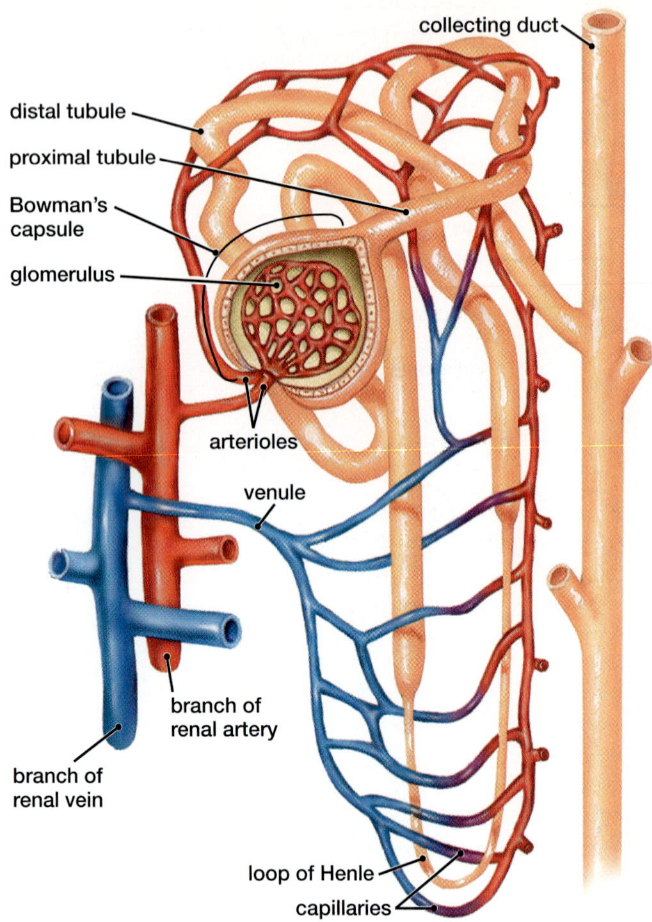

Figure 35-5 Individual nephron

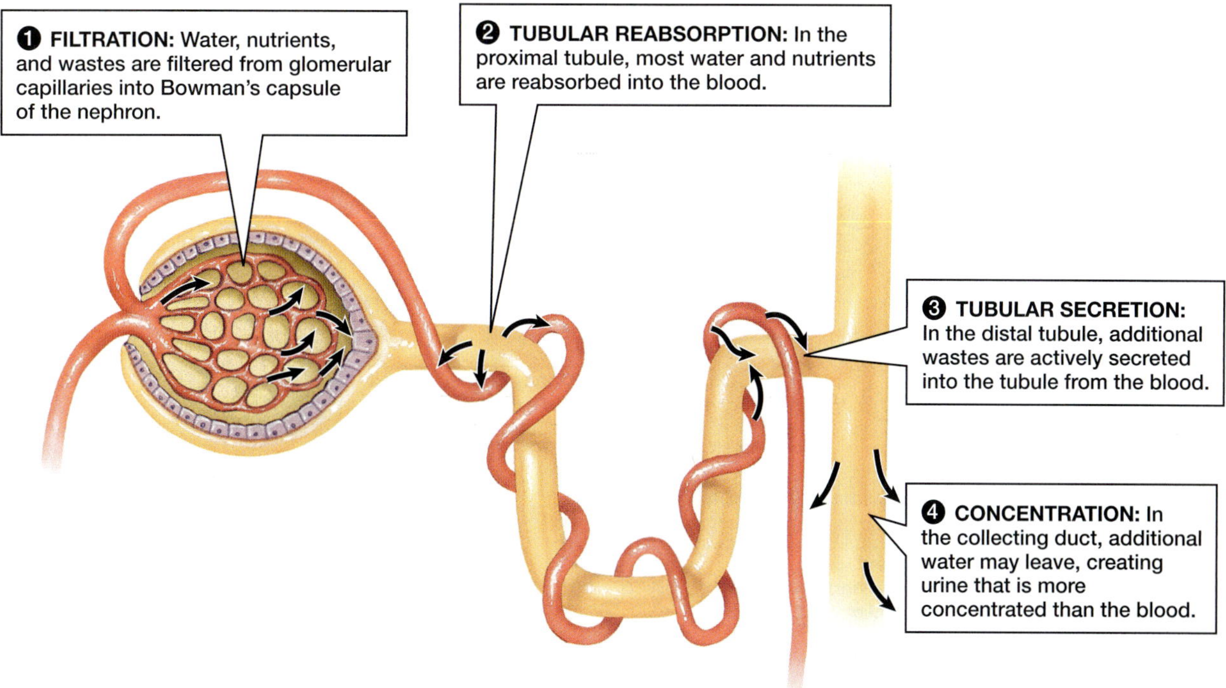

Figure 35-6 Urine formation

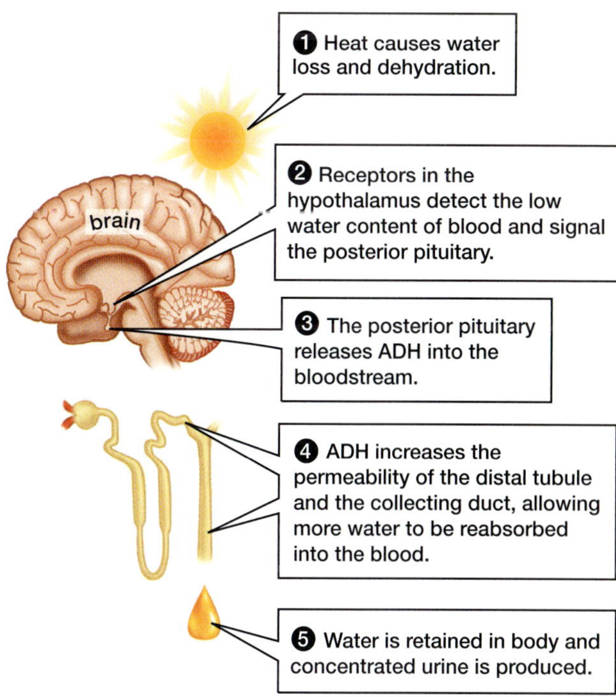

Figure 35-7 Regulation of water content of blood

(a) Freshwater fish (bluegill)

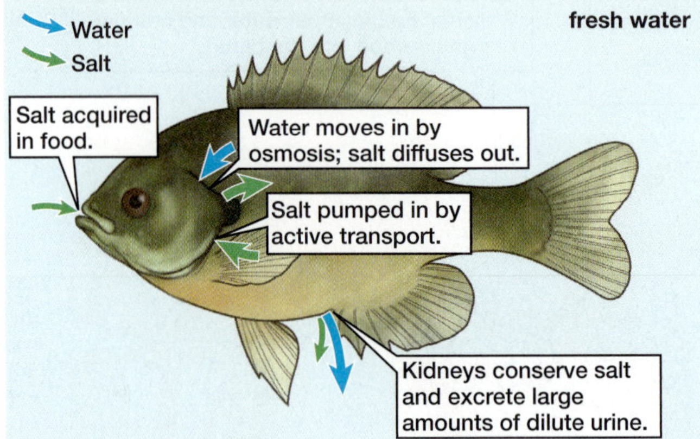

(b) Saltwater fish (garibaldi)

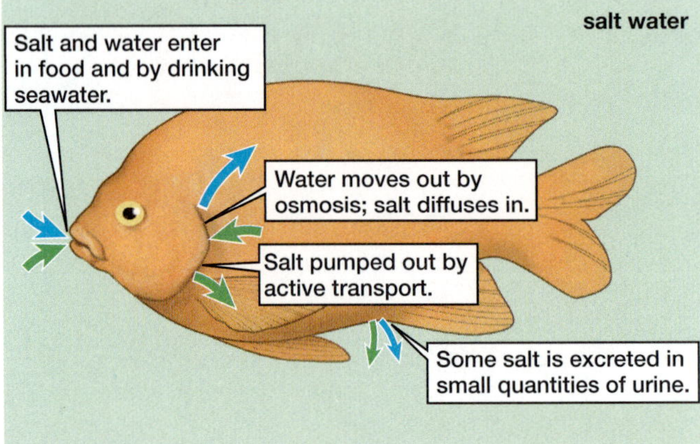

Figure 35-8 Osmoregulation freshwater fish

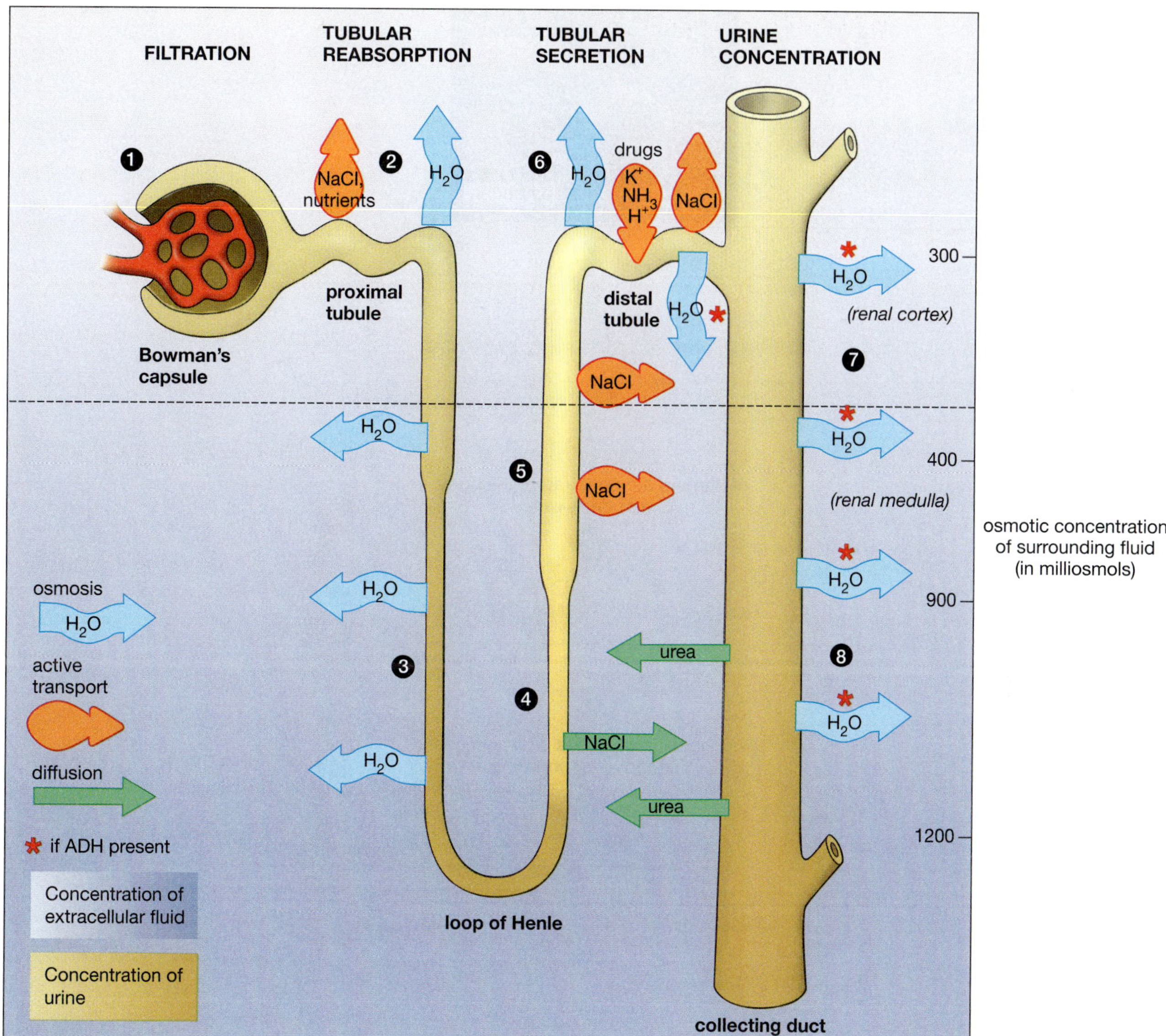

Figure E35-1 Details of urine formation

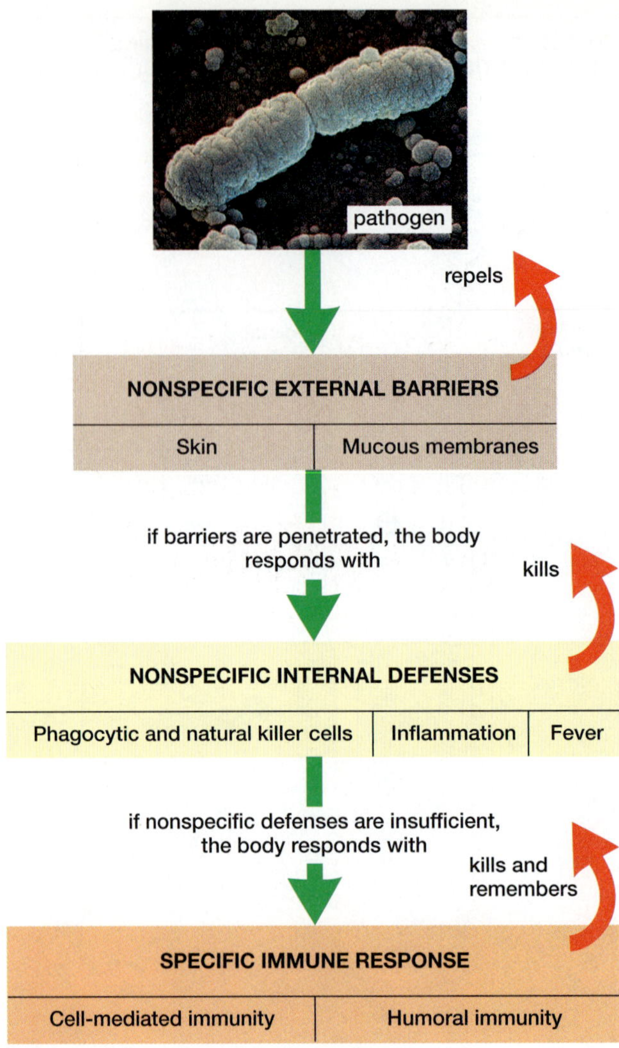

Figure 36-1 Levels of Defense

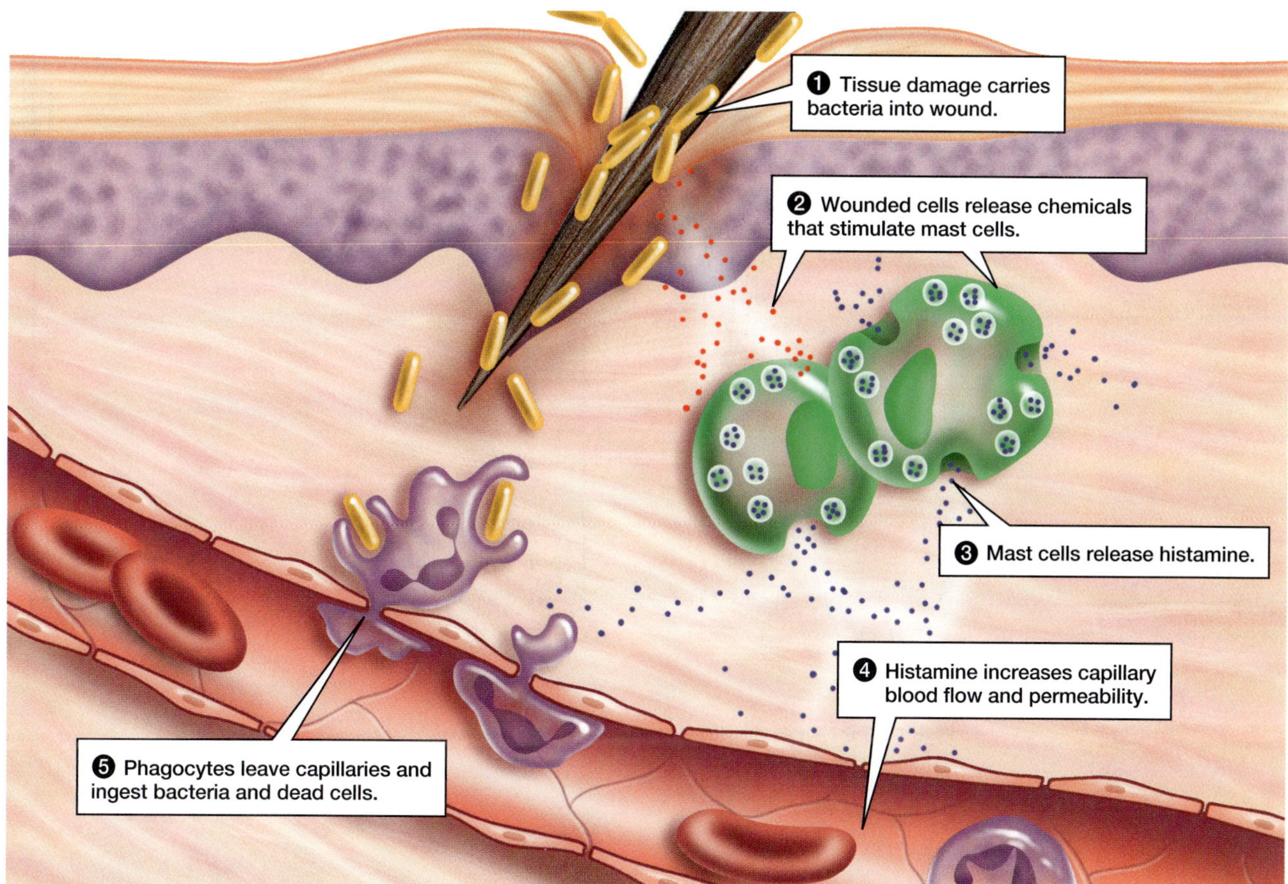

Figure 36-5 The inflammatory response

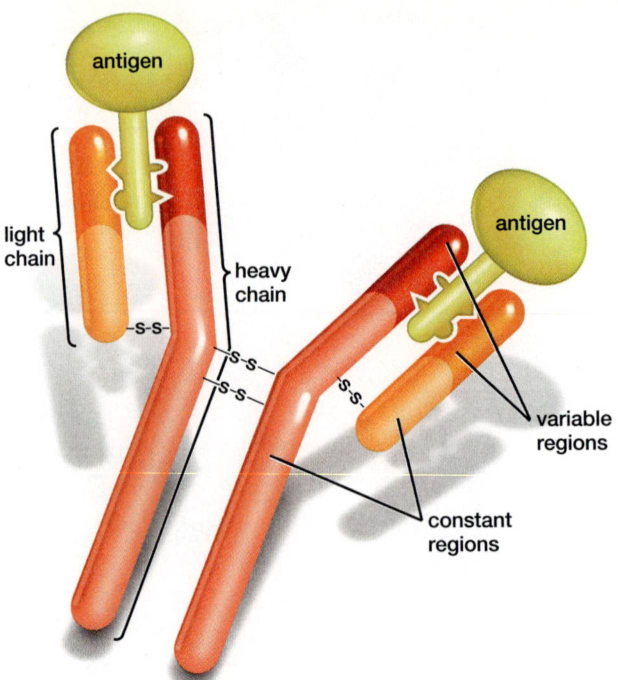

Figure 36-6 Antibody structure

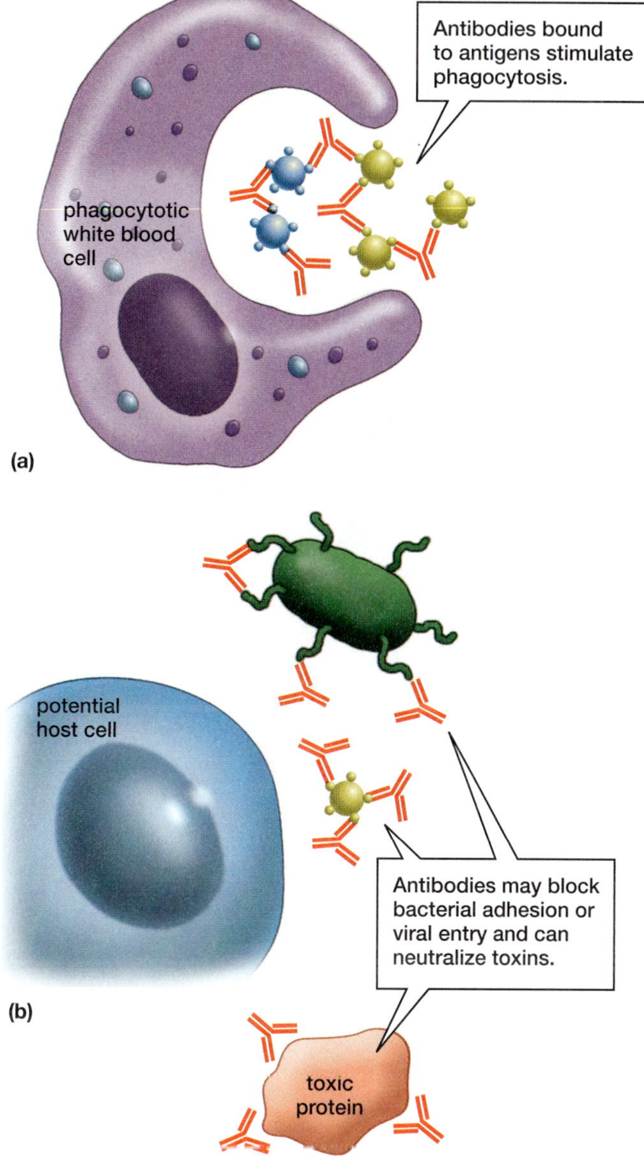

Figure 36-7 Examples of functions of antibodies

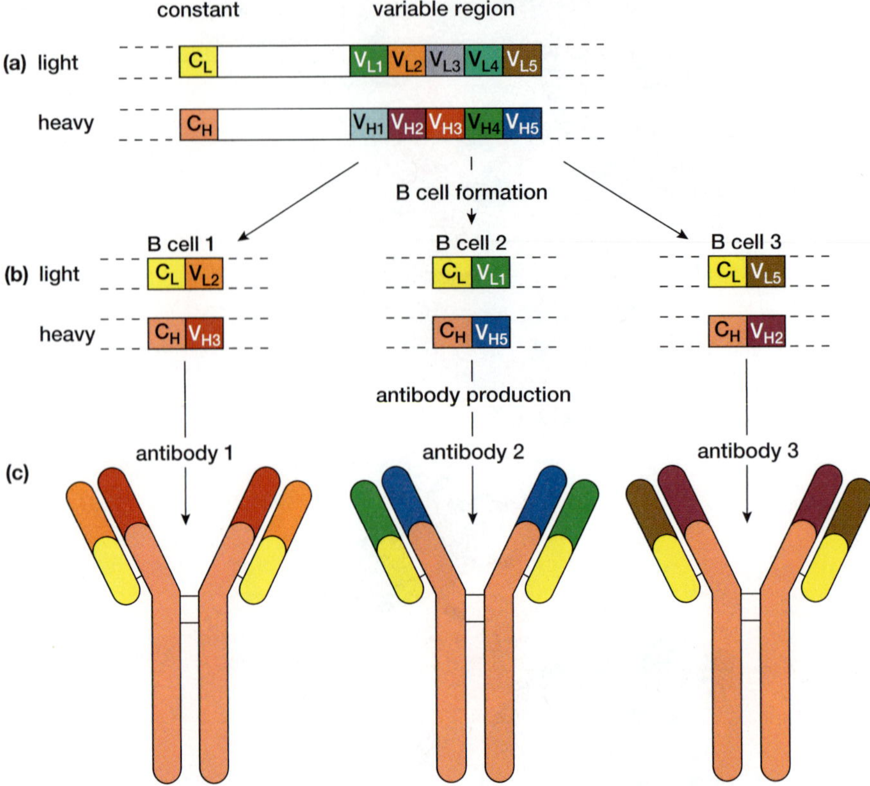

Figure 36-8 Recombination during construction

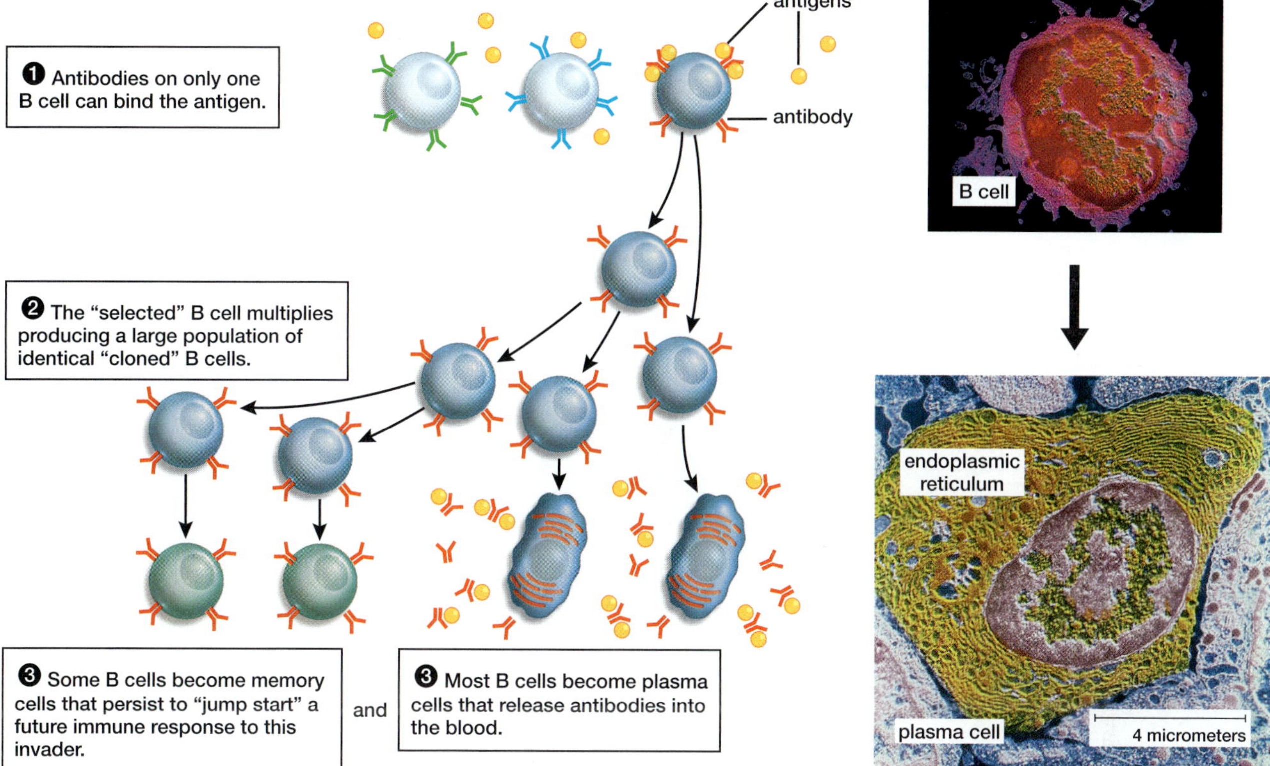

Figure 36-9 Clonal selection among B cells

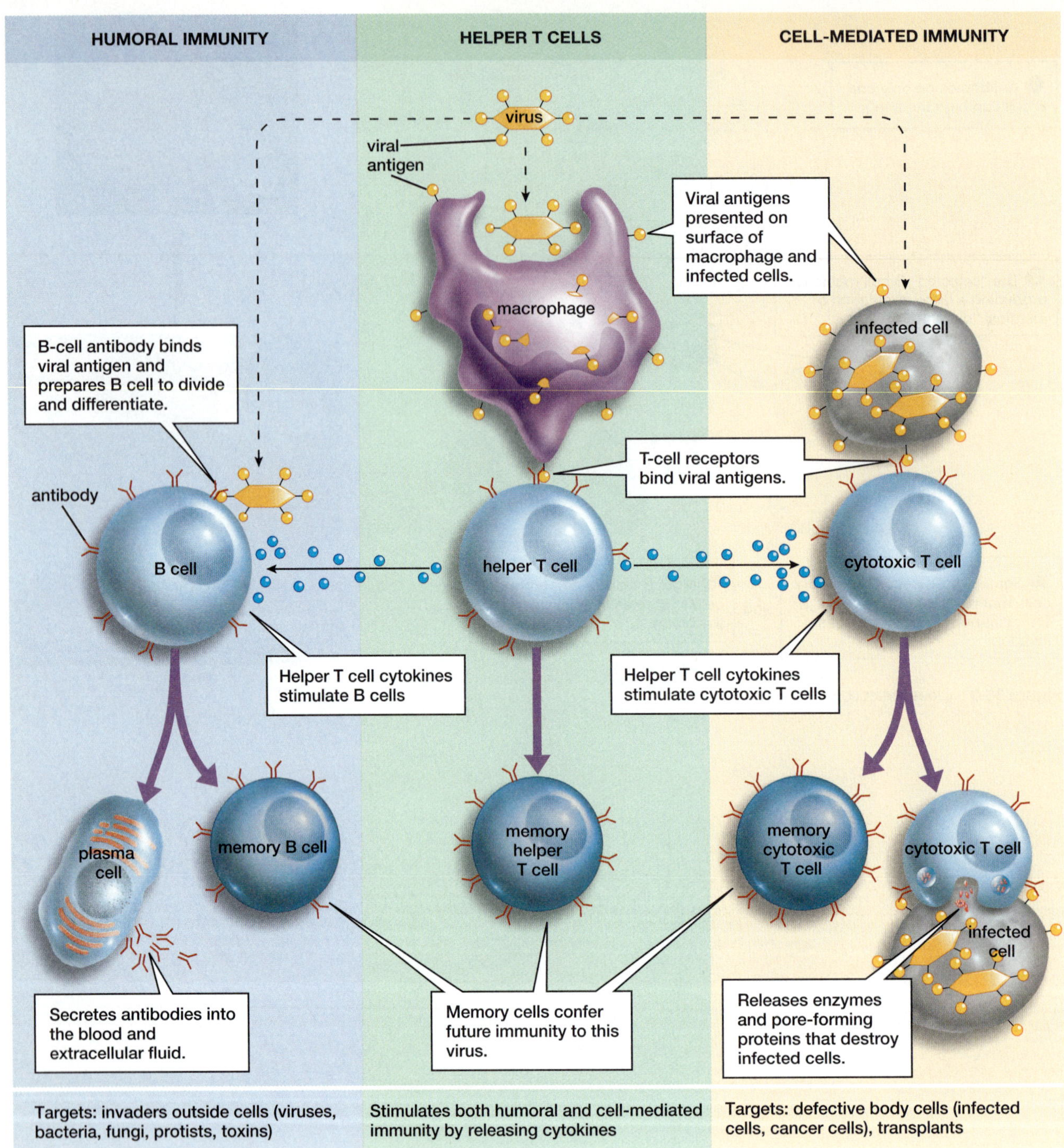

Figure 36-11 A summary of humoral and cell-mediated immune responses

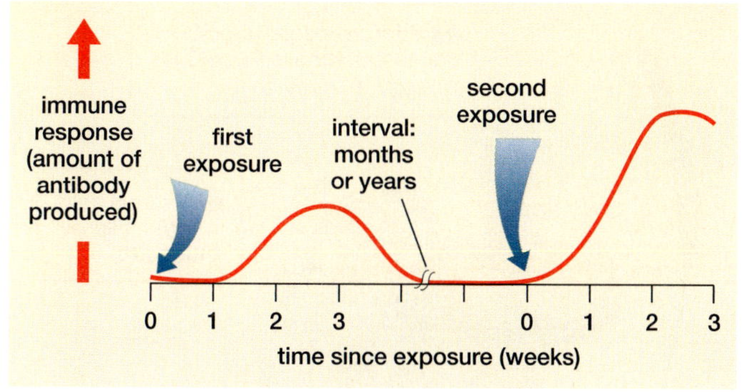

Figure 36-12 Memory and the immune response

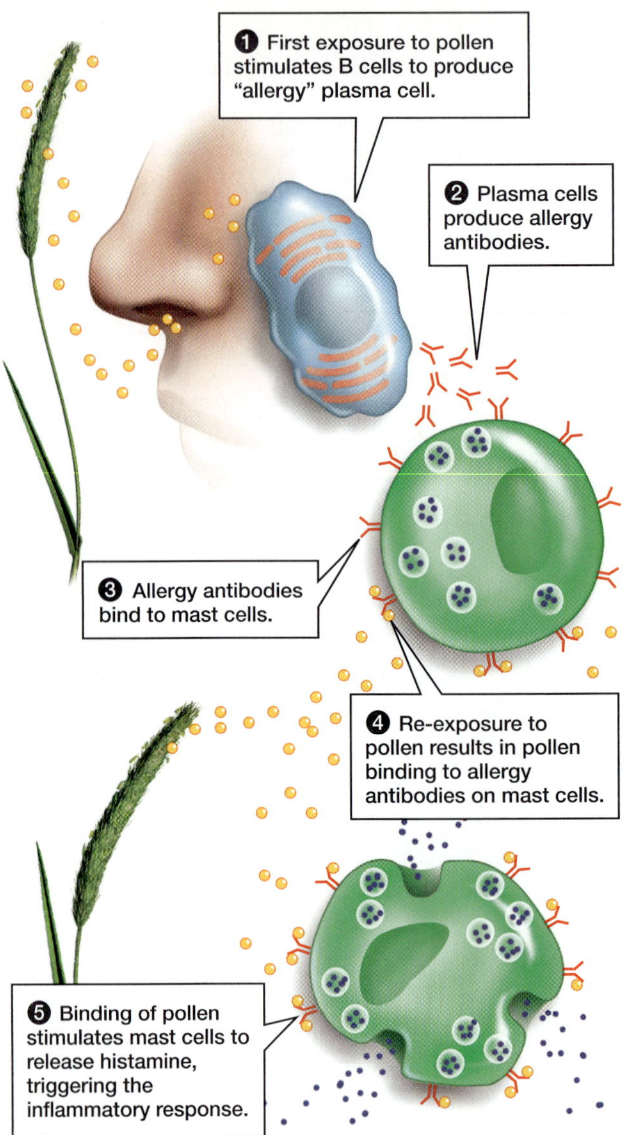

Figure 36-13 Allergic reactions

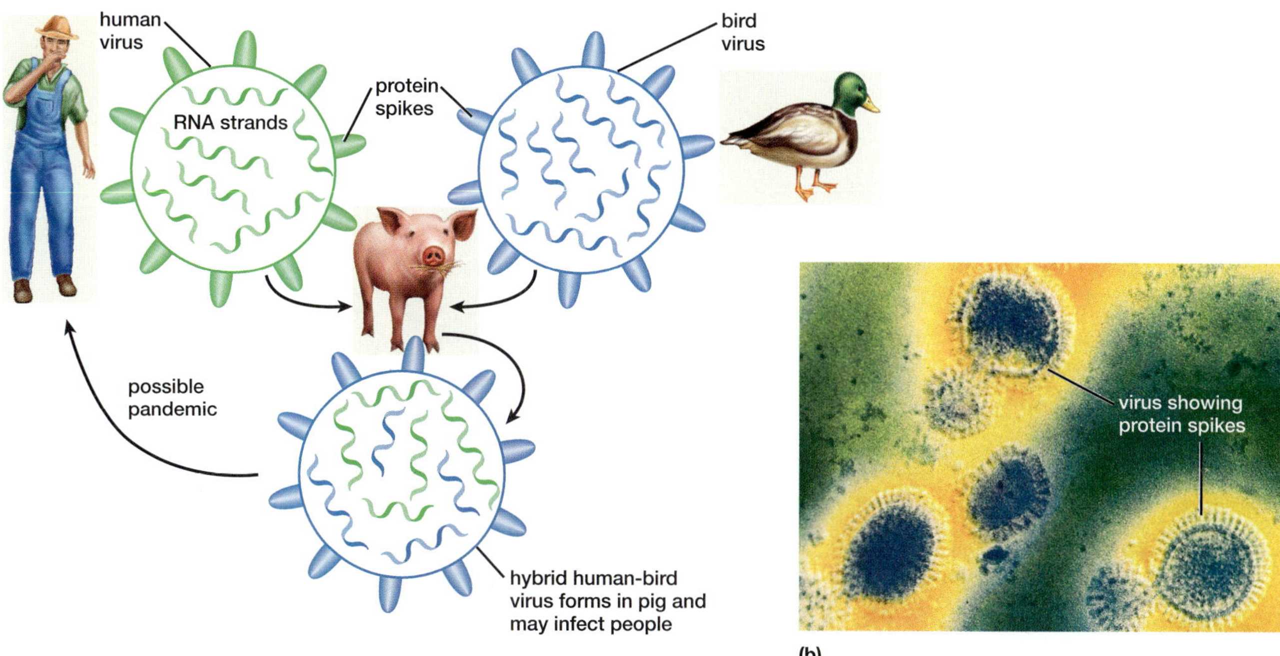

Figure 36-2 Flu virus

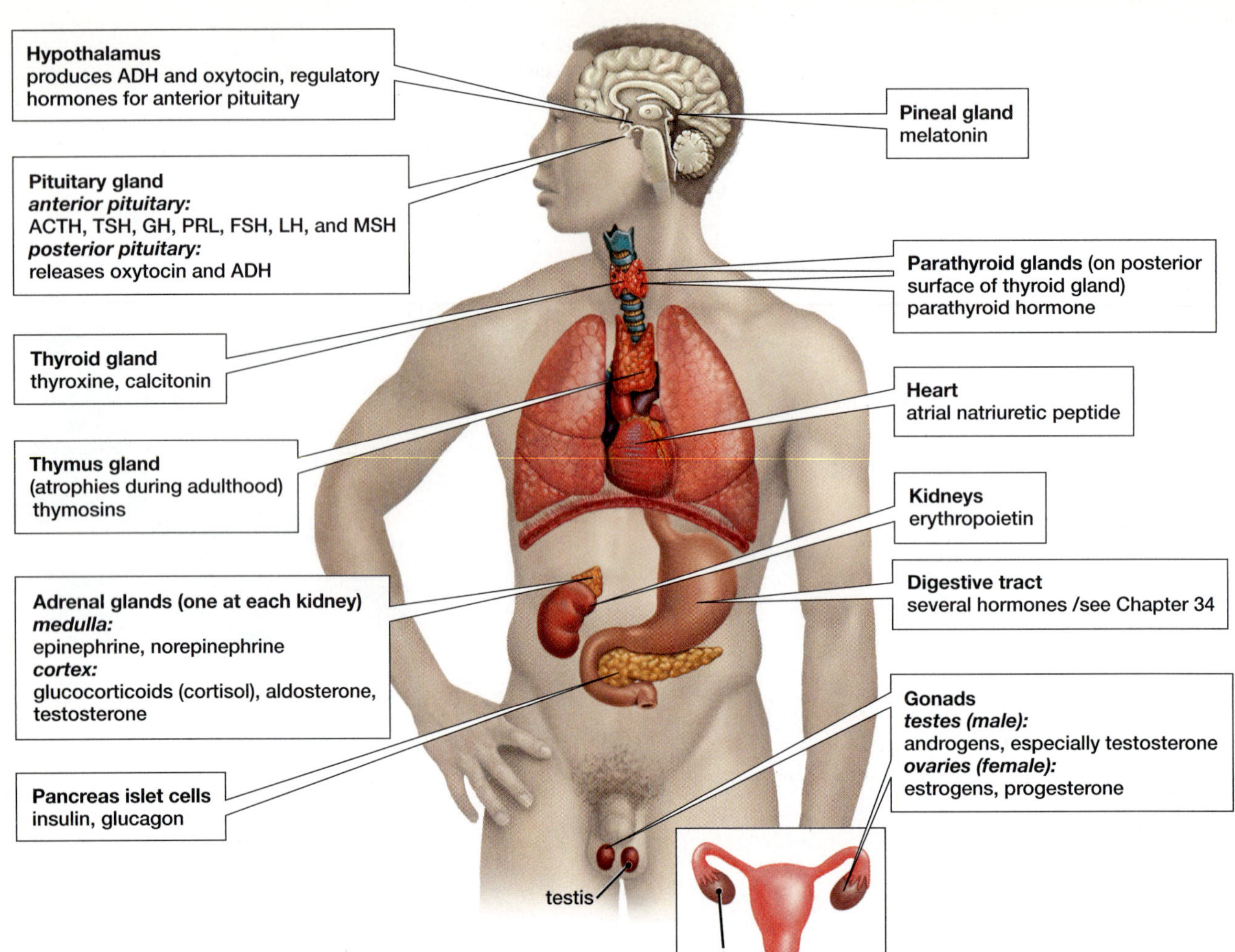

Figure 37-1 Major mammalian endocrine glands

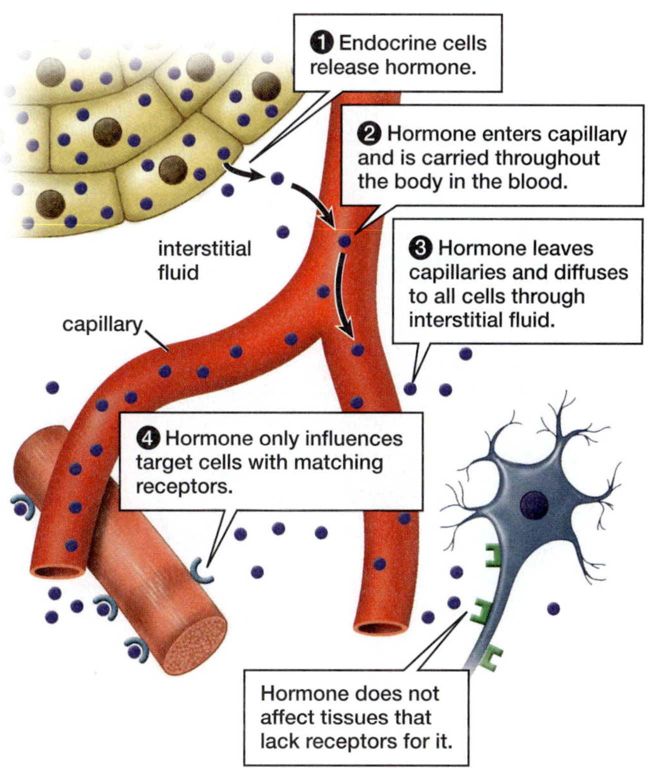

Figure 37-2 Endocrine glands, hormones and target cells

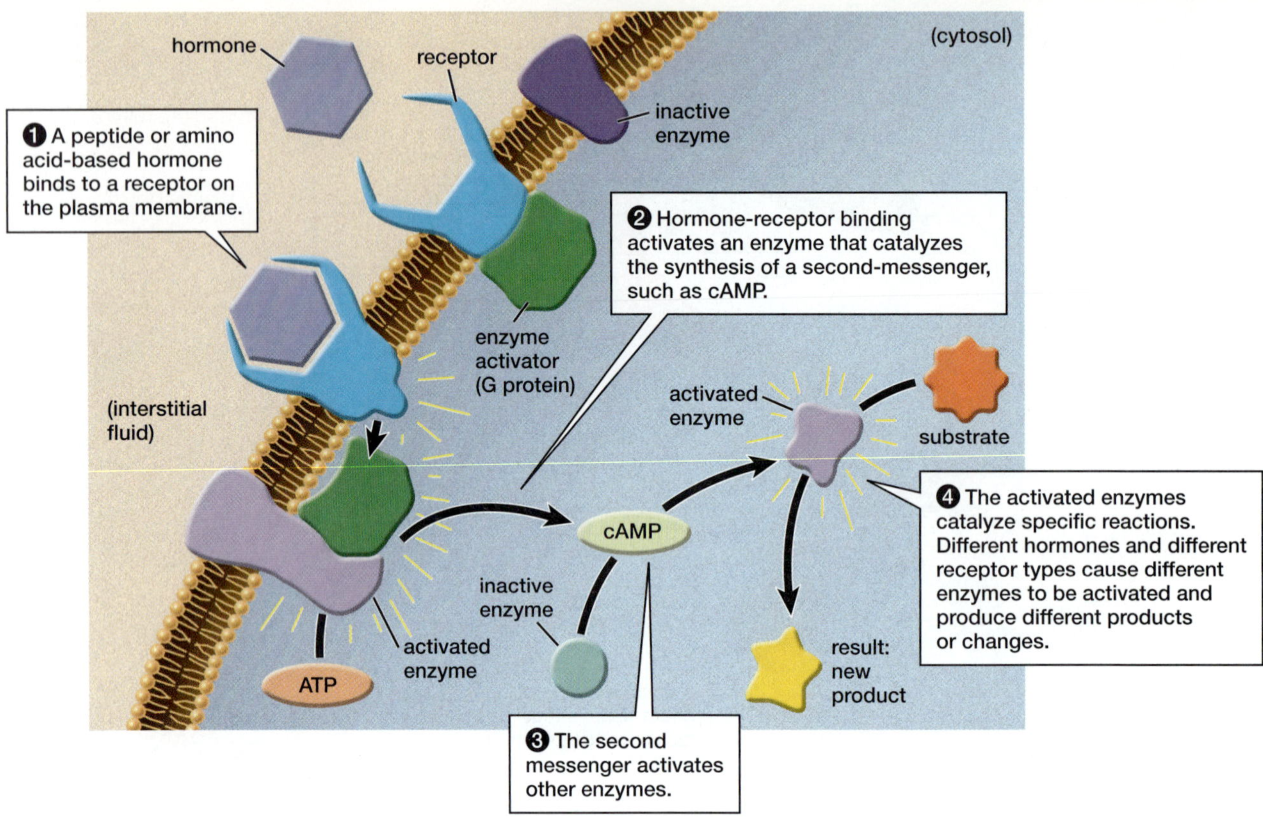

Figure 37-3 Peptide/amino acid hormone mechanism

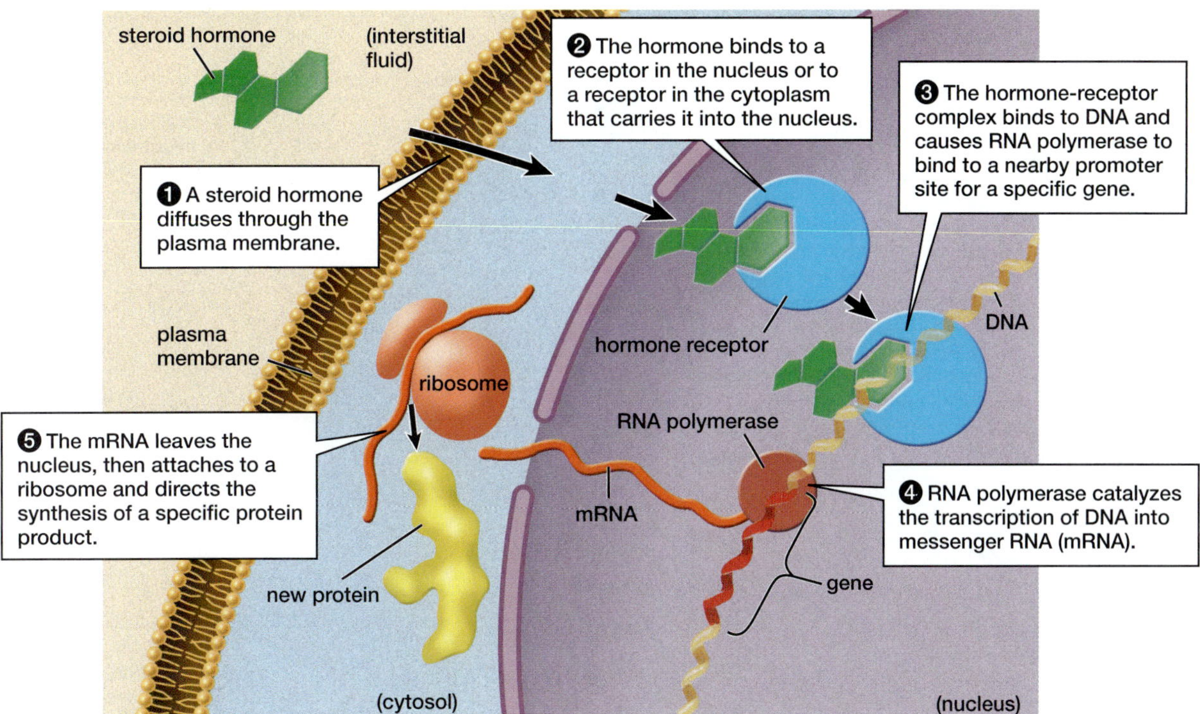

Figure 37-4 Steroid hormone mechanism

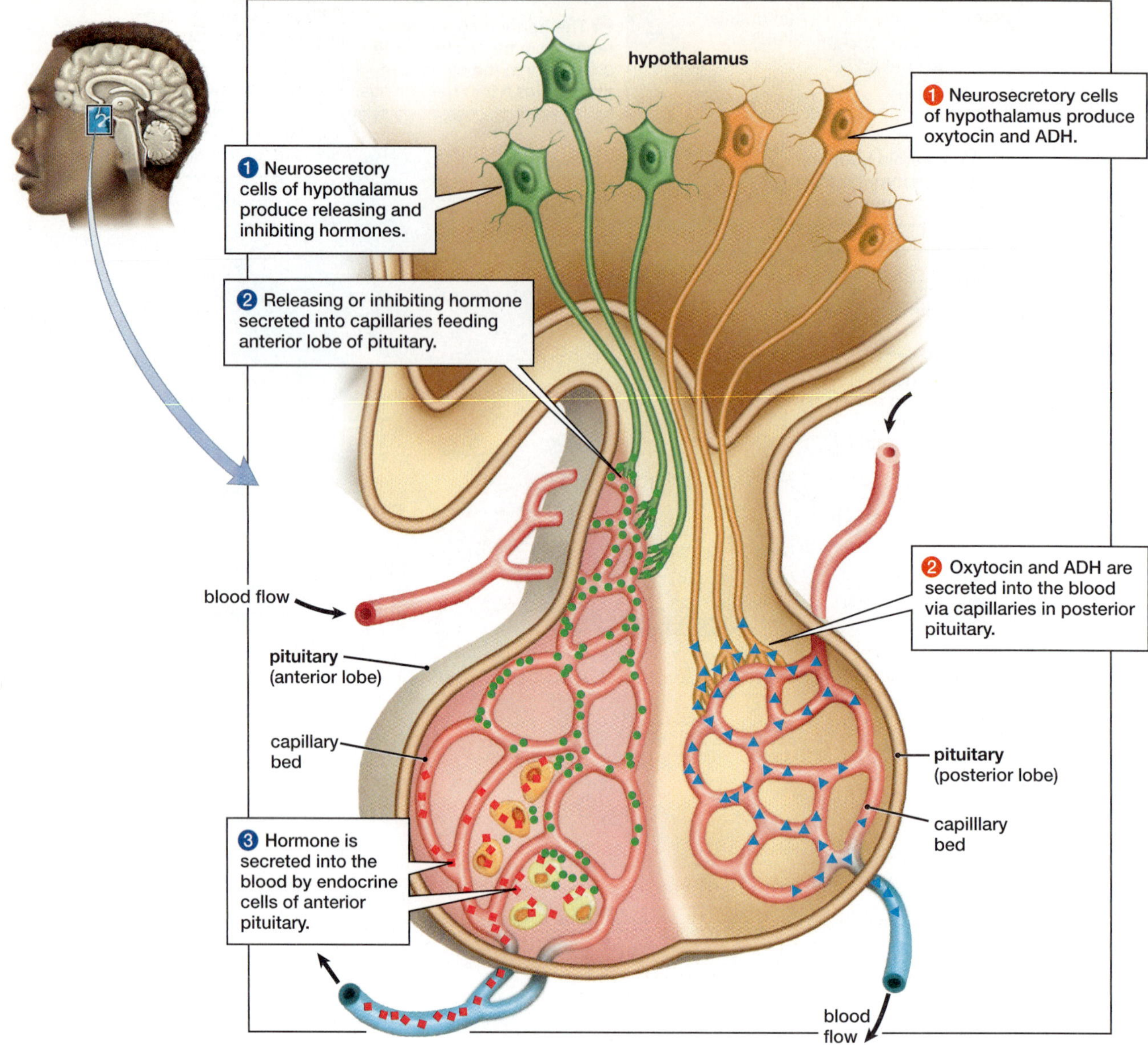

Figure 37-6 Hypothalamus controls pituitary

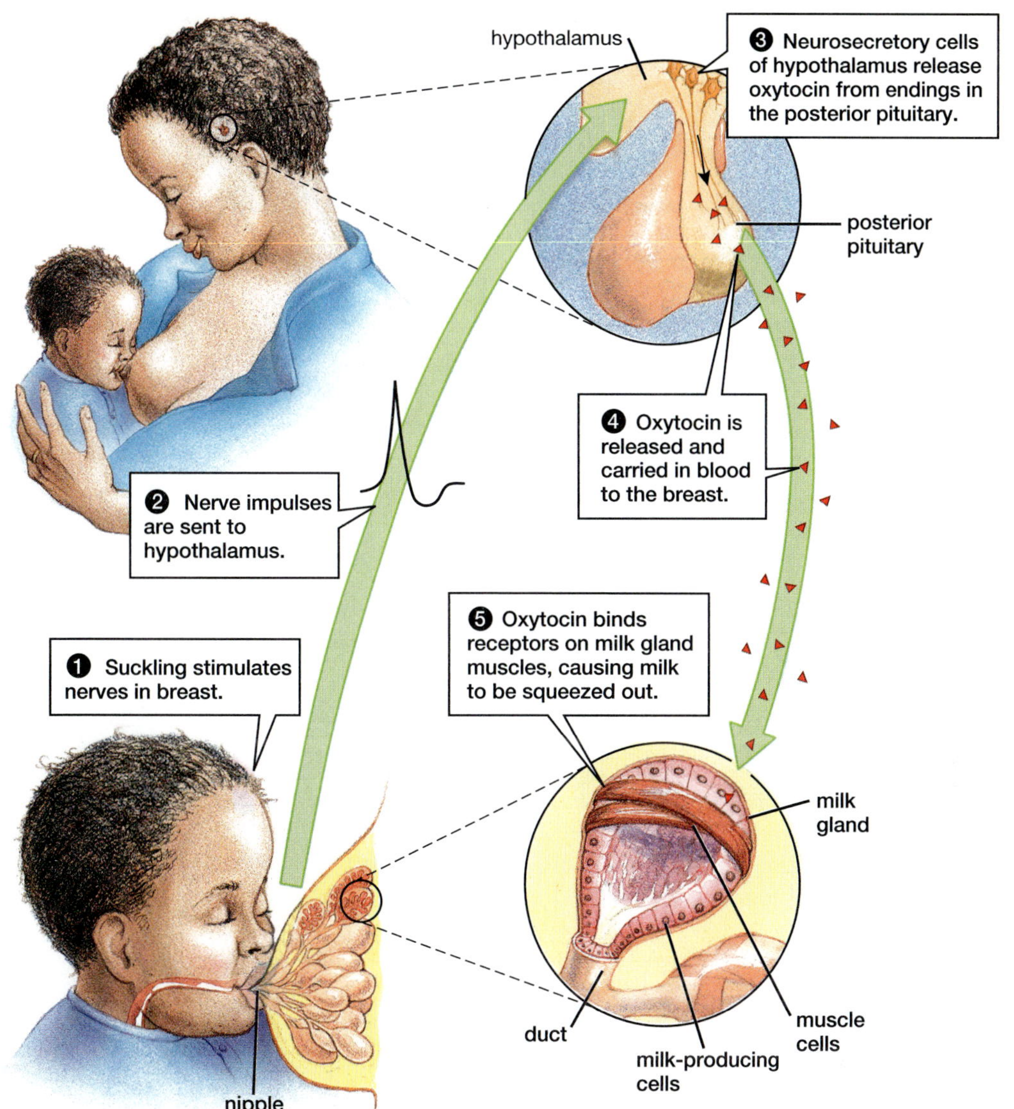

Figure 37-8 Control of lactation

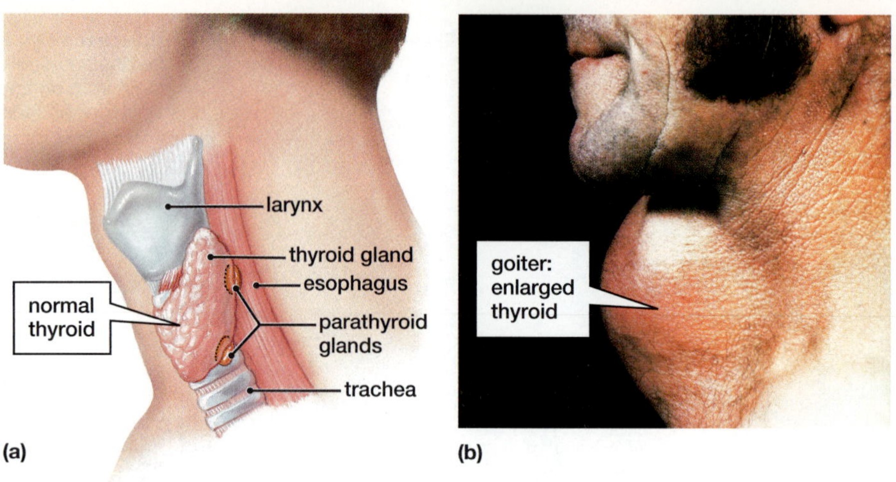

Figure 37-9 Thyroid

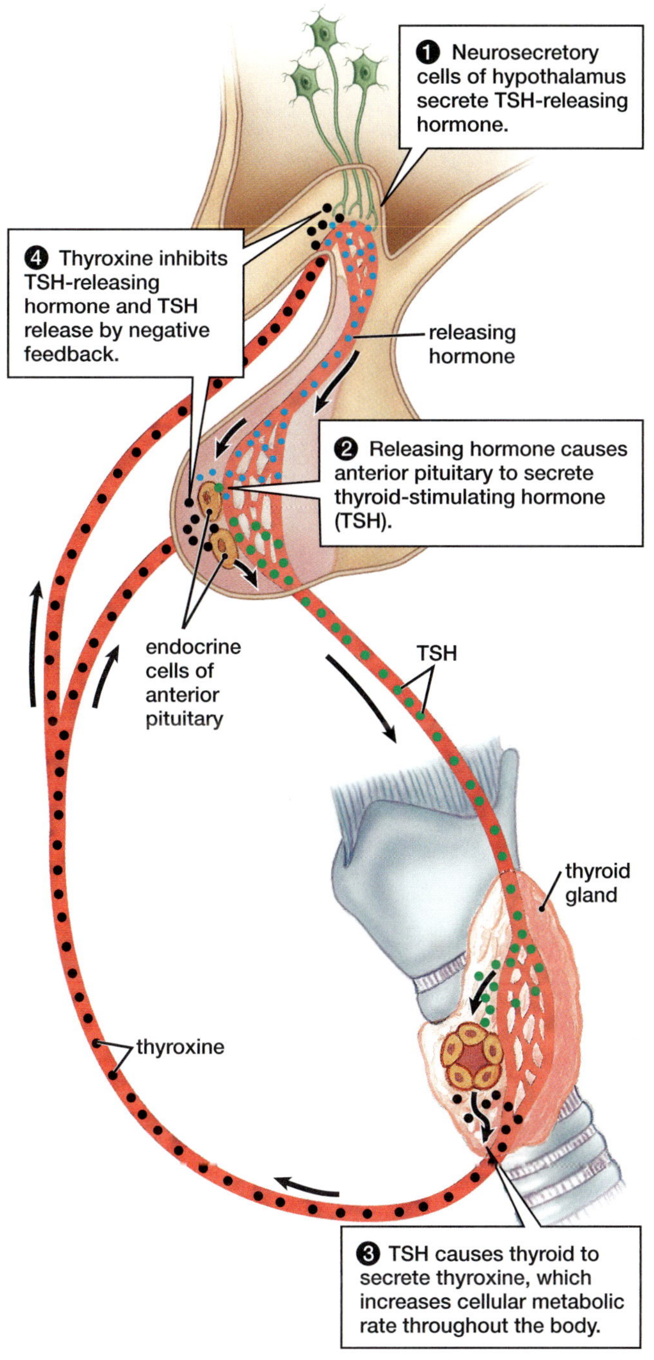

Figure 37-10 Feedback controls thyroxine release

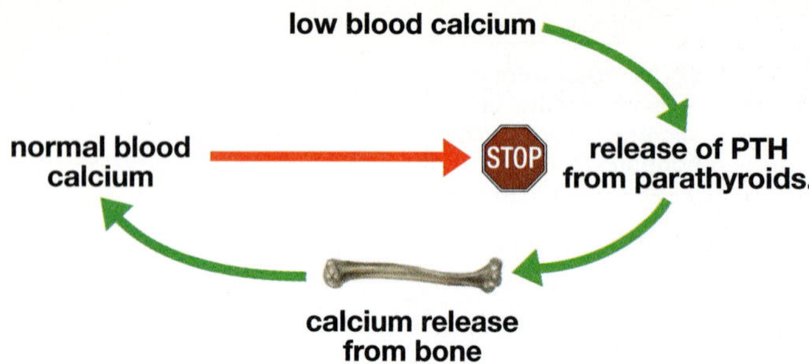

Figure 37-11 Feedback controls calcium

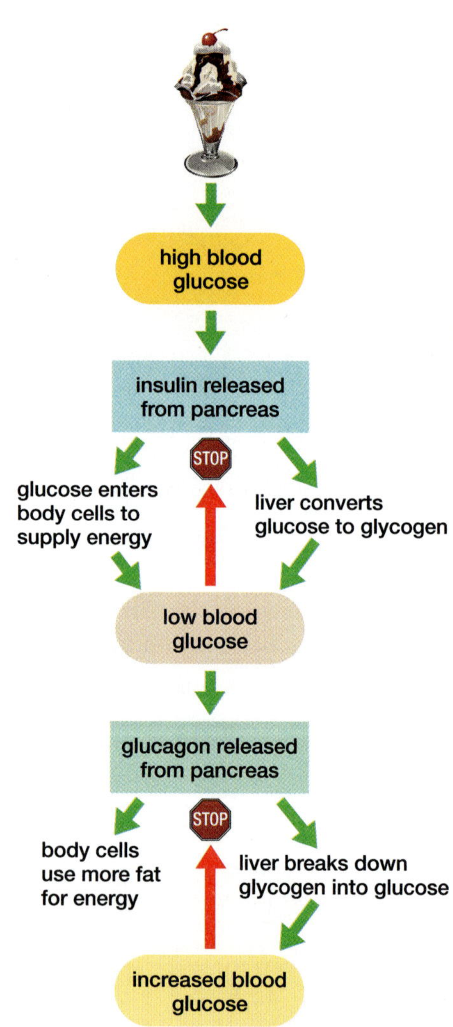

Figure 37-12 Control of insulin and glucagon

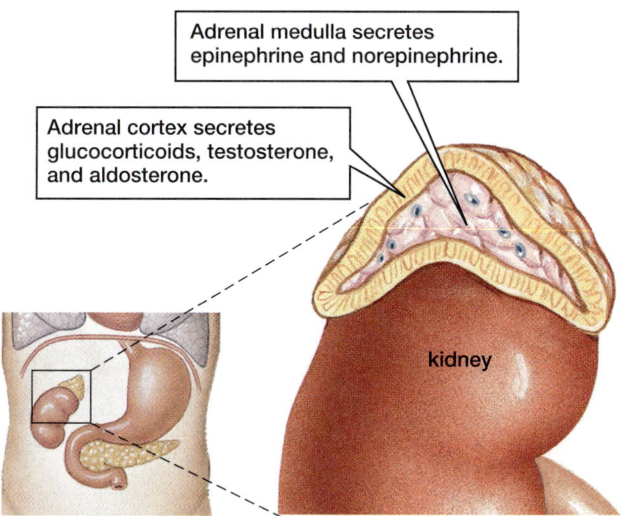

Figure 37-13 Adrenal gland

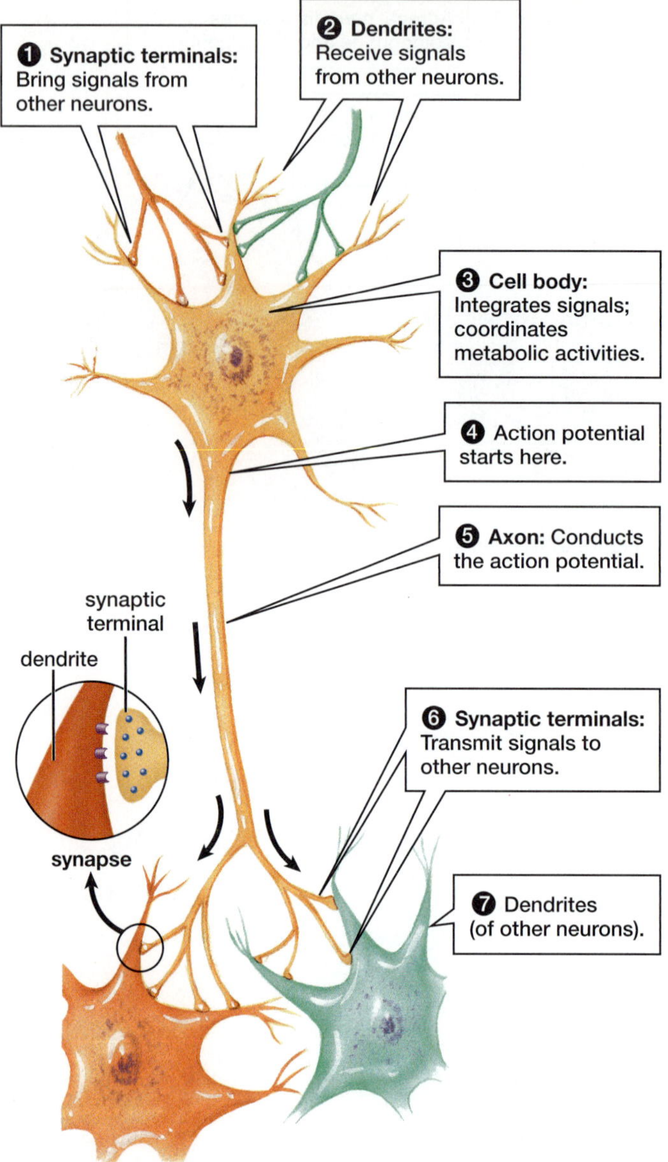

Figure 38-1 Neuron structure

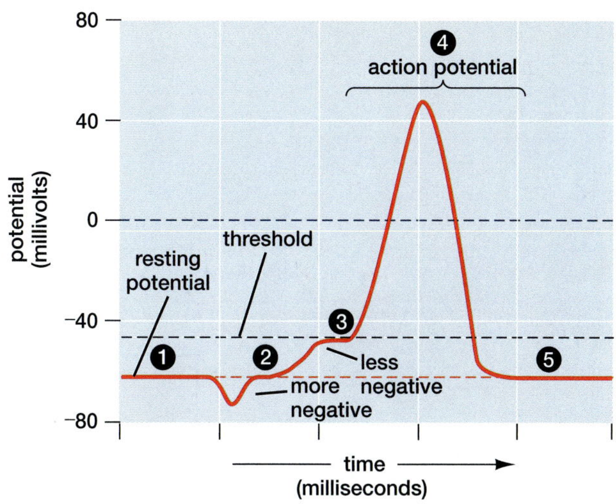

Figure 38-2 Action potential

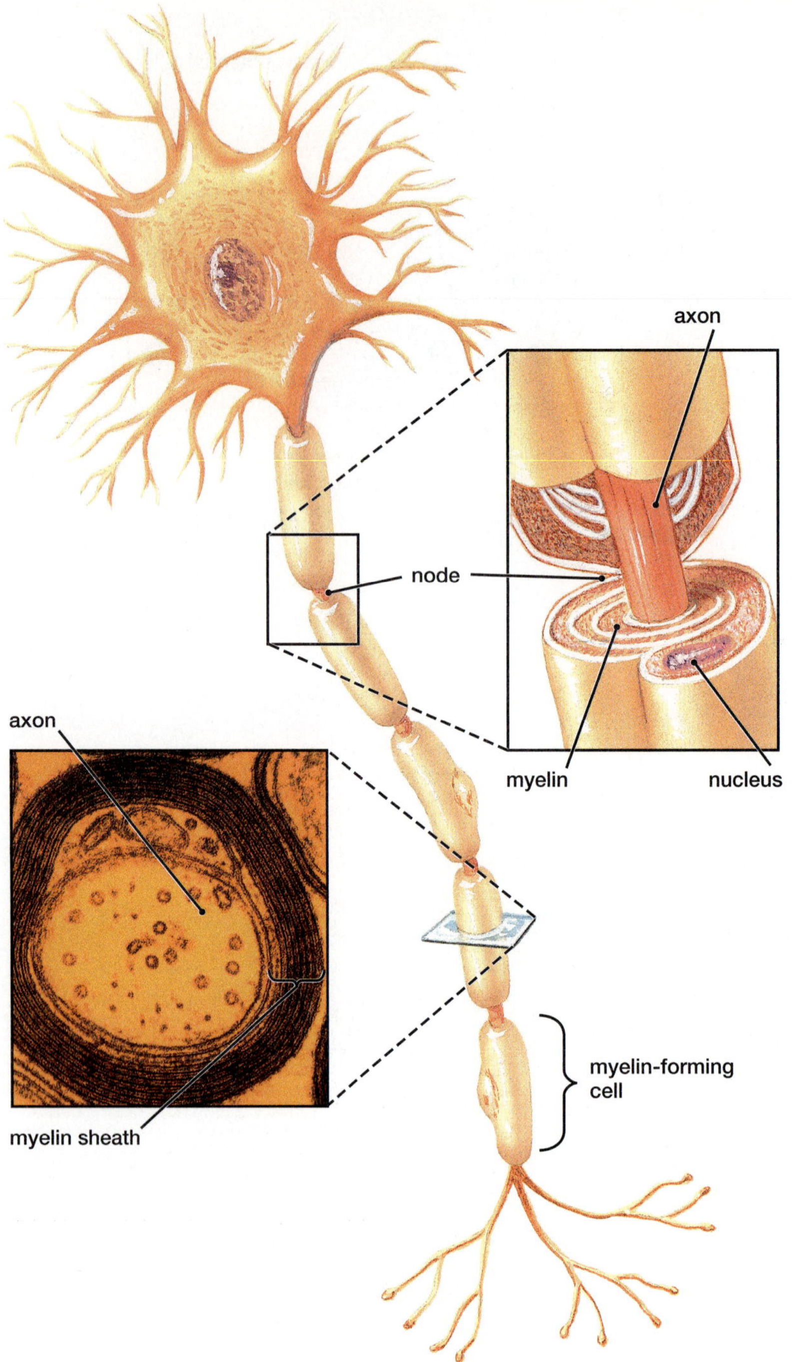

Figure 38-3 Myelinated axon

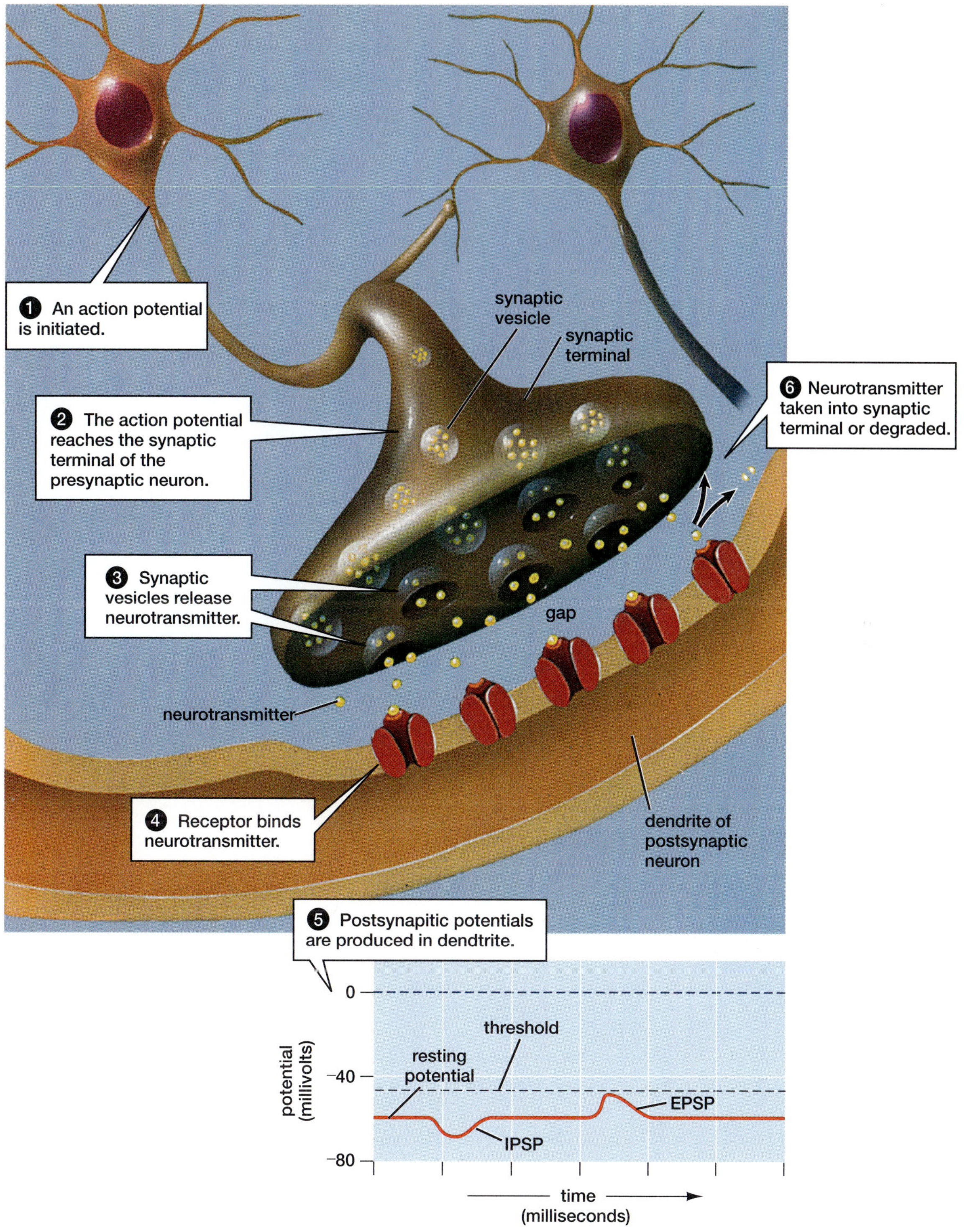

Figure 38-4 Synapse

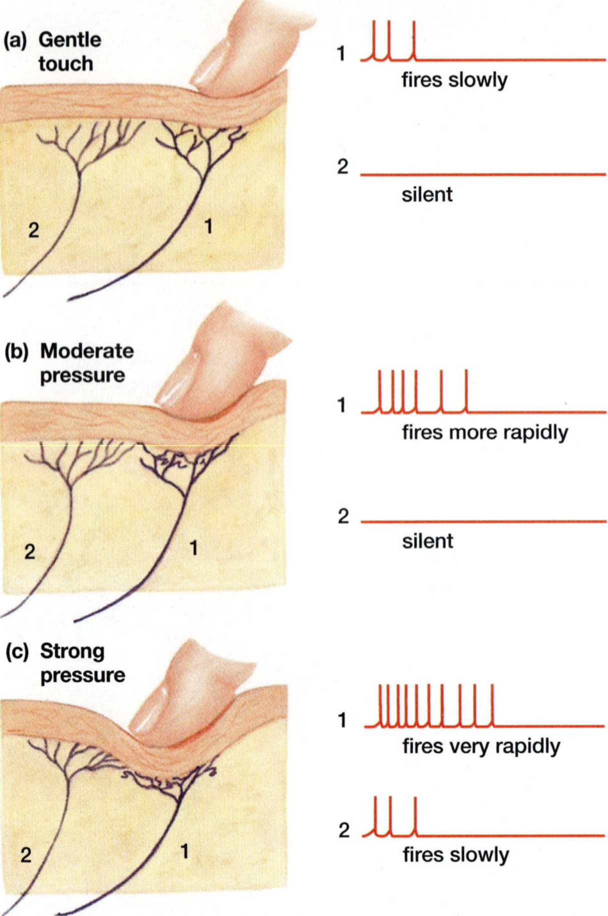

Figure 38-5 Signaling stimulus intensity

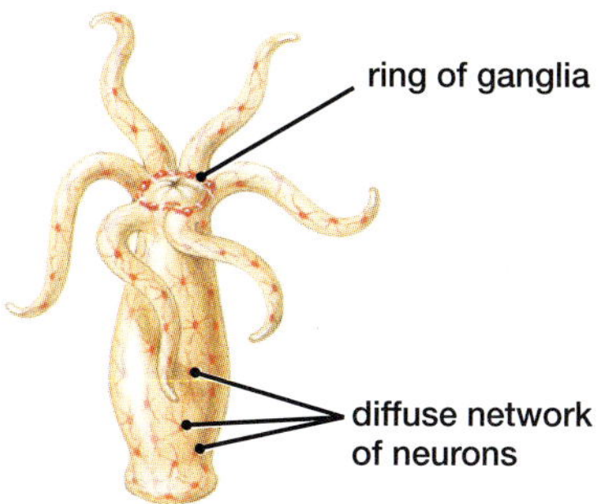

Figure 38-6a Nerve net in Hydra

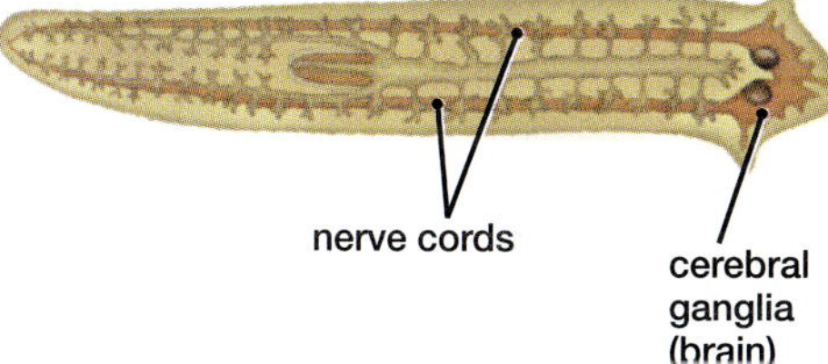

Figure 38-6b Nervous system in flatworm

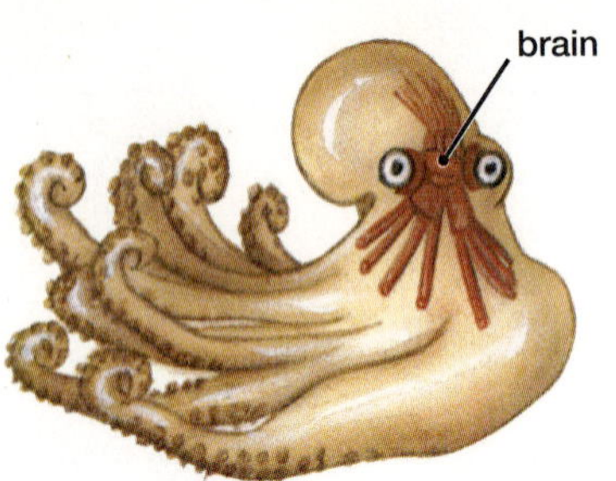

Figure 38-6c Nervous system in Octopus

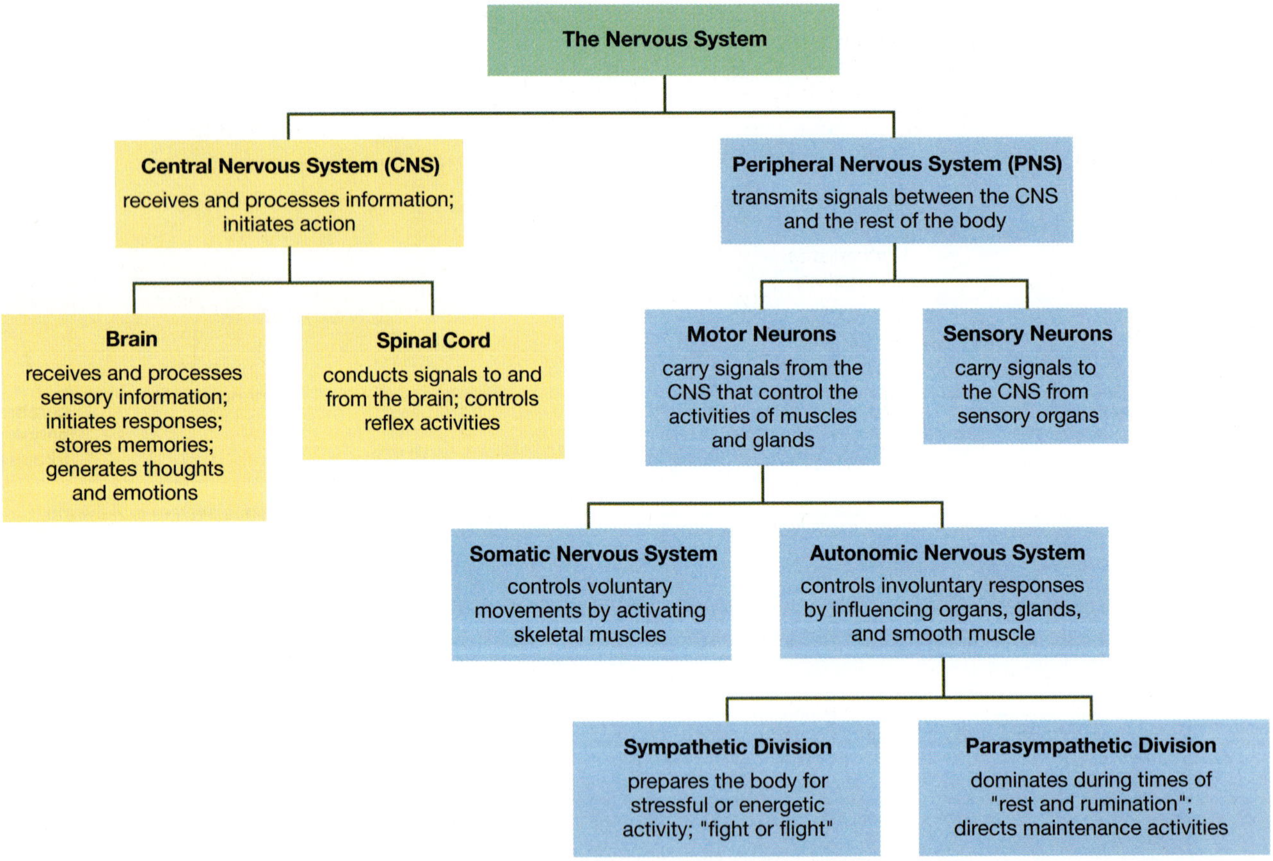

The Nervous System

- **Central Nervous System (CNS)**
  receives and processes information; initiates action
  - **Brain**
    receives and processes sensory information; initiates responses; stores memories; generates thoughts and emotions
  - **Spinal Cord**
    conducts signals to and from the brain; controls reflex activities
- **Peripheral Nervous System (PNS)**
  transmits signals between the CNS and the rest of the body
  - **Motor Neurons**
    carry signals from the CNS that control the activities of muscles and glands
    - **Somatic Nervous System**
      controls voluntary movements by activating skeletal muscles
    - **Autonomic Nervous System**
      controls involuntary responses by influencing organs, glands, and smooth muscle
      - **Sympathetic Division**
        prepares the body for stressful or energetic activity; "fight or flight"
      - **Parasympathetic Division**
        dominates during times of "rest and rumination"; directs maintenance activities
  - **Sensory Neurons**
    carry signals to the CNS from sensory organs

Figure 38-7 Organization of vertebrate nervous system

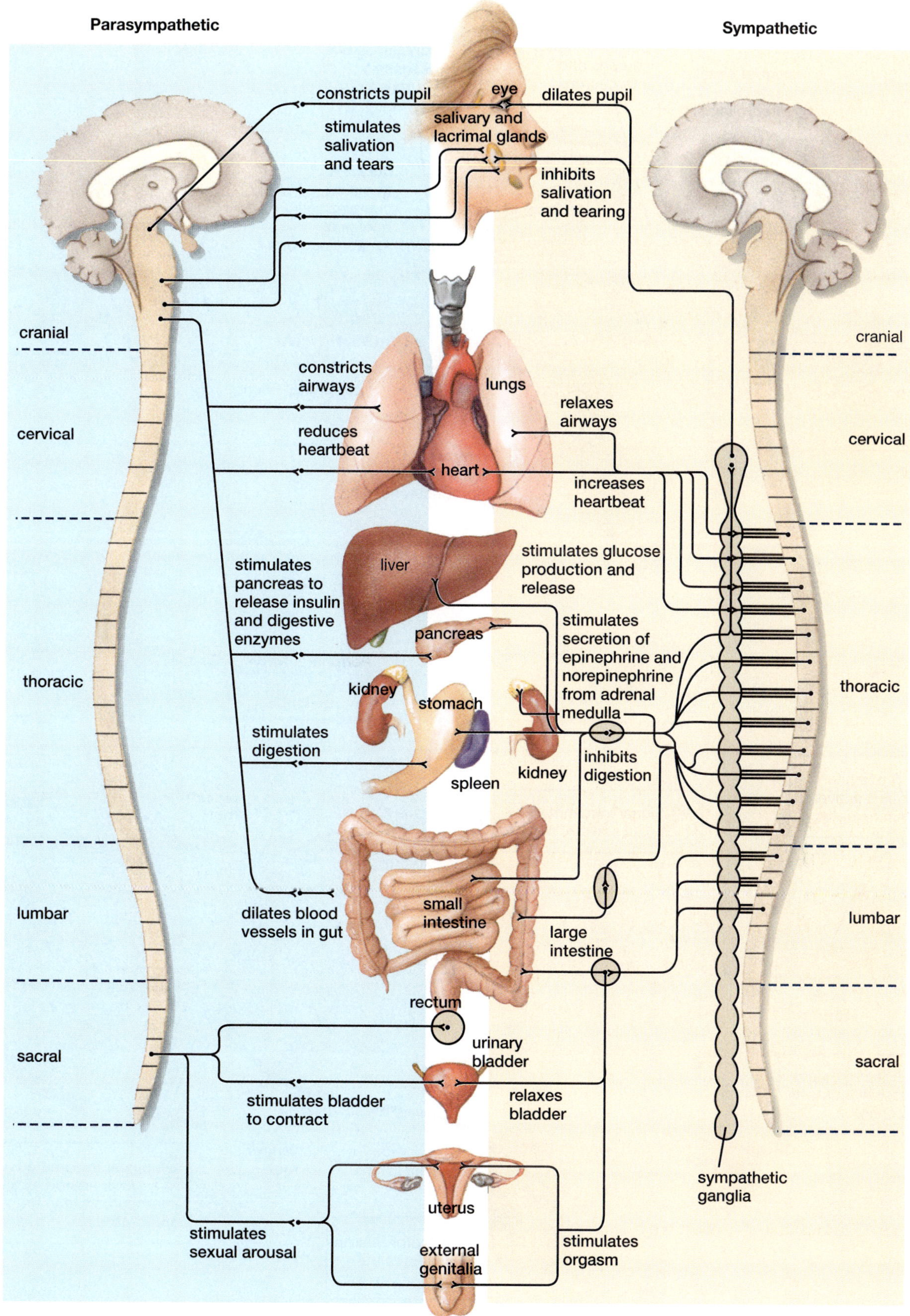

Figure 38-8 Autonomic nervous system

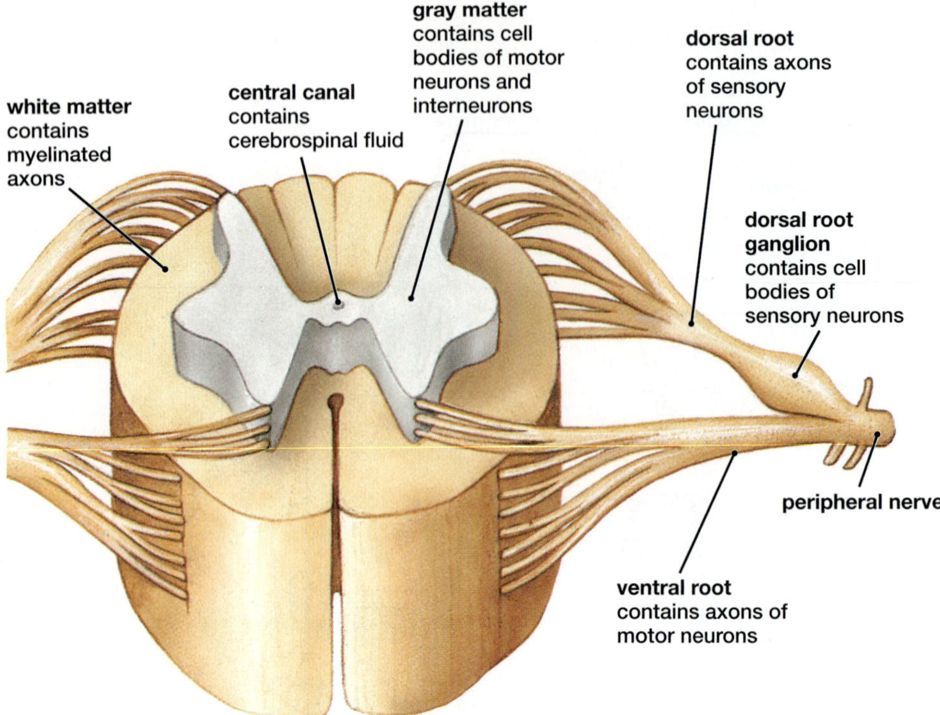

Figure 38-9 Spinal cord

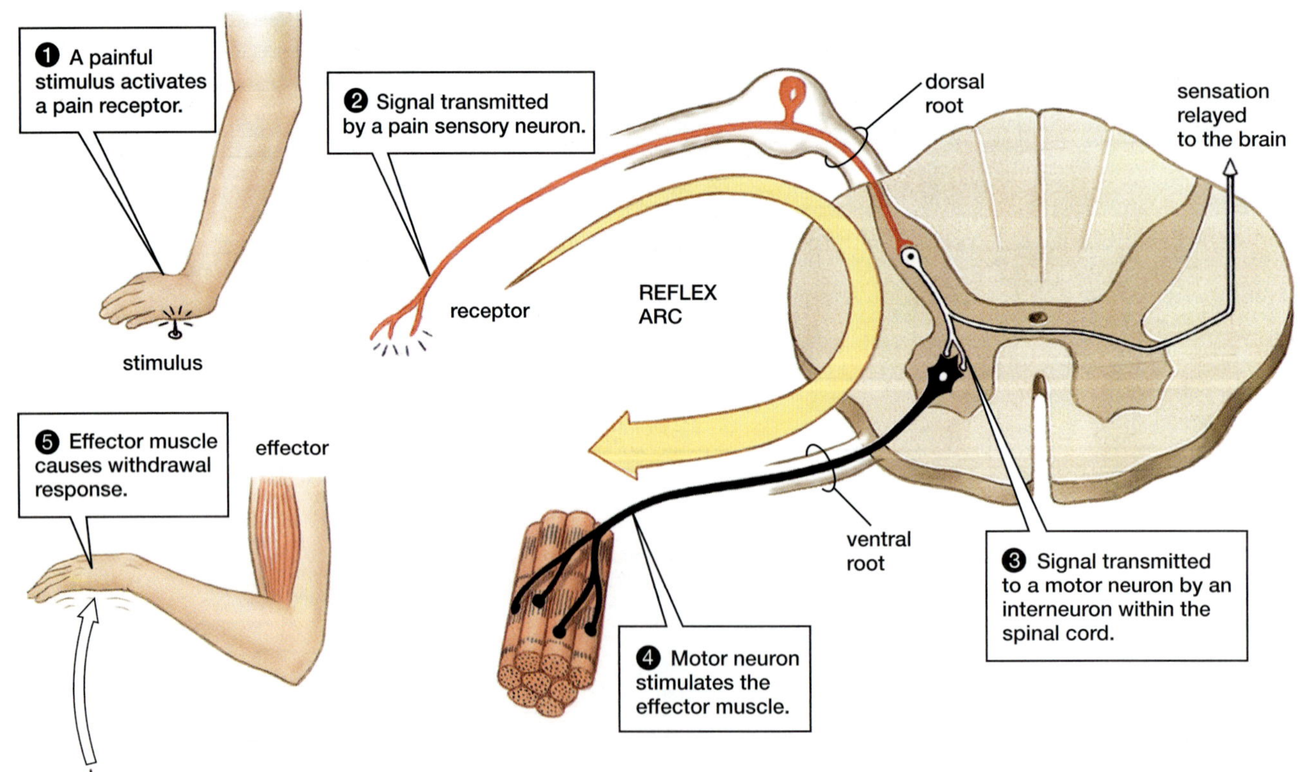

Figure 38-10 Pain-withdrawal reflex

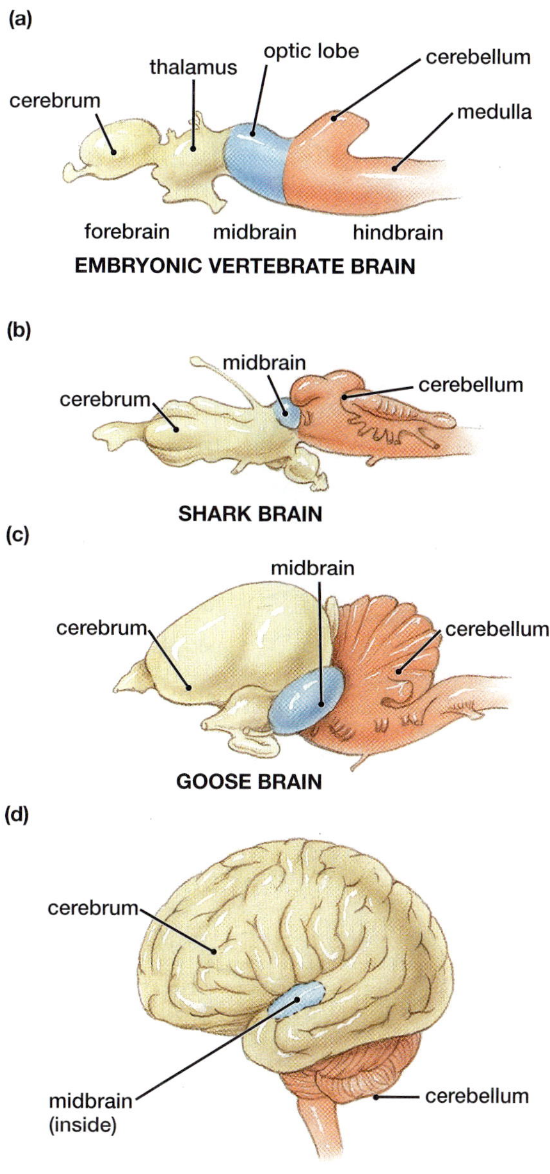

Figure 38-11 Vertebrate brains

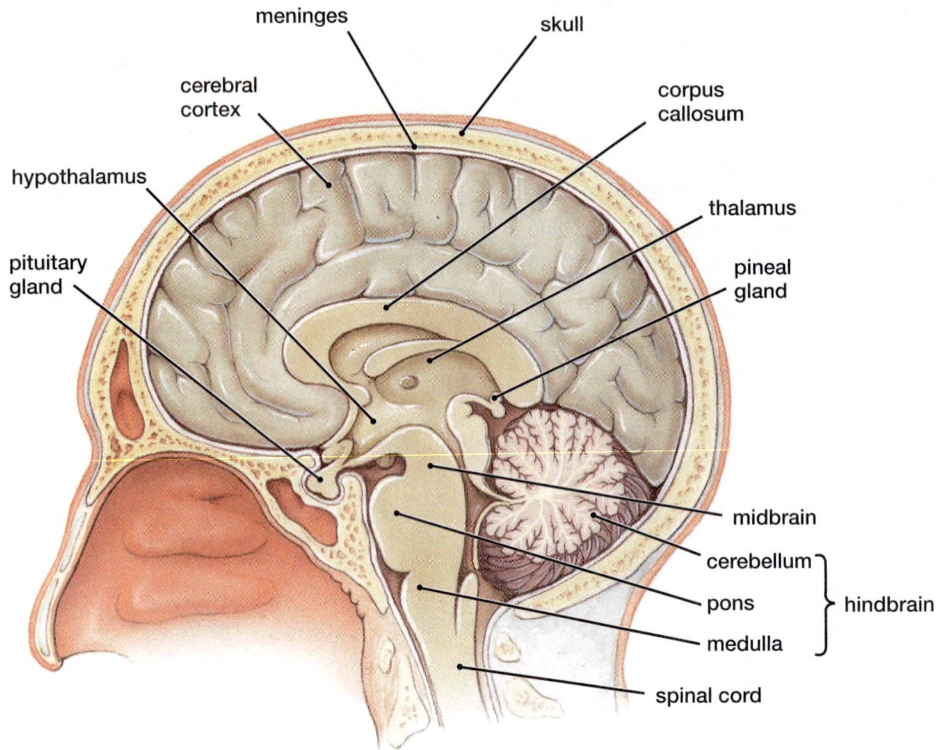

Figure 38-12 Longitudinal section of human brain

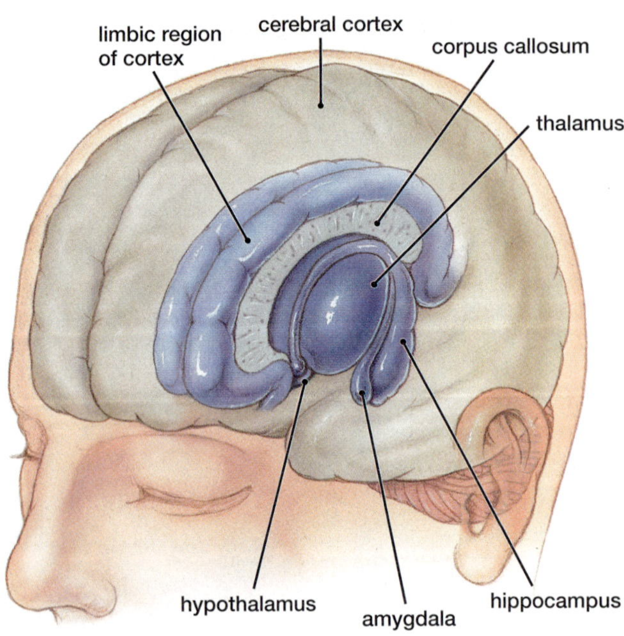

Figure 38-13 Limbic system

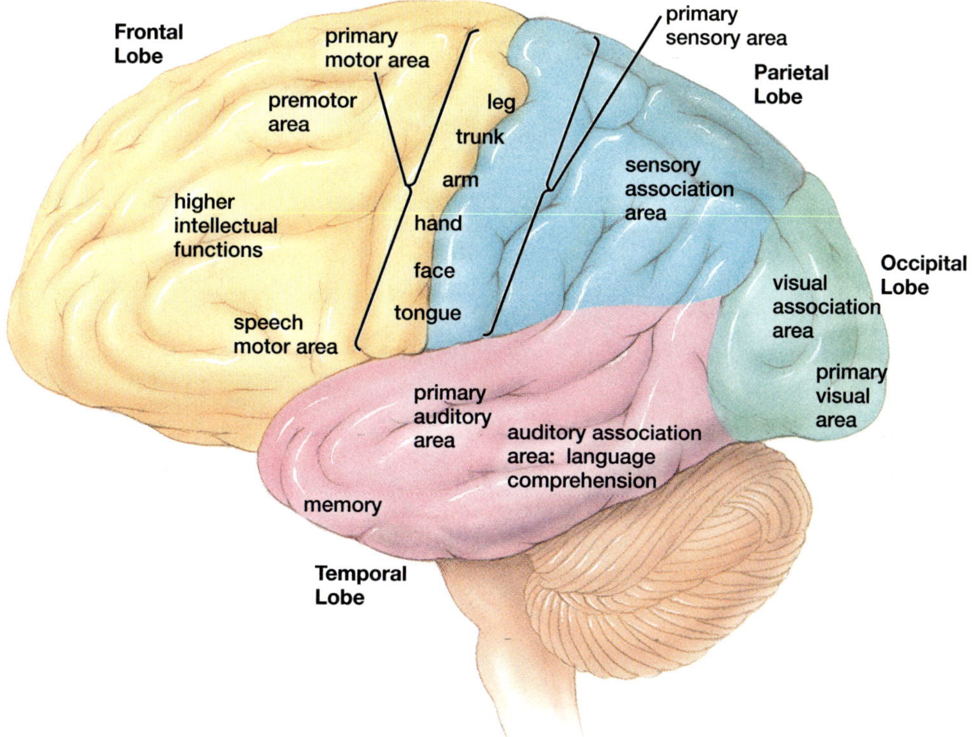

Figure 38-14 Human cerebral cortex

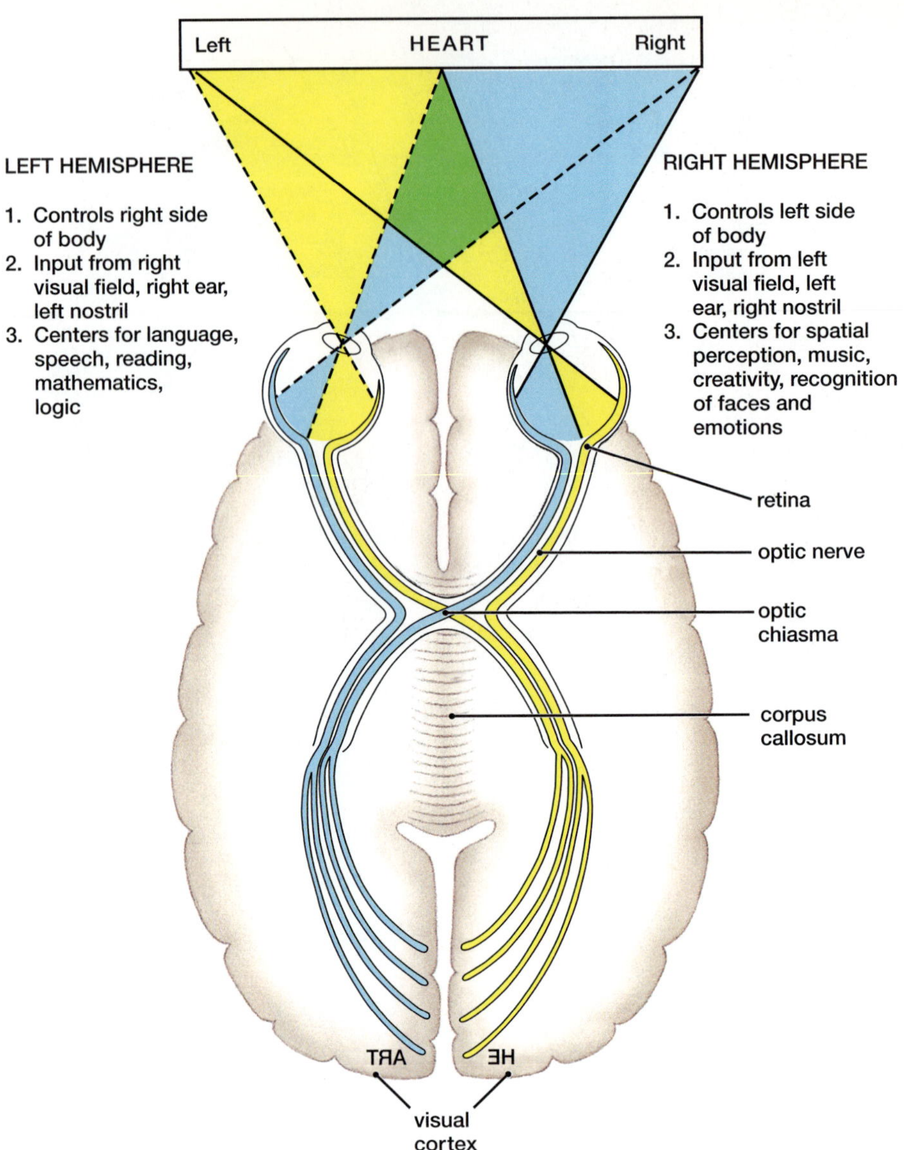

Figure 38-15 Visual field projection to cortex

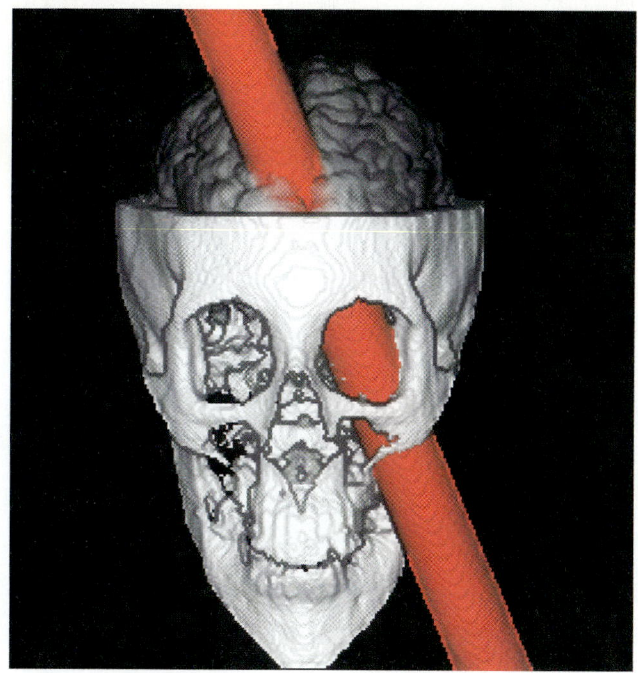

Figure 38-16 Reconstruction of Phineas Gage skull

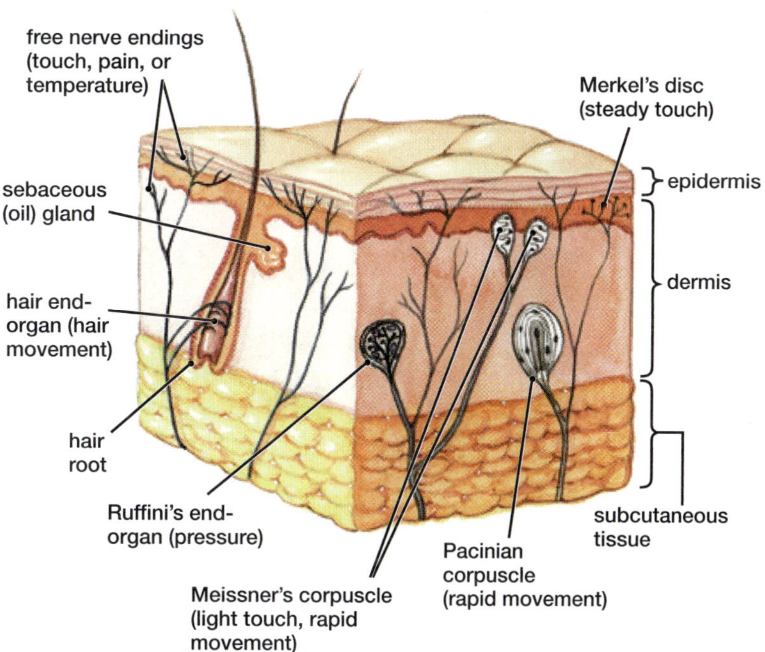

Figure 38-17 Skin senses

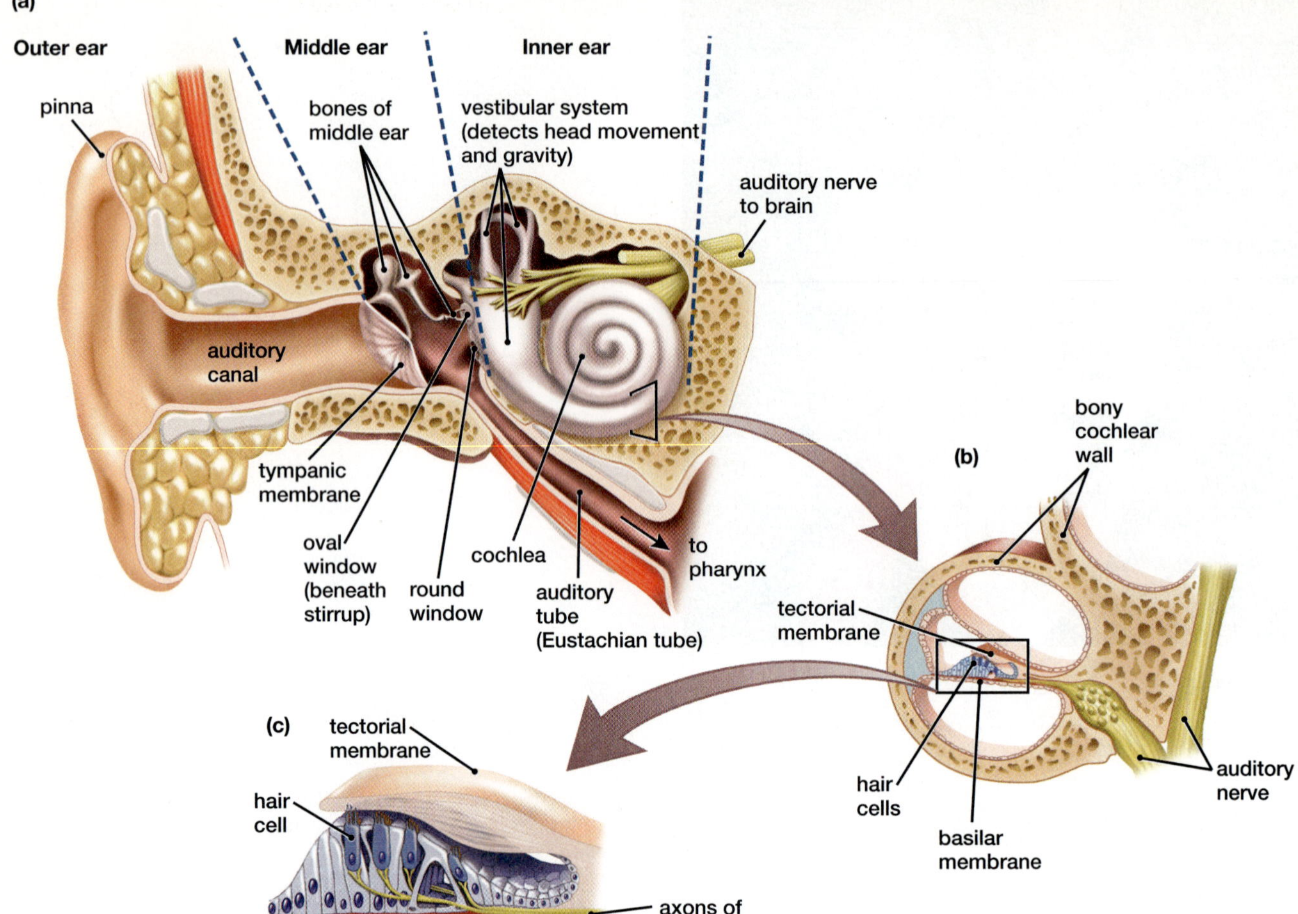

Figure 38-18 Ear

(a) Anatomy of the human eye

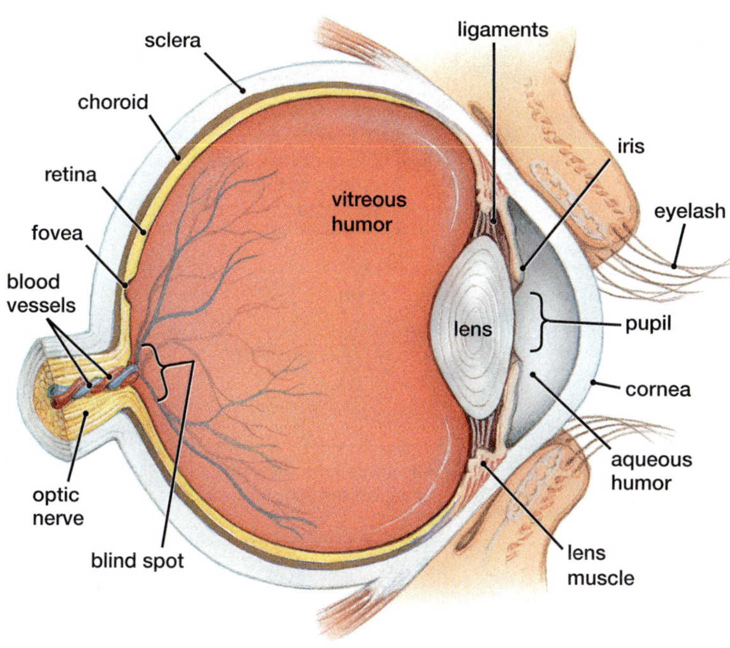

(b) Layers of the retina

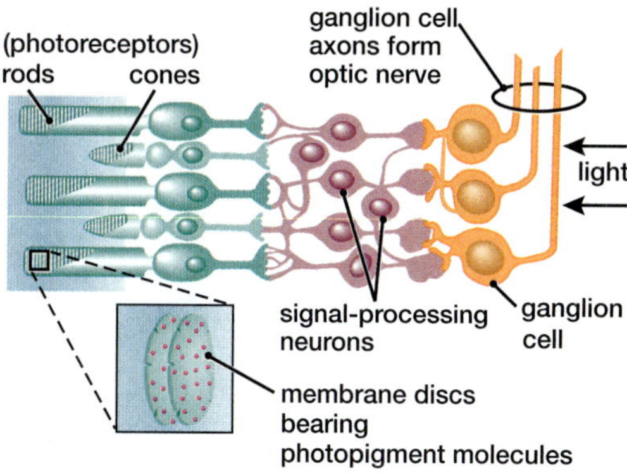

Figure 38-21 Eye

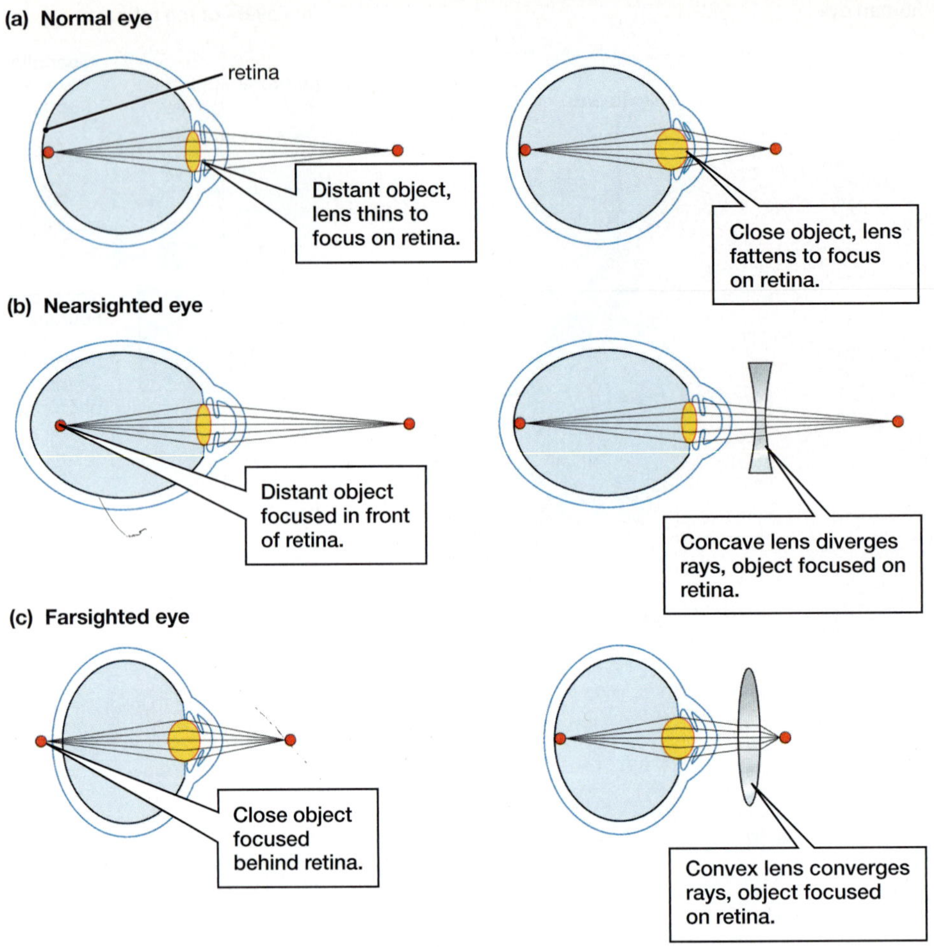

(a) **Normal eye**

retina

Distant object, lens thins to focus on retina.

Close object, lens fattens to focus on retina.

(b) **Nearsighted eye**

Distant object focused in front of retina.

Concave lens diverges rays, object focused on retina.

(c) **Farsighted eye**

Close object focused behind retina.

Convex lens converges rays, object focused on retina.

Figure 38-22 Eye focusing

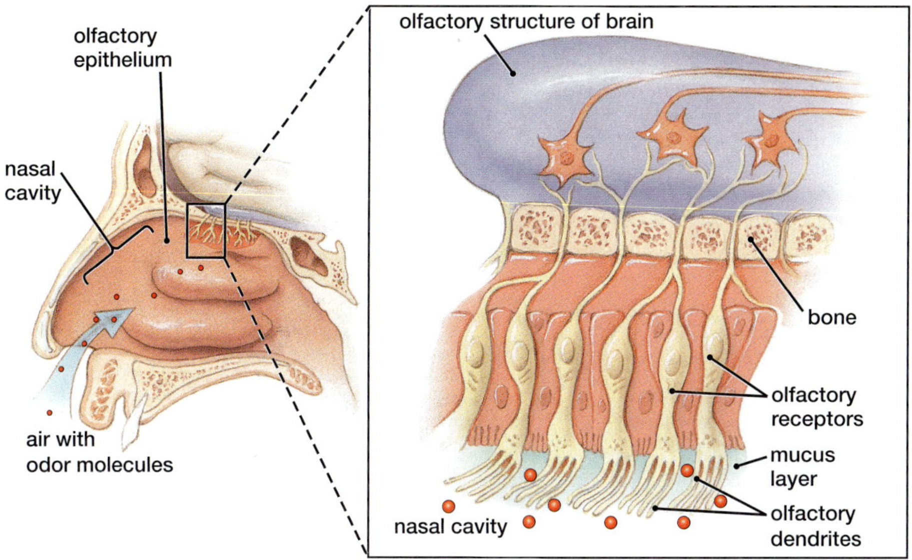

Figure 38-26 Olfaction

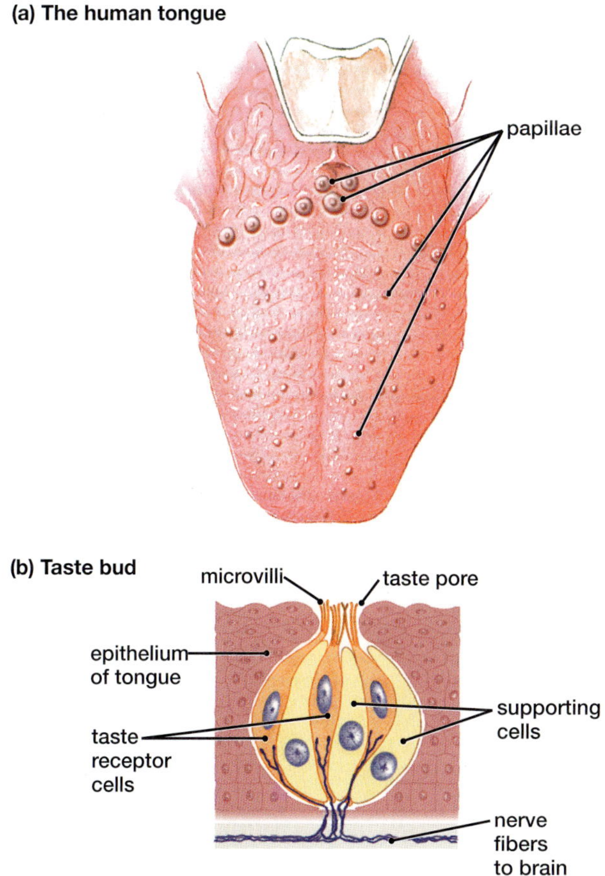

Figure 38-27 Taste

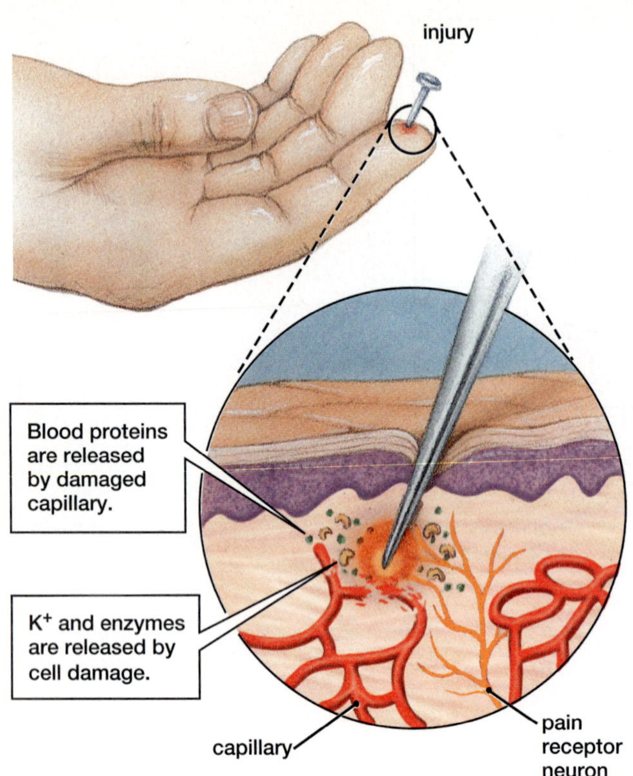

Figure 38-28 Pain reception

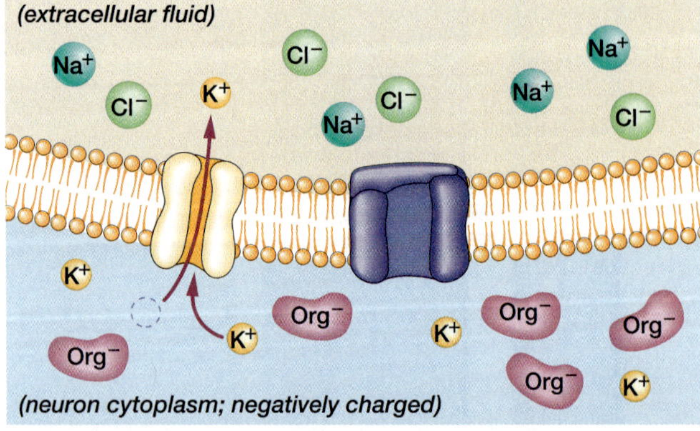

Figure E38-1 Resting potential ion movement

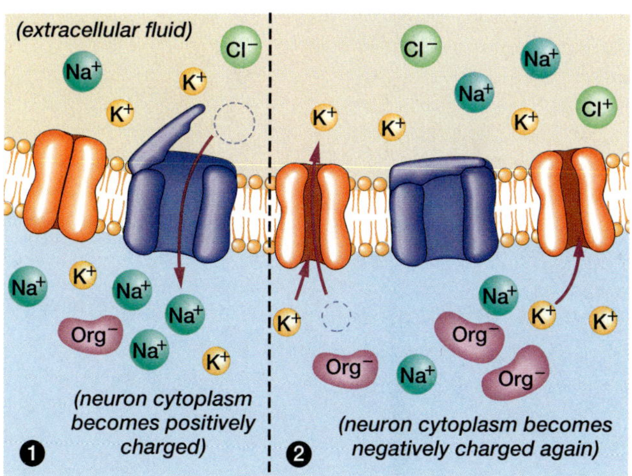

Figure E38-2 Action potential ion movement

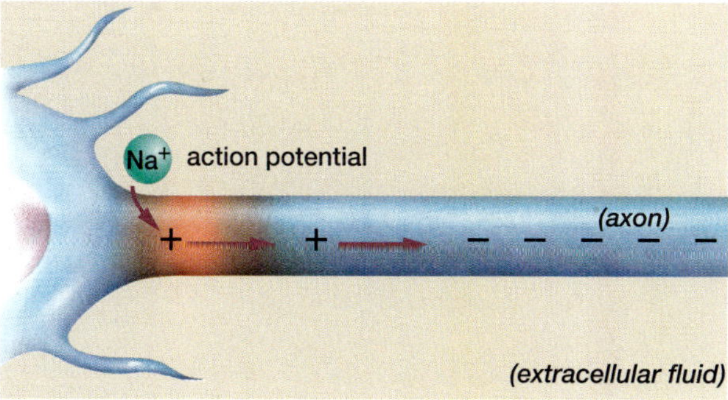

Figure E38-3 Action potential conduction, part 1

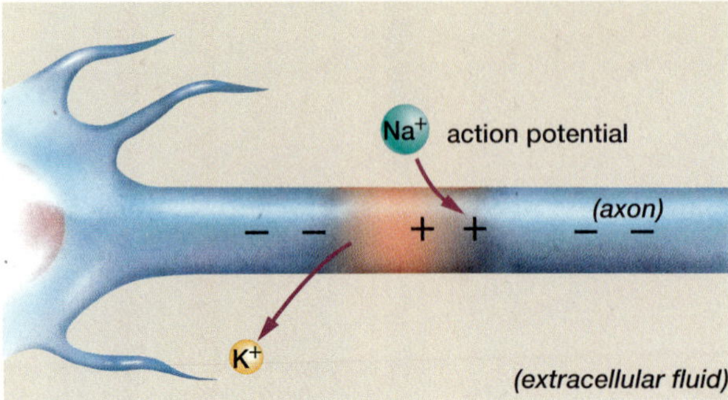

Figure E38-4 Action potential conduction, part 2

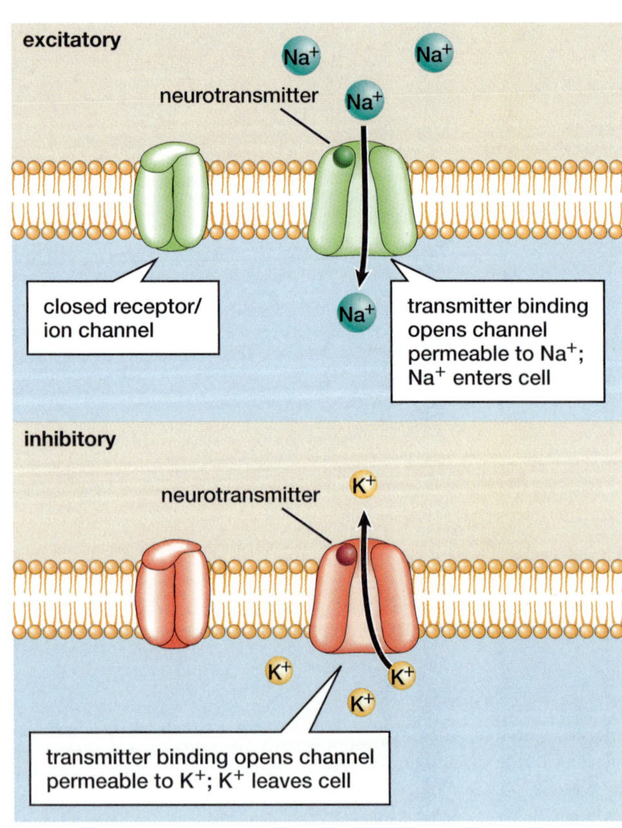

Figure E38-5 Ion channel action in synapses

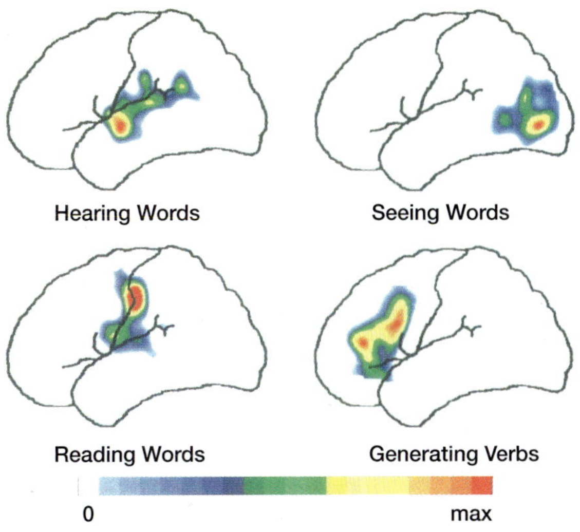

Figure E38-7 PET scan of language areas

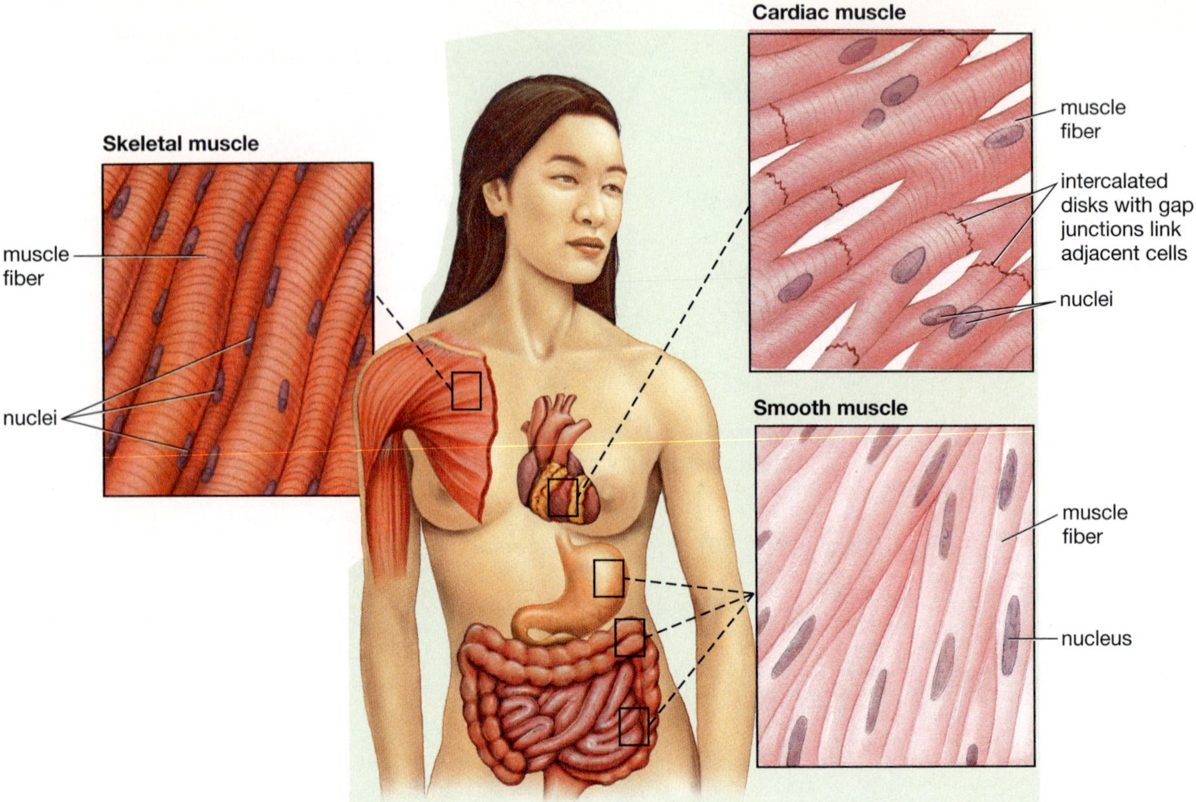

Table 39-1 Skeletal muscle

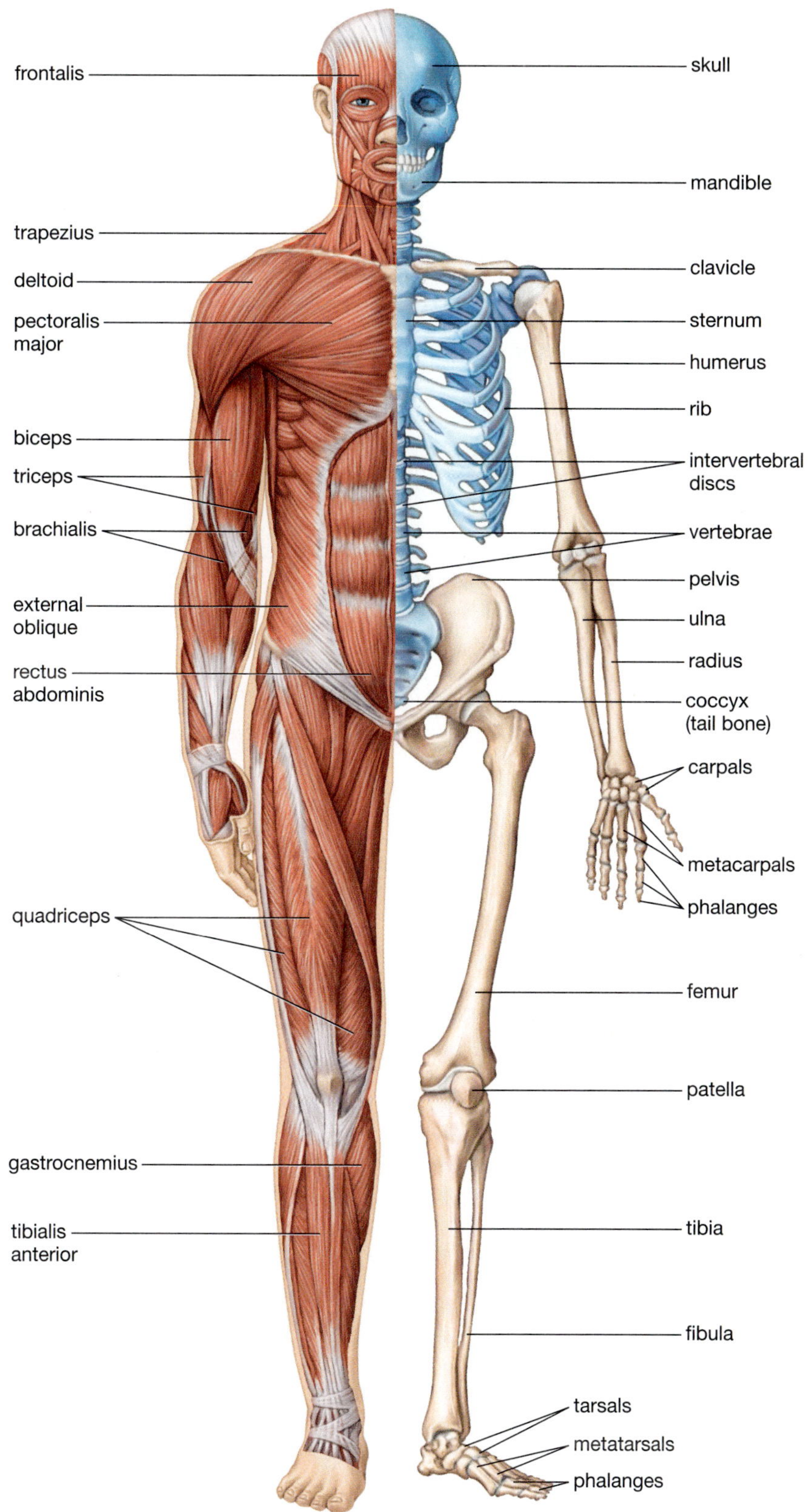

Figure 39-1 Skeleton and muscles

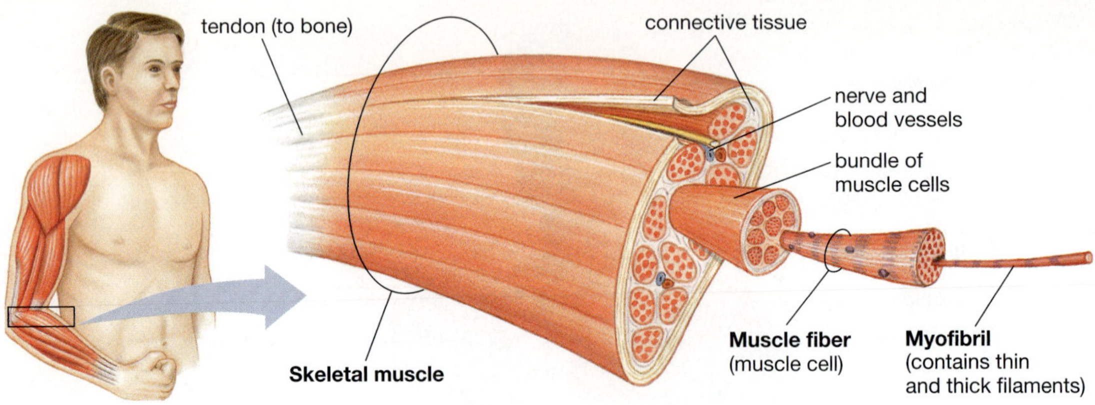

Figure 39-2 Skeletal muscle structure

(a) Cross section of fiber

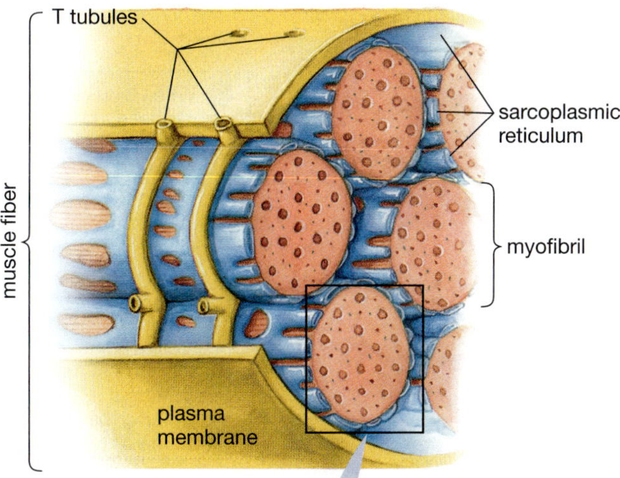

(b) Myofibril and sarcomere

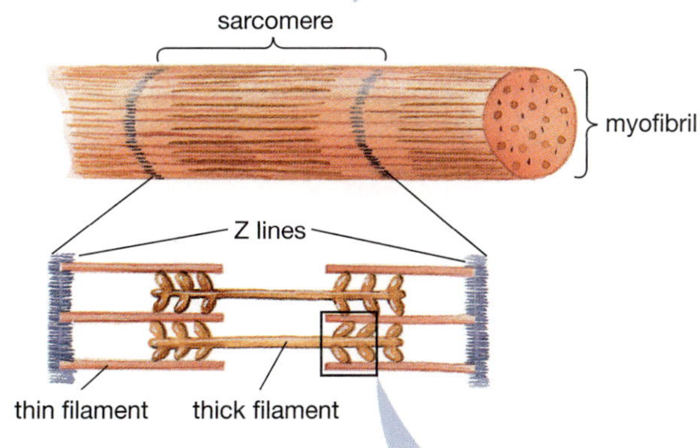

(c) Thick and thin filaments

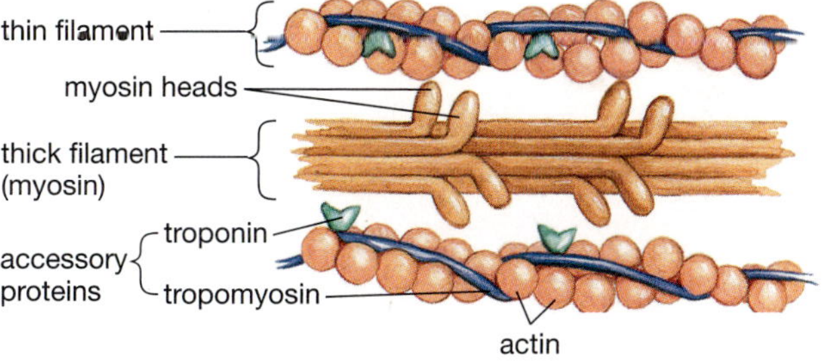

Figure 39-3 Muscle fiber structure

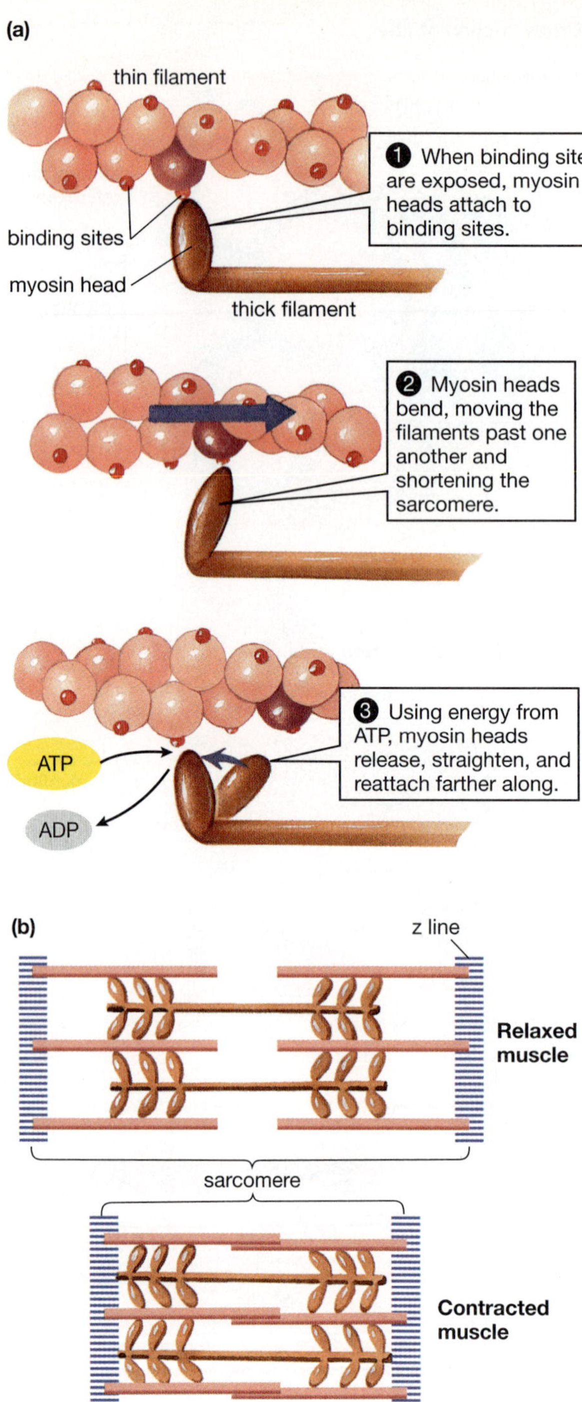

Figure 39-4 Muscle movement

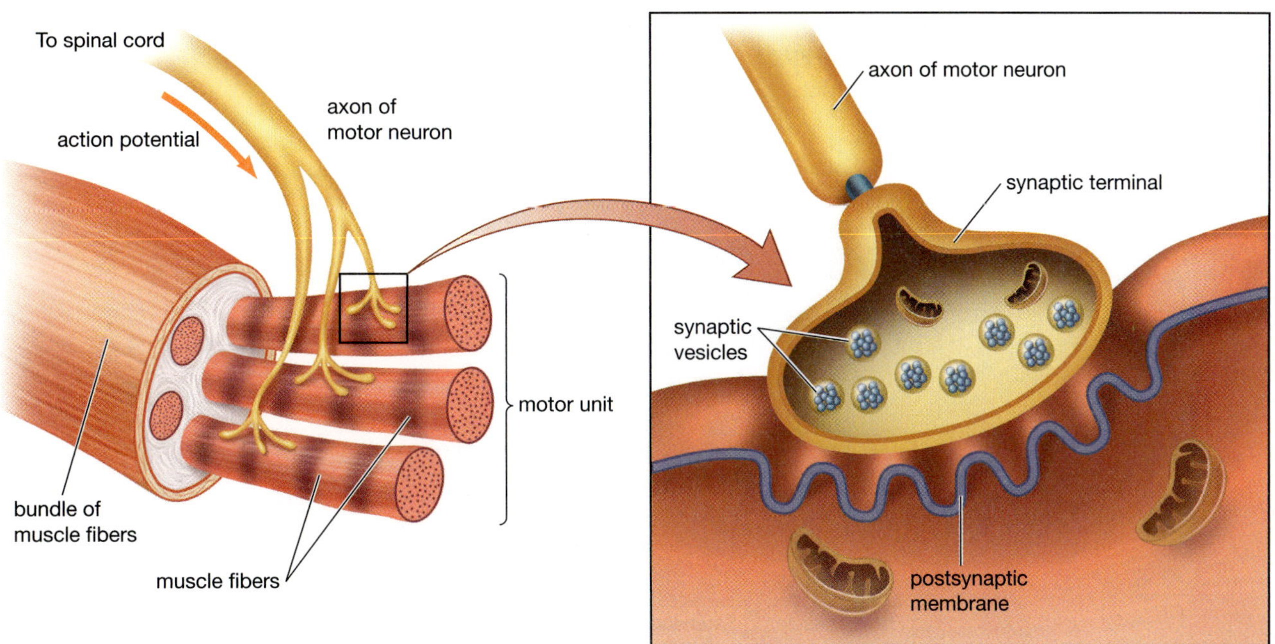

Figure 39-5 Motor unit and neuromuscular junction

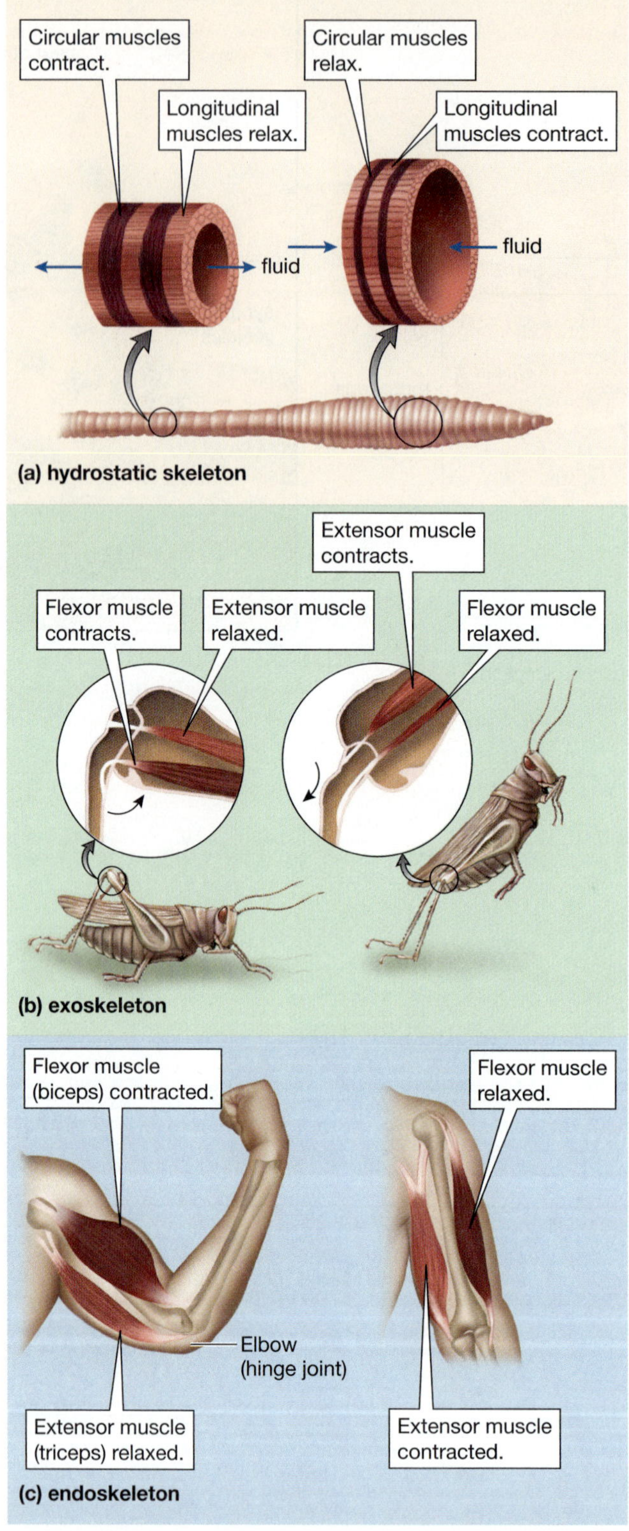

Figure 39-8 Antagonistic muscles in earthworm, insect and human arm

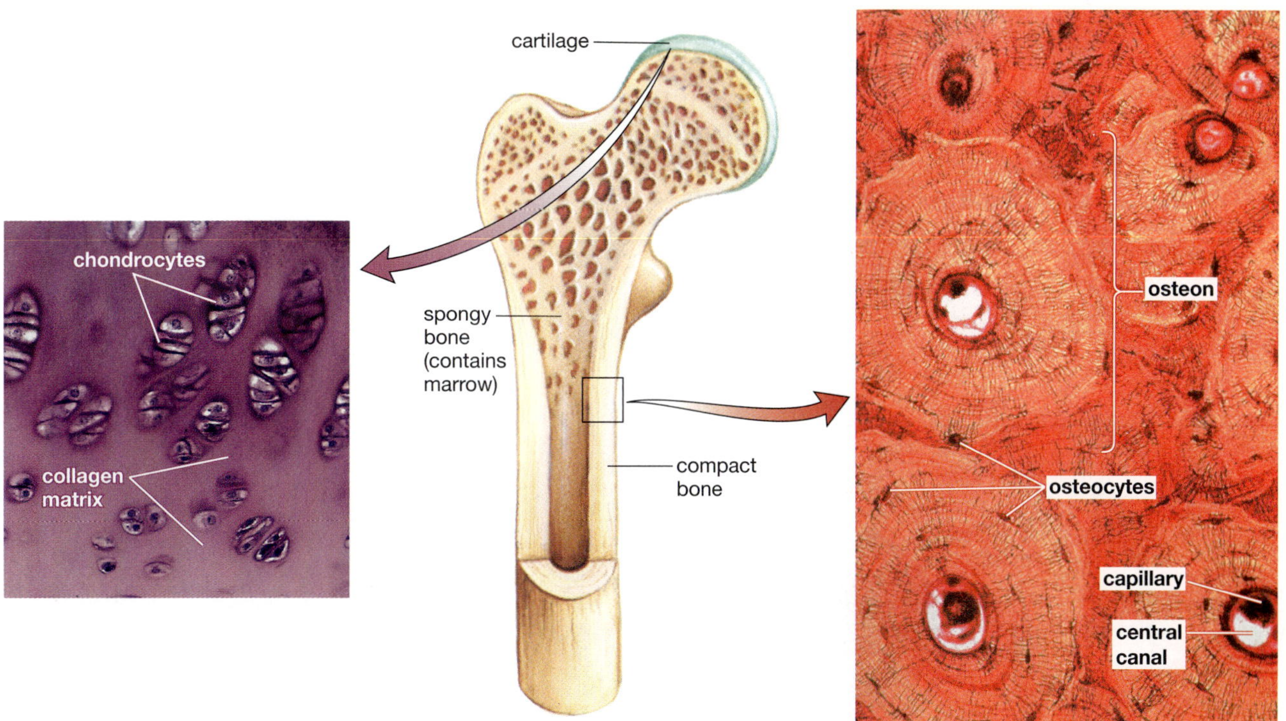

Figure 39-10 Cartilage and bone

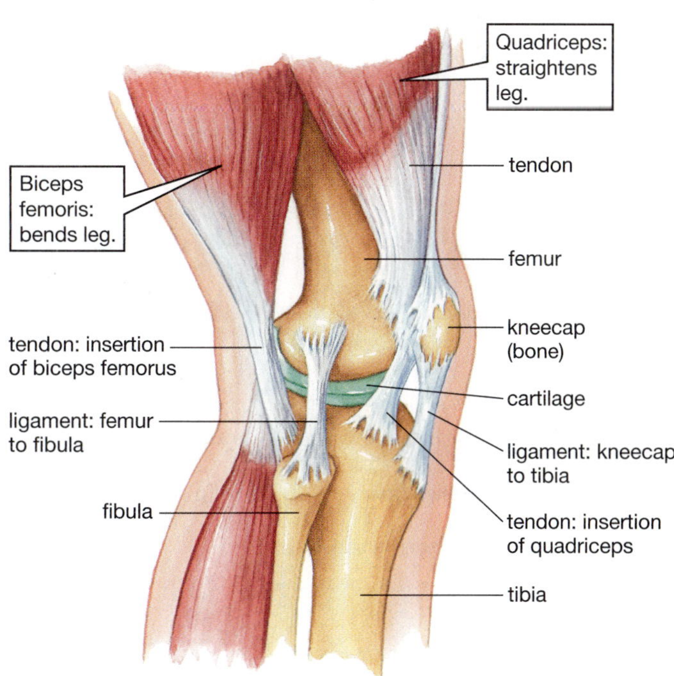

Figure 39-11 Human knee

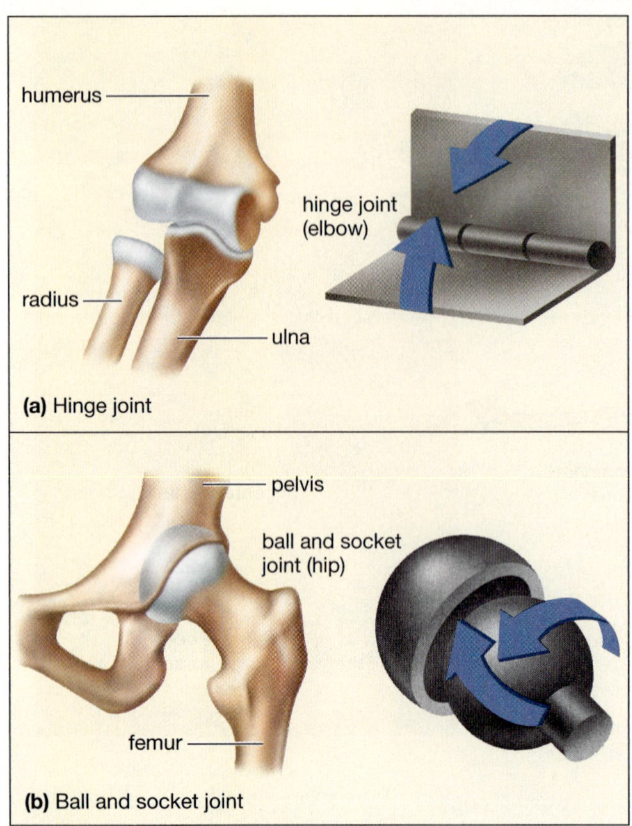

Figure 39-12 Hinge and ball and socket joints

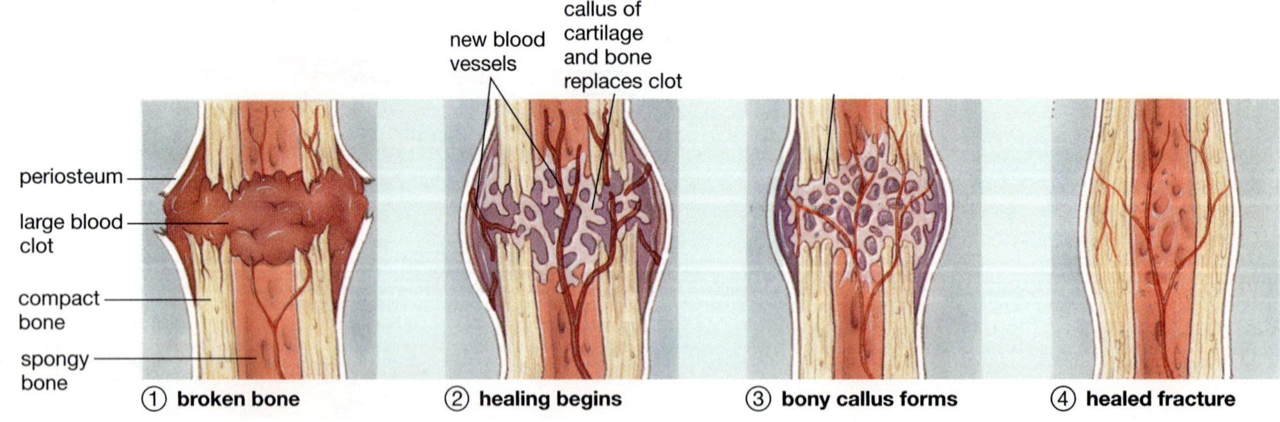

Figure E39-1 Bone healing series

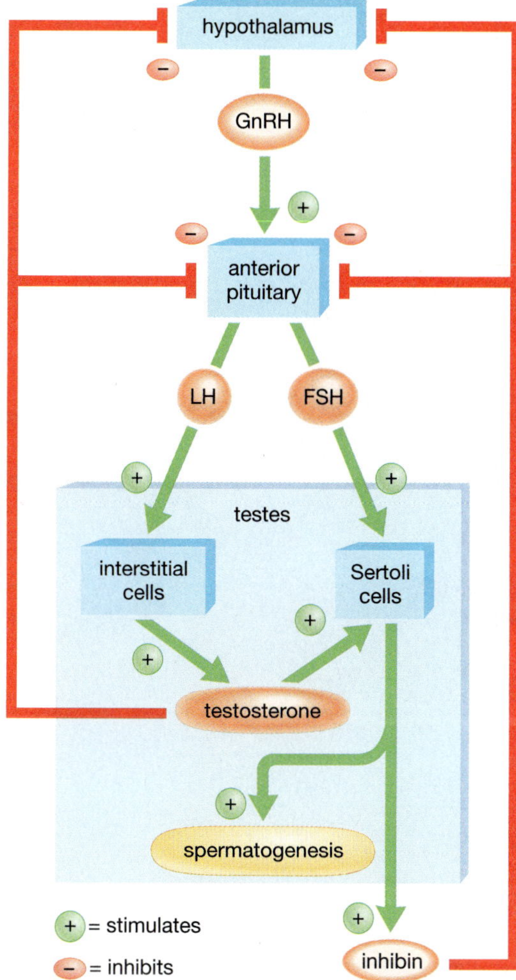

Figure 40-15 Hormonal control of spermatogenesis

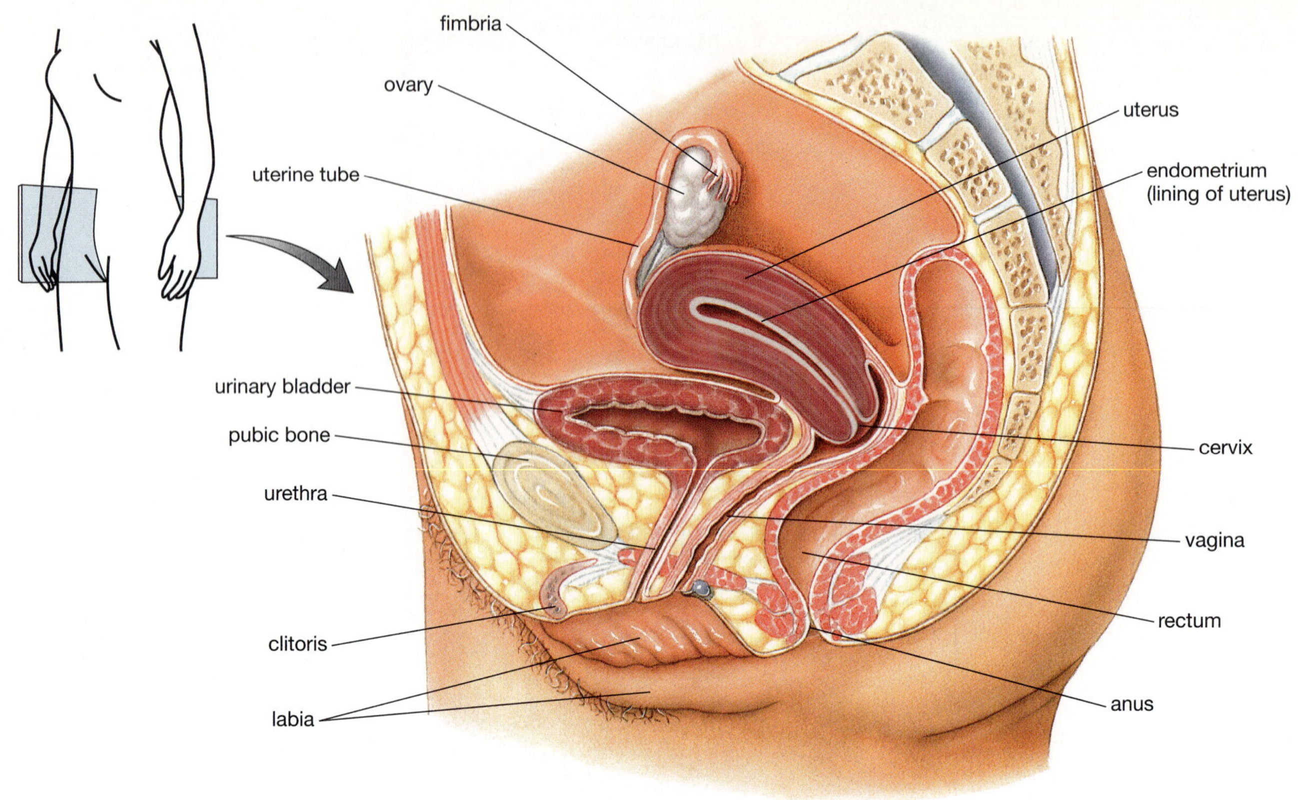

Figure 40-16 Human female reproductive tract

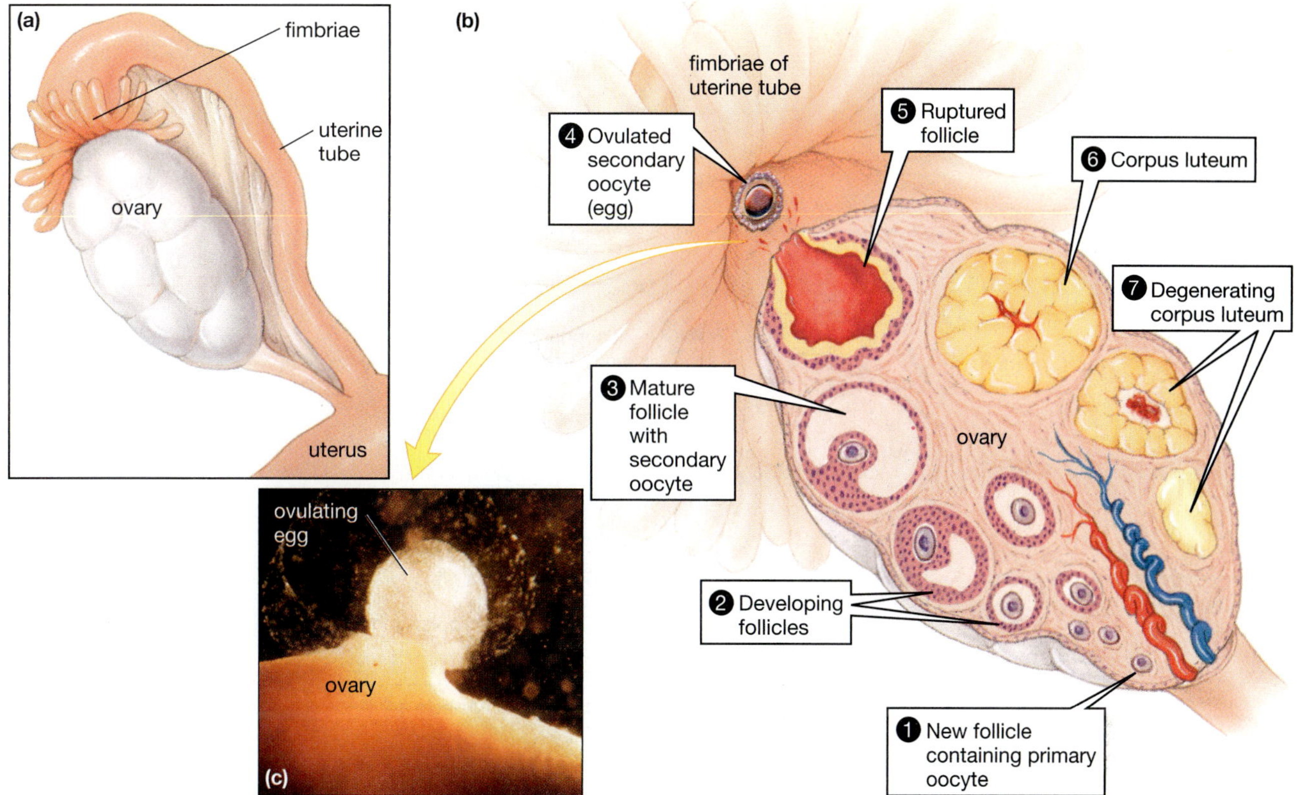

Figure 40-17 Structures involved in egg formation

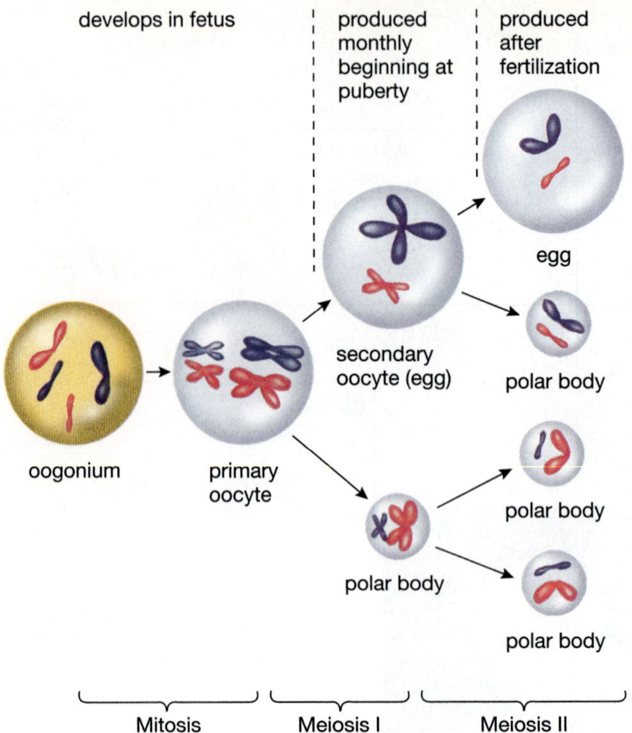

Figure 40-18 Meiosis produces egg cells

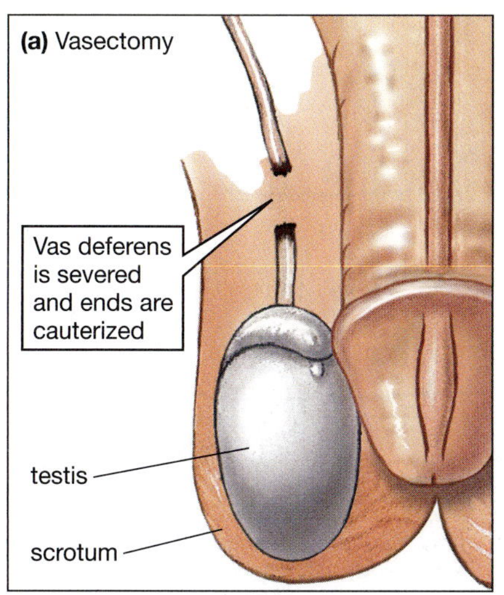

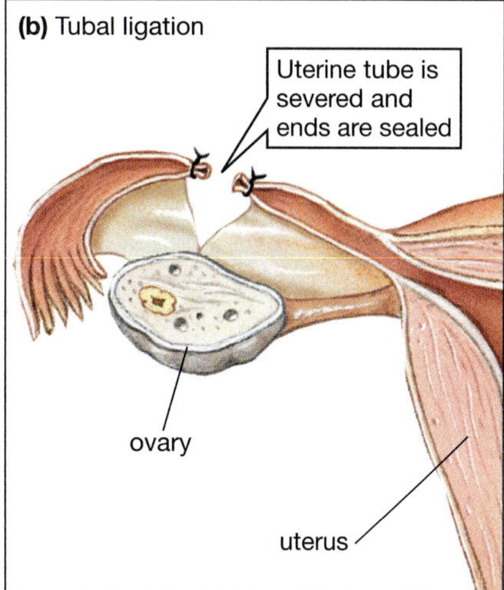

Figure 40-21 Sterilization

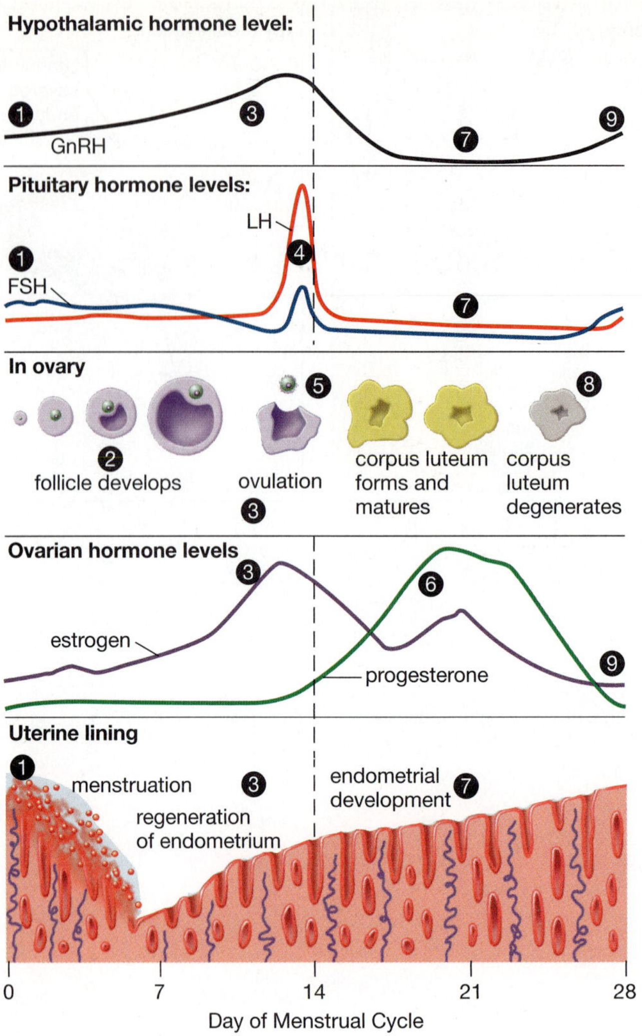

Figure E40-1 Menstrual cycle

Figure 41-1 Indirect development

ectoderm mesoderm endoderm

(a) The blastula just before gastrulation. The three embryonic tissue types have not yet formed. Colors indicate the future fate of the cells after they begin differentiating in the gastrula.

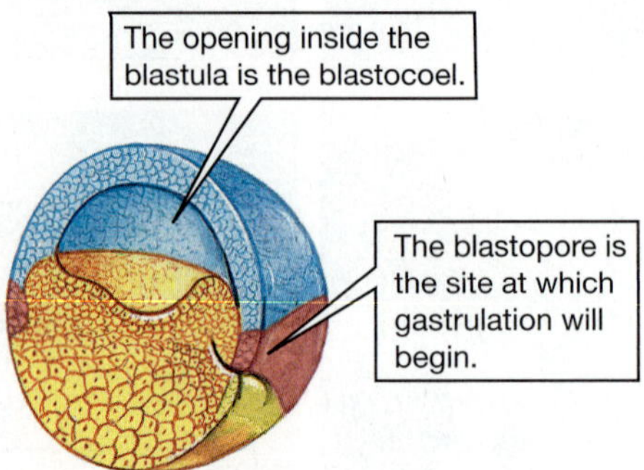

(b) Cells migrate at the start of gastrulation. Cells migrating in will form the endoderm and mesoderm layers of the gastrula; the cells remaining on the surface will form ectoderm.

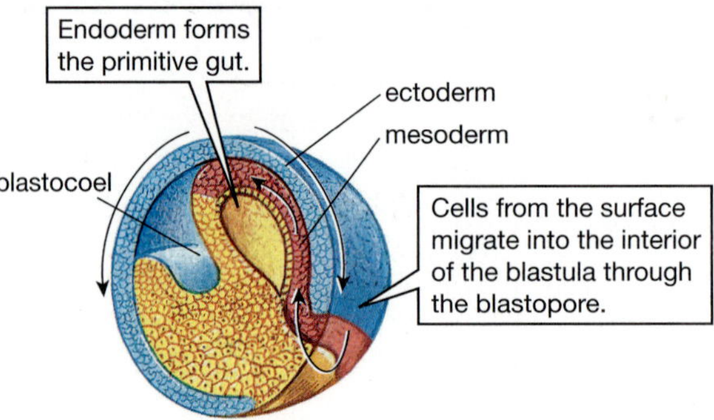

(c) Mesoderm differentiates.

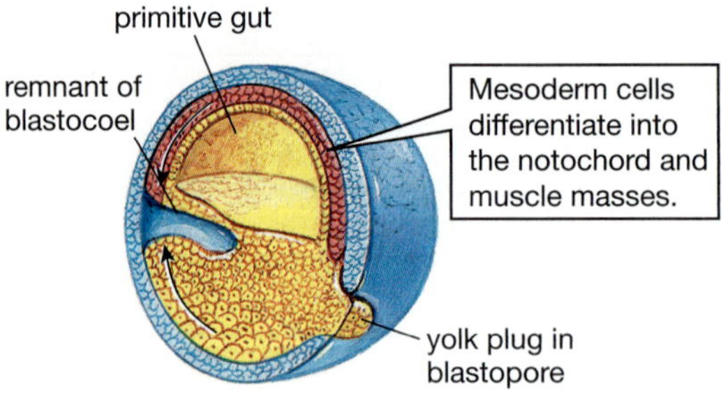

Figure 41-3 Blastula to gastrula

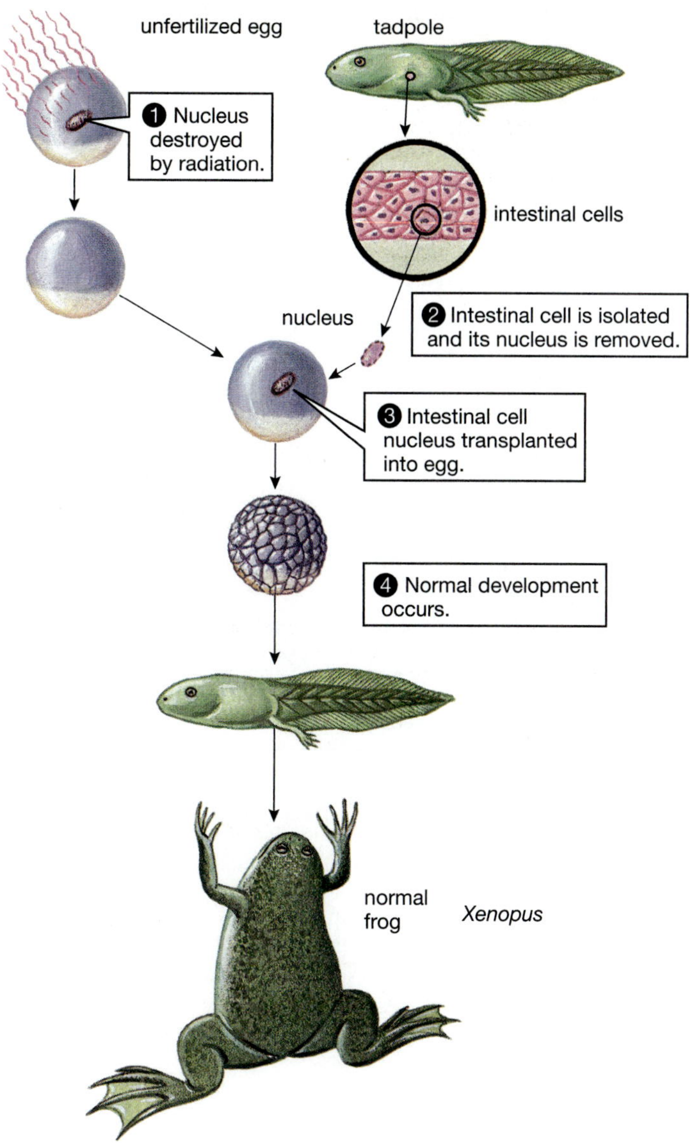

Figure 41-5 Frog differentiation

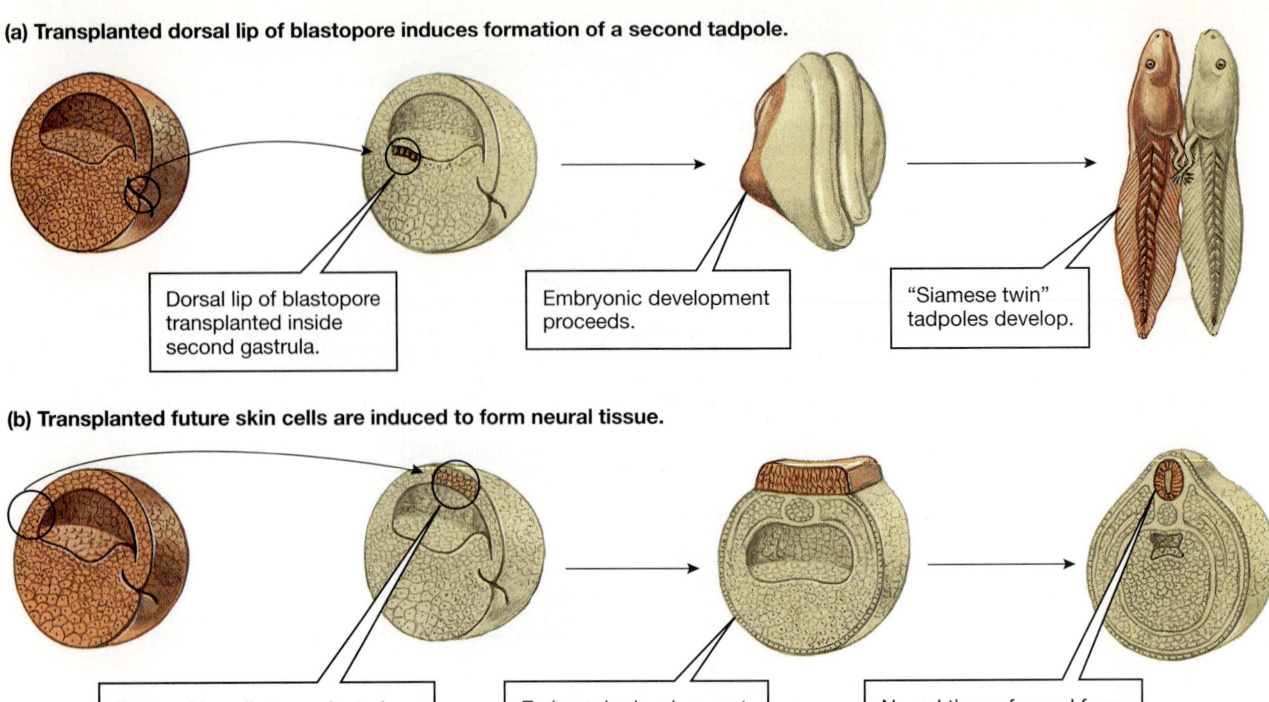

Figure 41-6 Induction

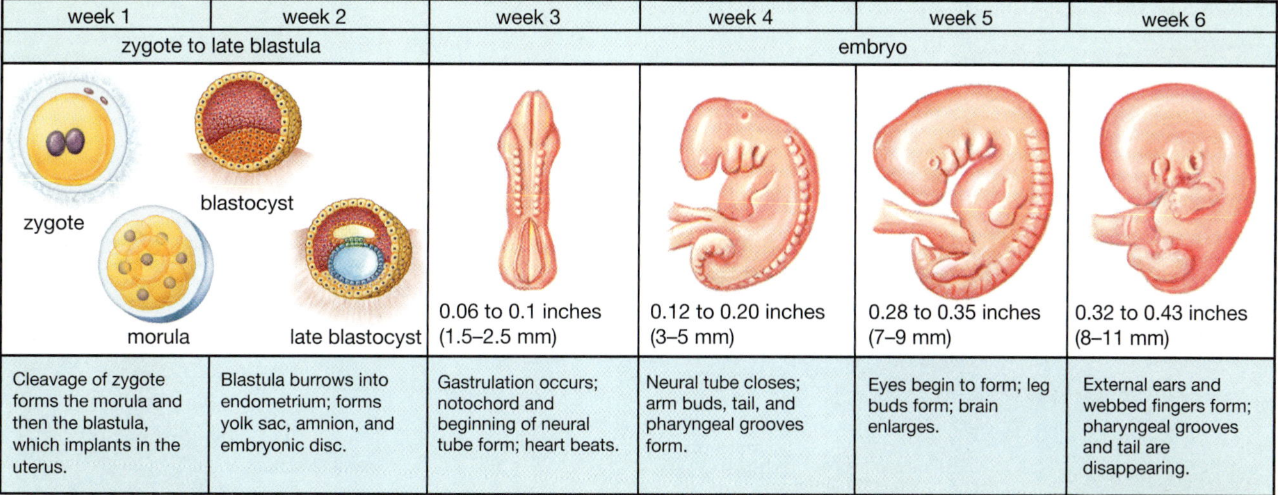

week 7	week 8	week 10	week 12	week 16
embryo		fetus		
0.67 to 0.79 inches (1.7–2.0 cm)	0.90 to 1.10 inches (2.3–2.8 cm)	1.25–1.75 inches (3.2–4.4 cm)	2–3 inches (5–7.6 cm)	4–5 inches (10.2–12.7 cm)
Webbed toes form; bones begin to stiffen; back straightens; eyelids begin forming.	All major organs and male genitals begin to form; arms can bend; fingers are distinct. Facial features and outer ears take shape.	After 8 weeks; the embryo is called a fetus. Red blood cells form; toes separate; eyelids have developed; major brain parts are present; hands can form fists.	Neck is well-defined; all organs are present; male and female genitals are present; arms and legs move; teeth begin to form; heartbeat can be detected electronically.	Sucking and swallowing movements occur; liver and pancreas begin functioning. Body has grown relative to the head; major organs continue developing. Mother may feel movement; weight is about 5 oz.

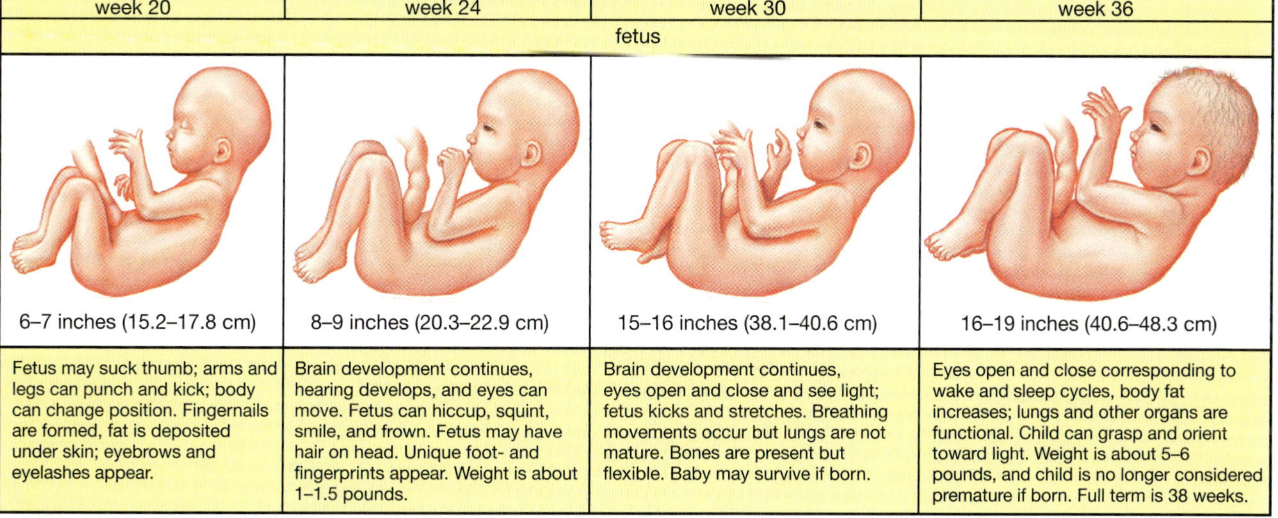

Figure 41-8 Development

(a) The first week

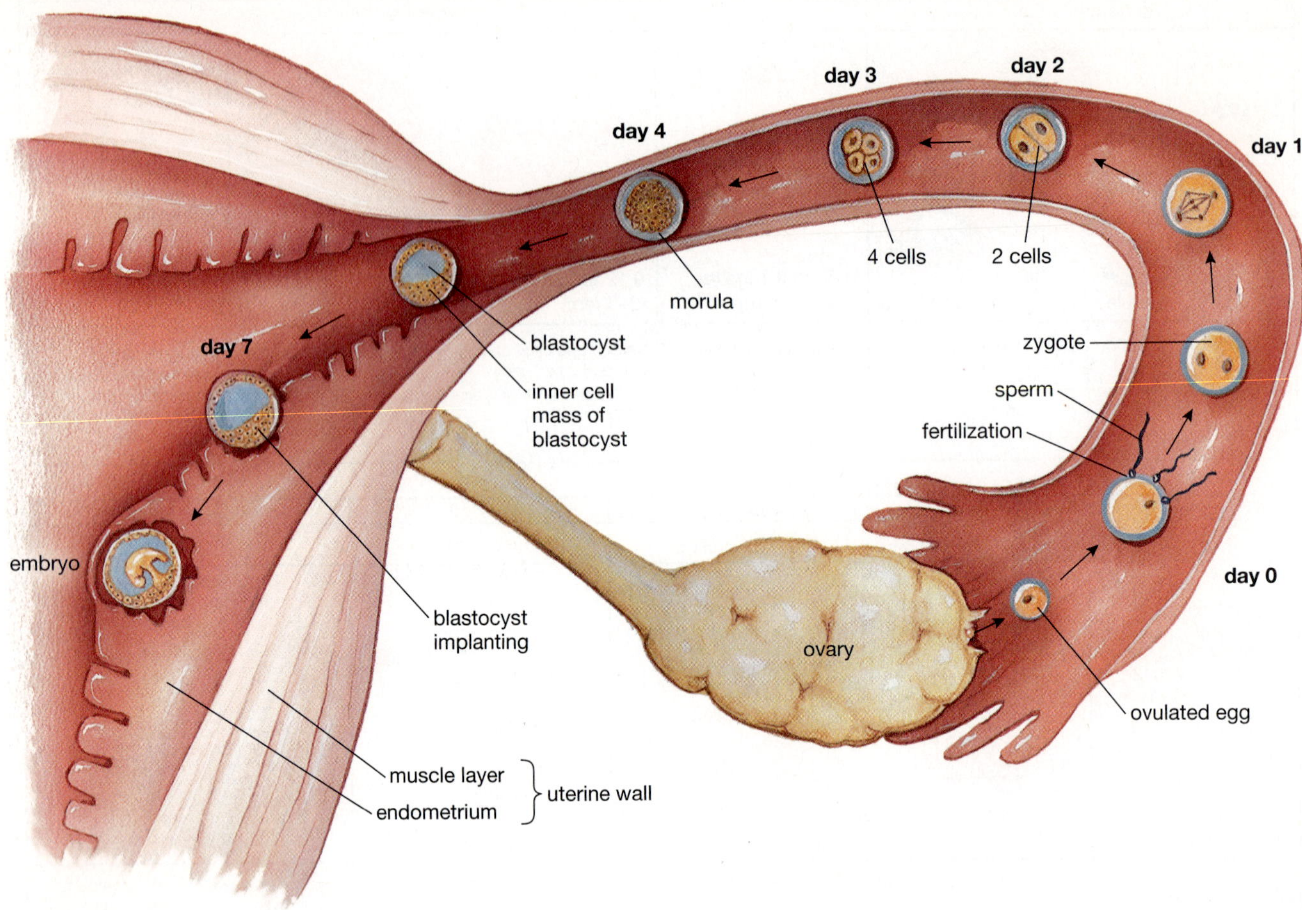

Figure 41-9 Journey of egg

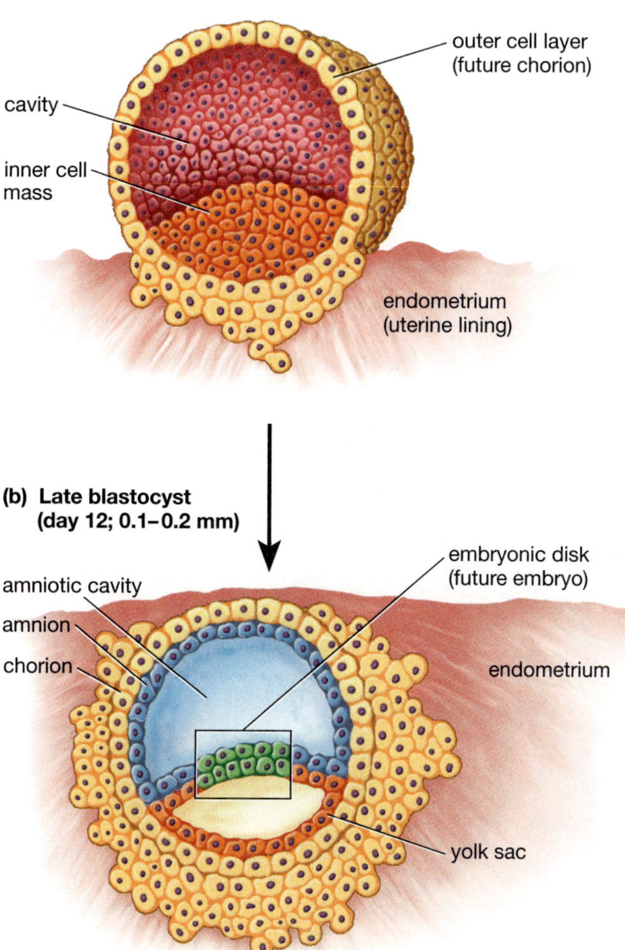

Figure 41-10 Implantation

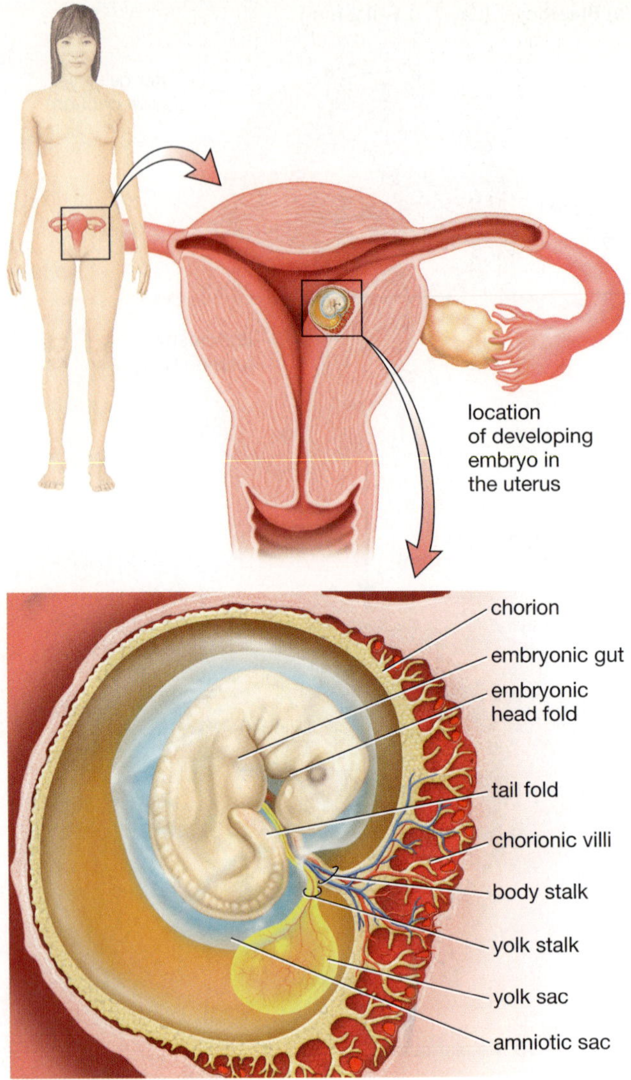

Figure 41-11 4 week embryo

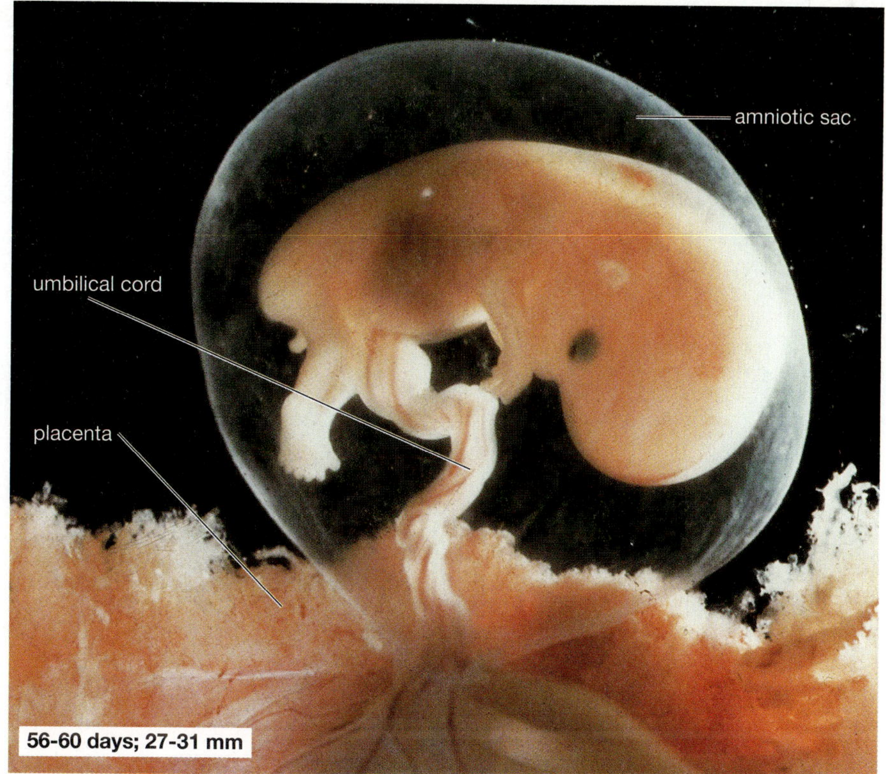

Figure 41-13 8 week embryo

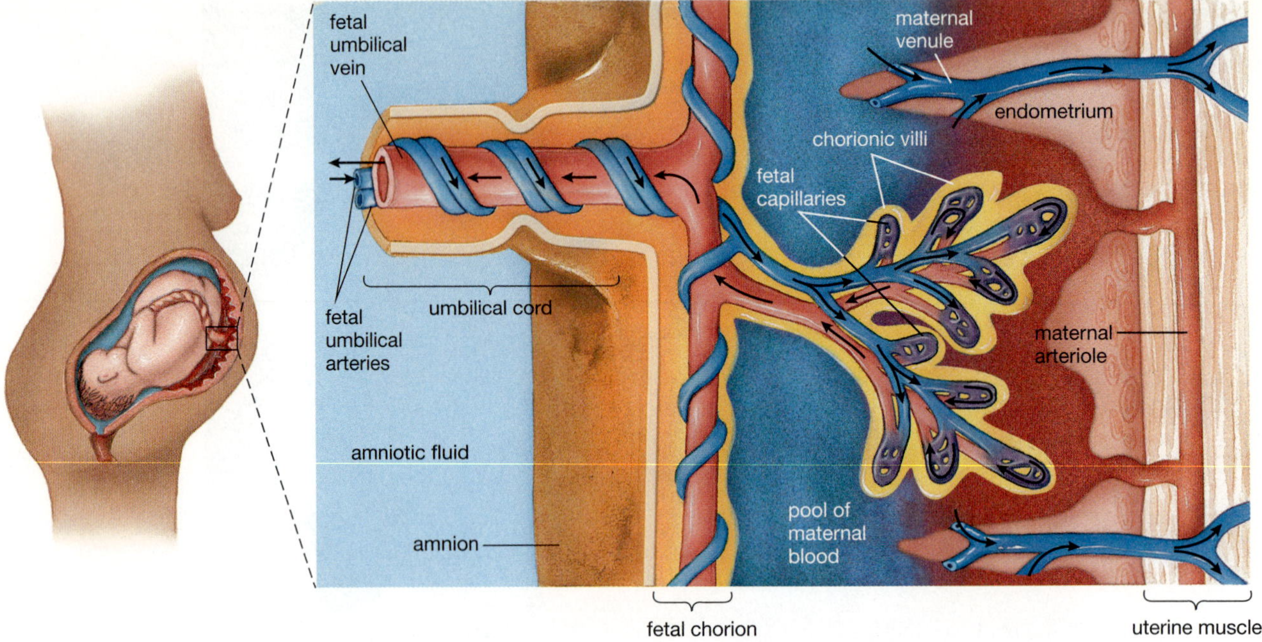

Figure 41-14 Placenta

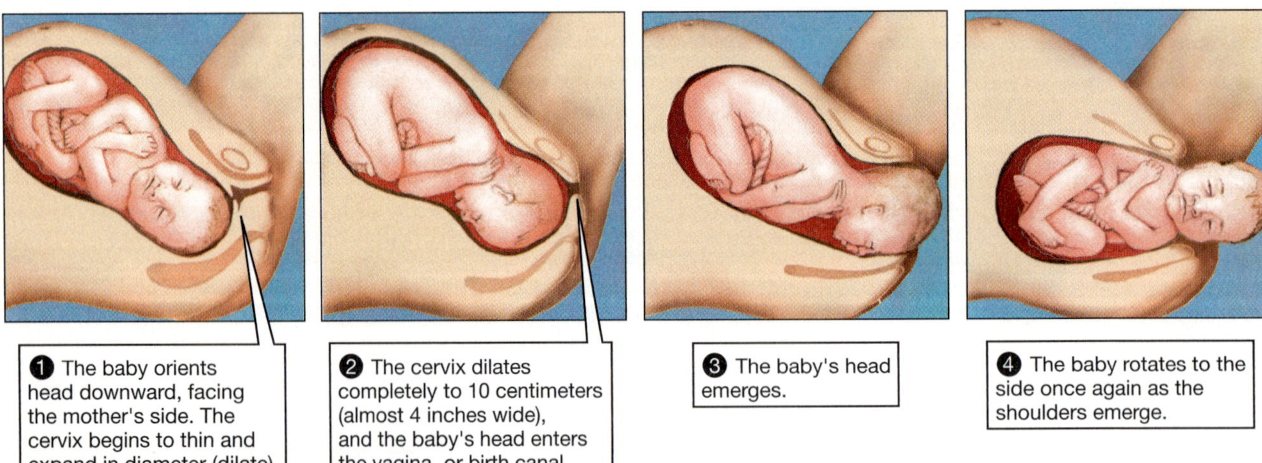

Figure 41-15 Birth

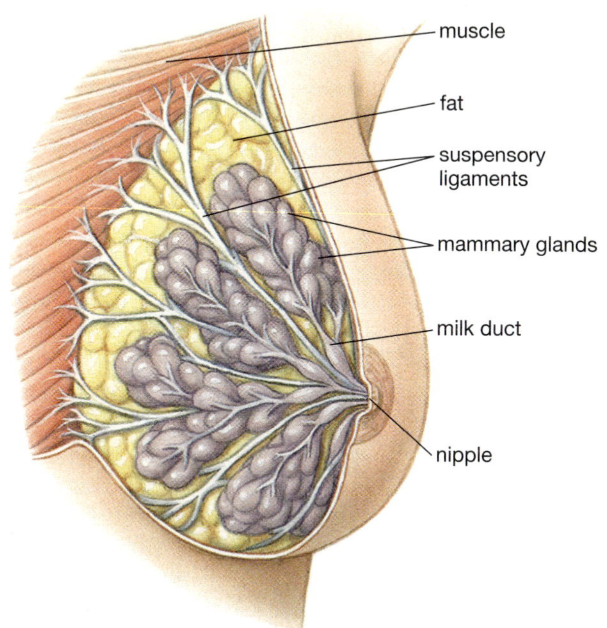

Figure 41-16 Breast

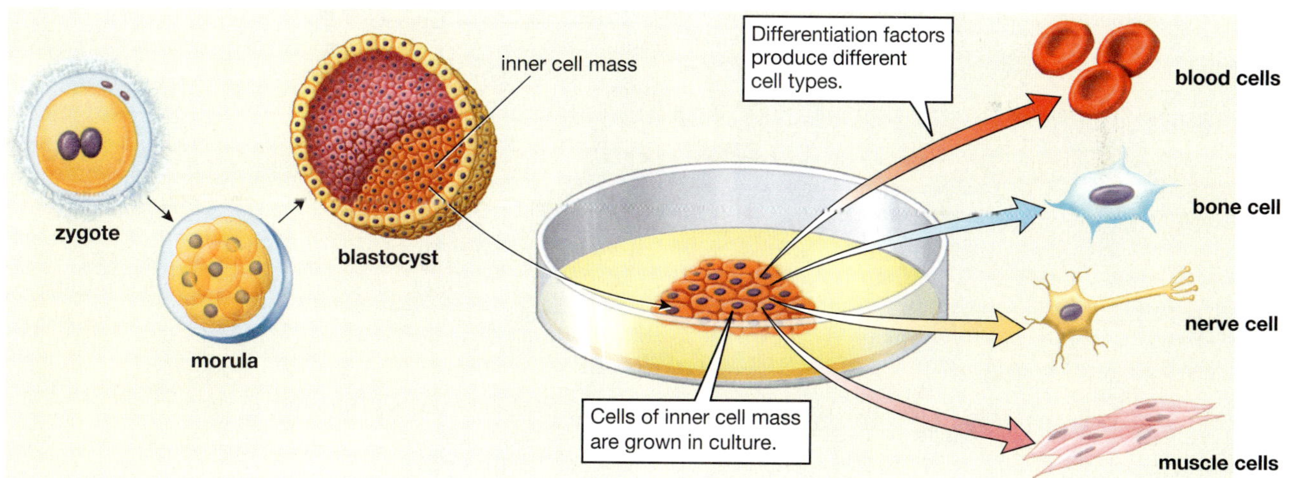

Figure E41-1 Stem cell culture

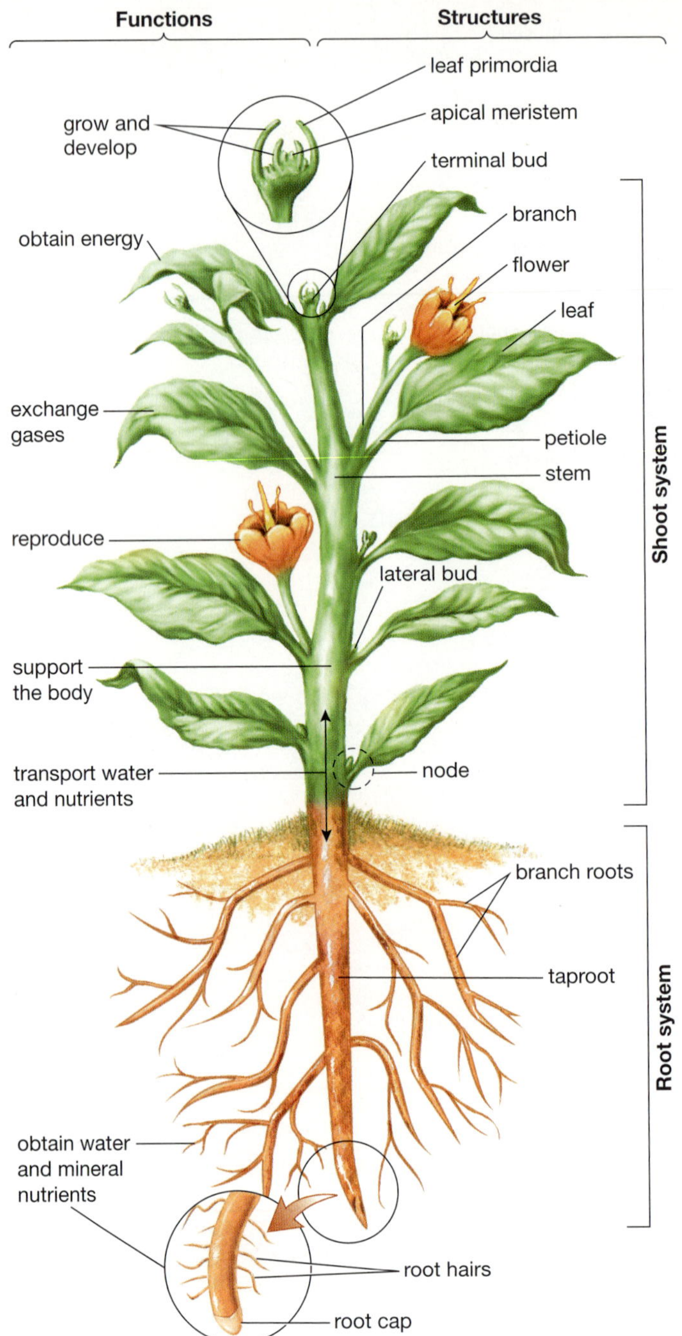

Figure 42-1 Flowering plant structure

Figure 42-2 Monocots and dicots

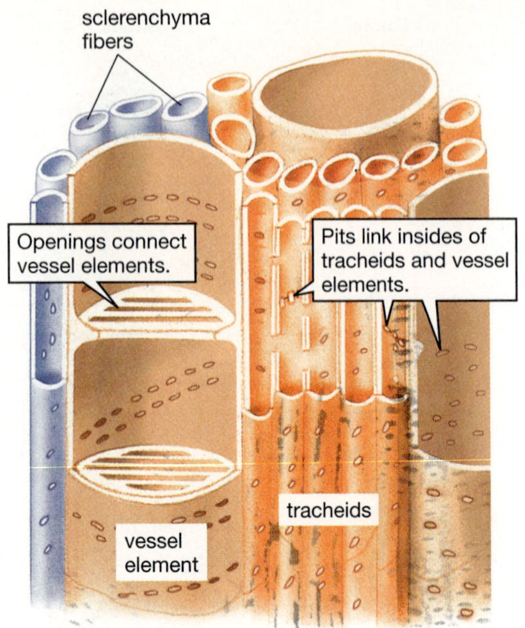

Figure 42-6a Xylem

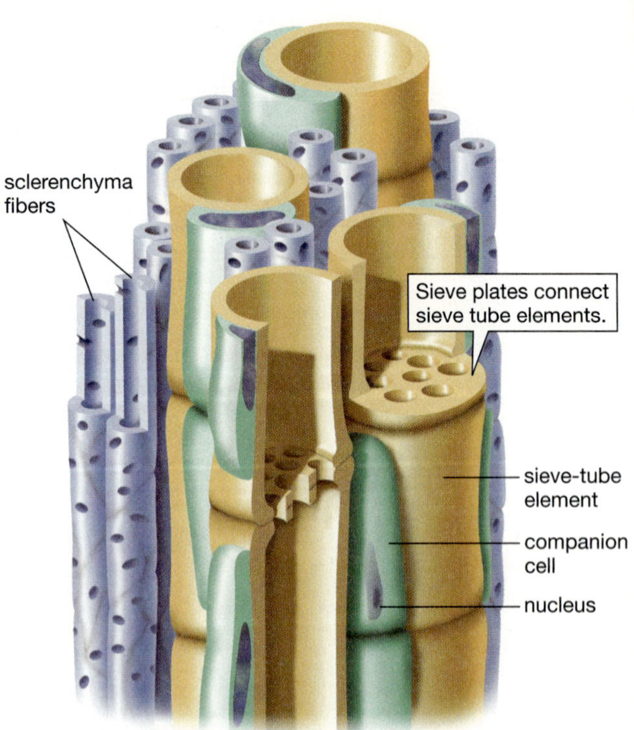

Figure 42-7a Phloem

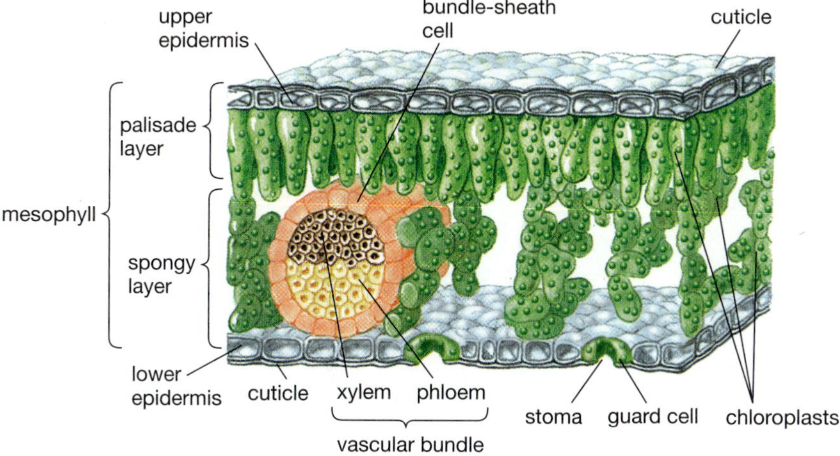

Figure 42-8a Leaf structure

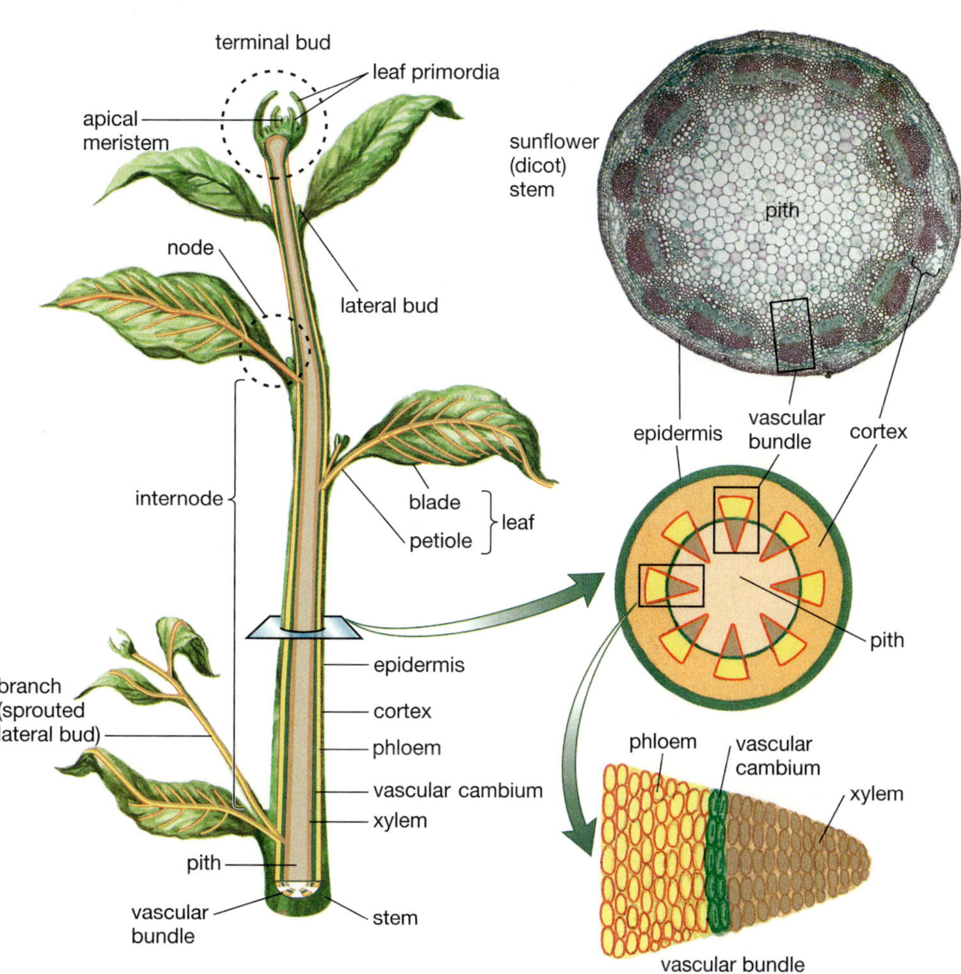

Figure 42-9 Dicot stem

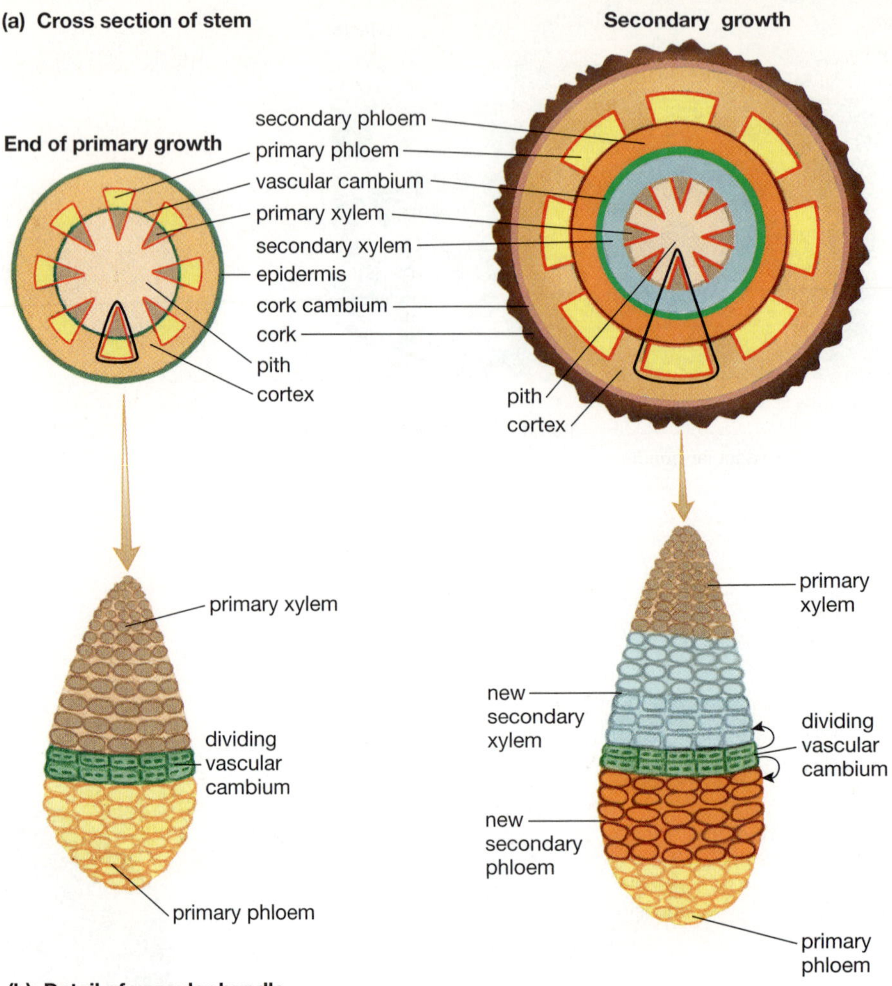

Figure 42-11 Secondary growth in stems

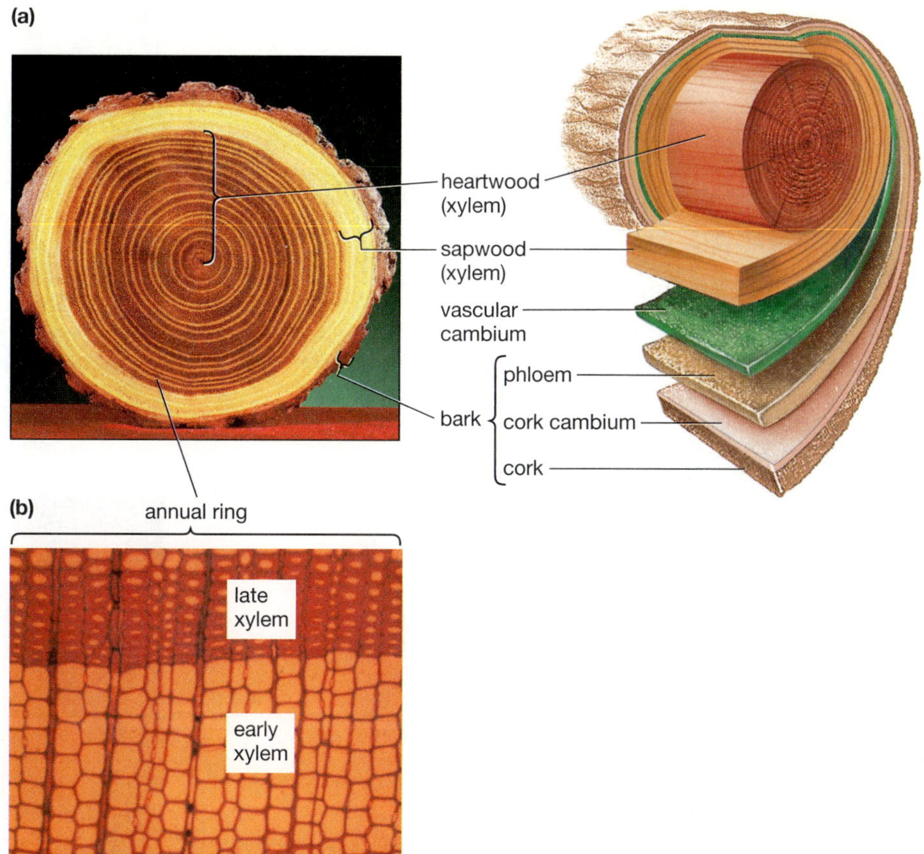

Figure 42-12 Annual rings

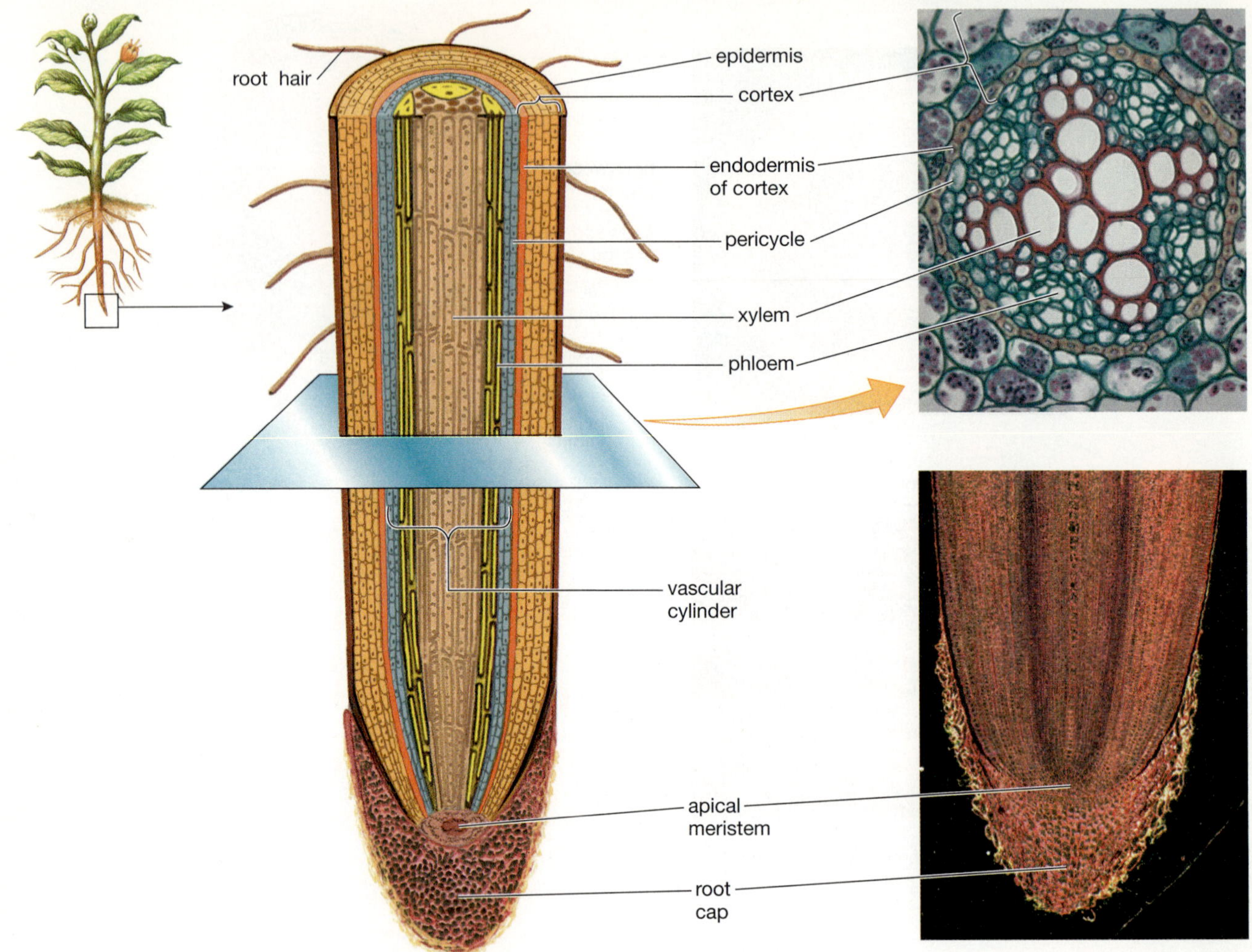

Figure 42-15 Root structure

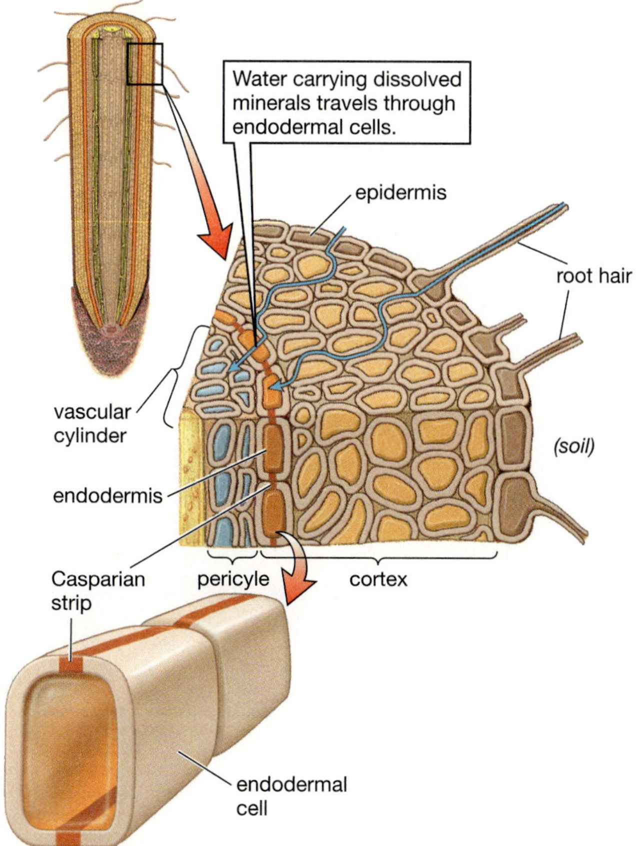

Figure 42-18 Casparian strip

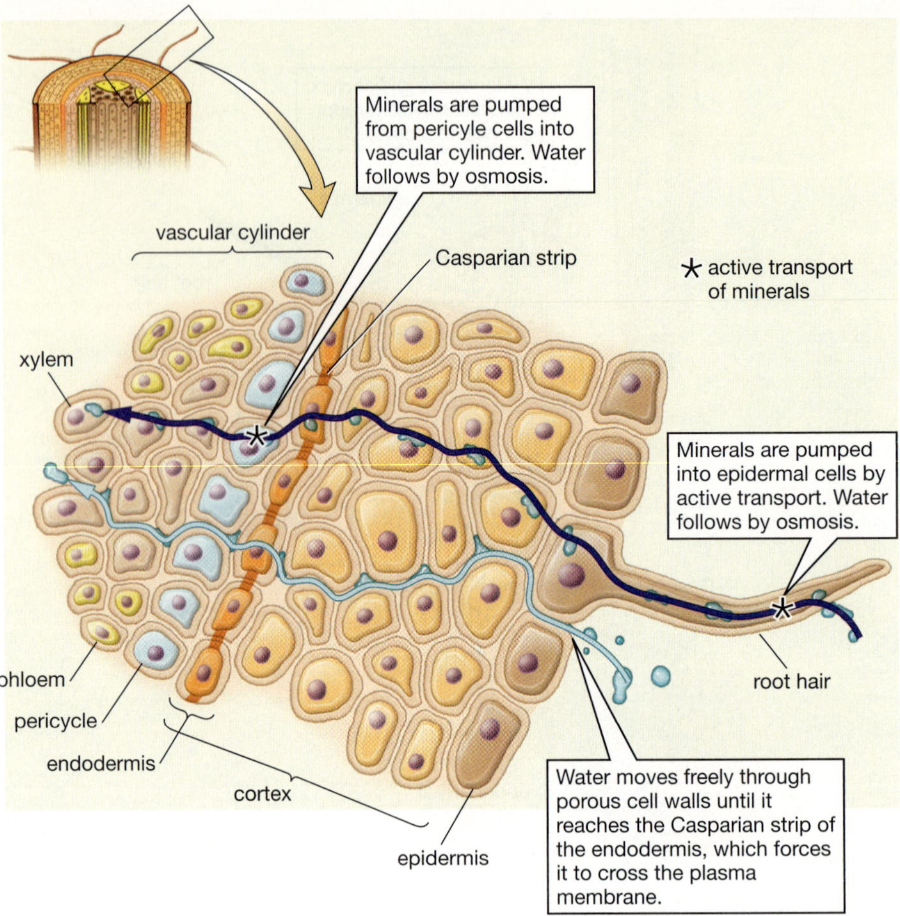

Figure E42-1 Mineral/water up

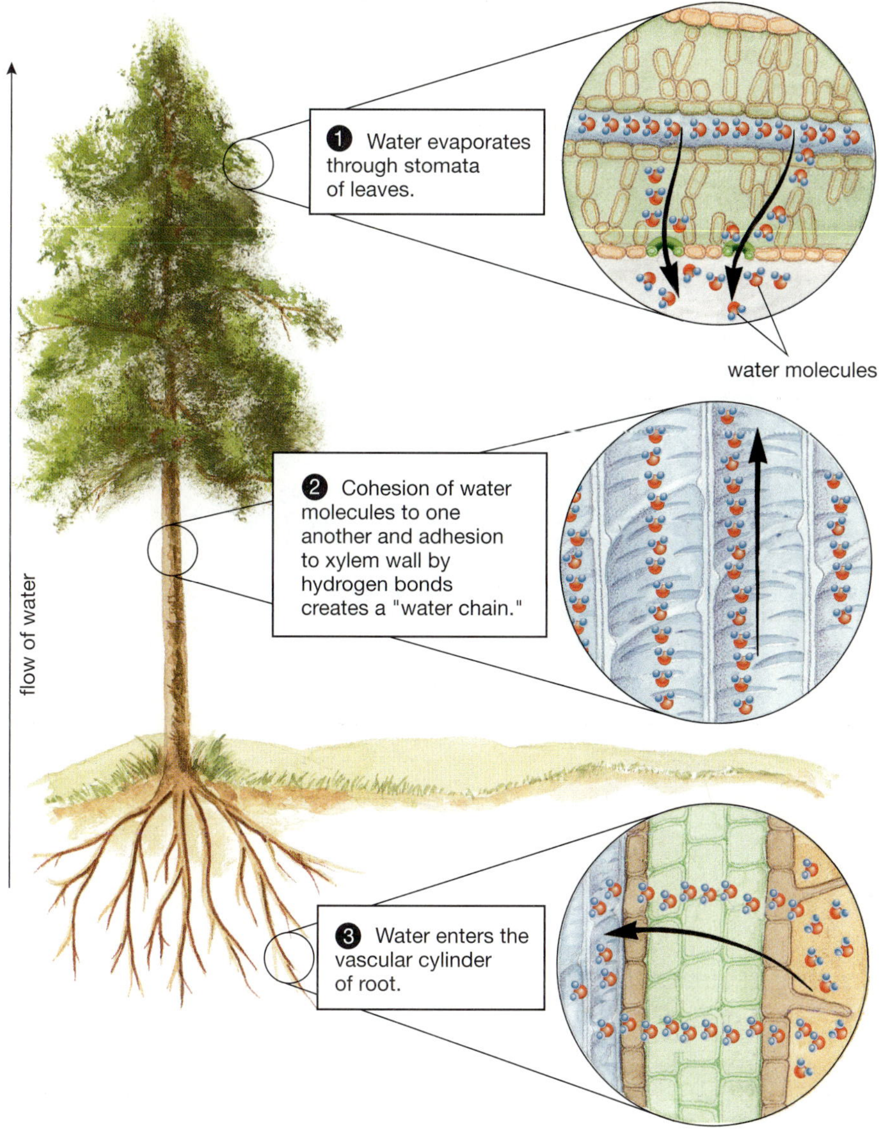

Figure 42-21 Cohesion tension

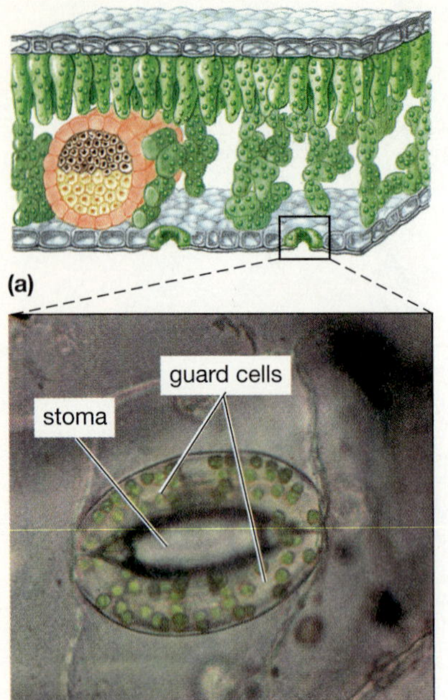

Figure 42-22 Stomata

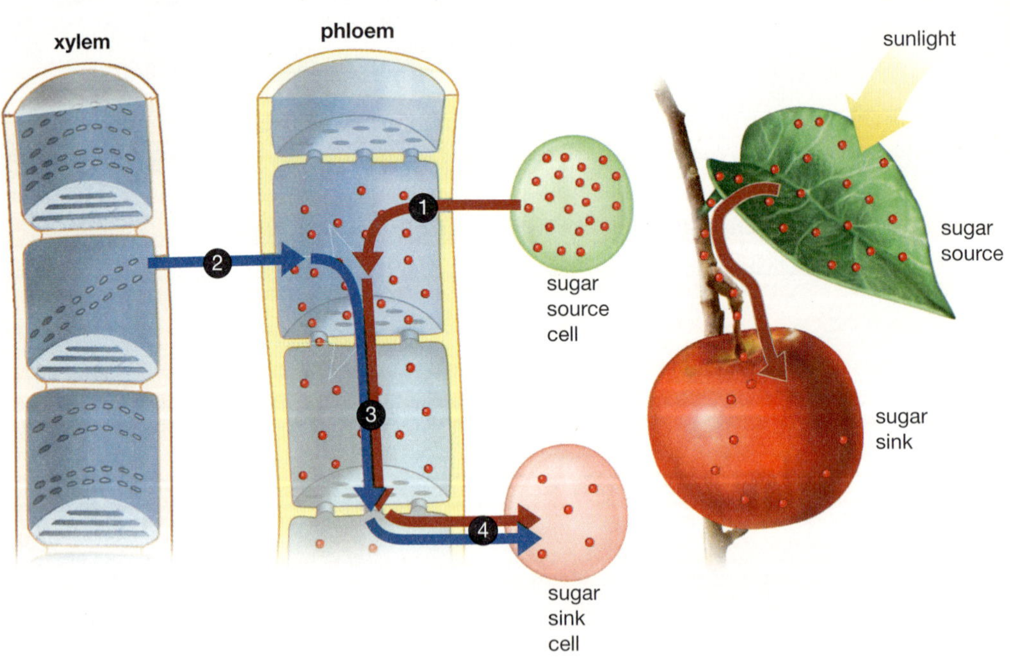

Figure 42-24 Pressure-flow

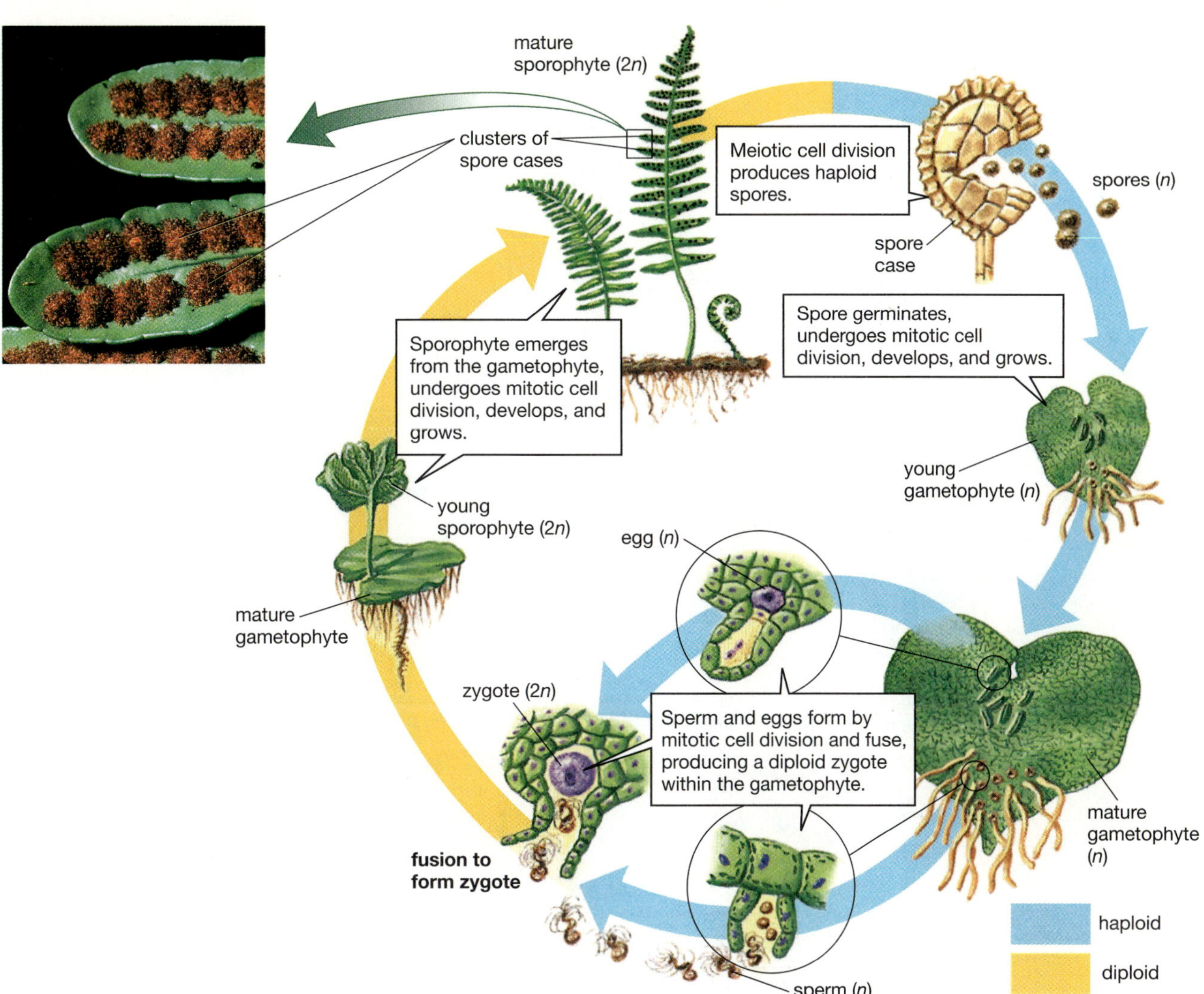

Figure 43-2 Fern life cycle

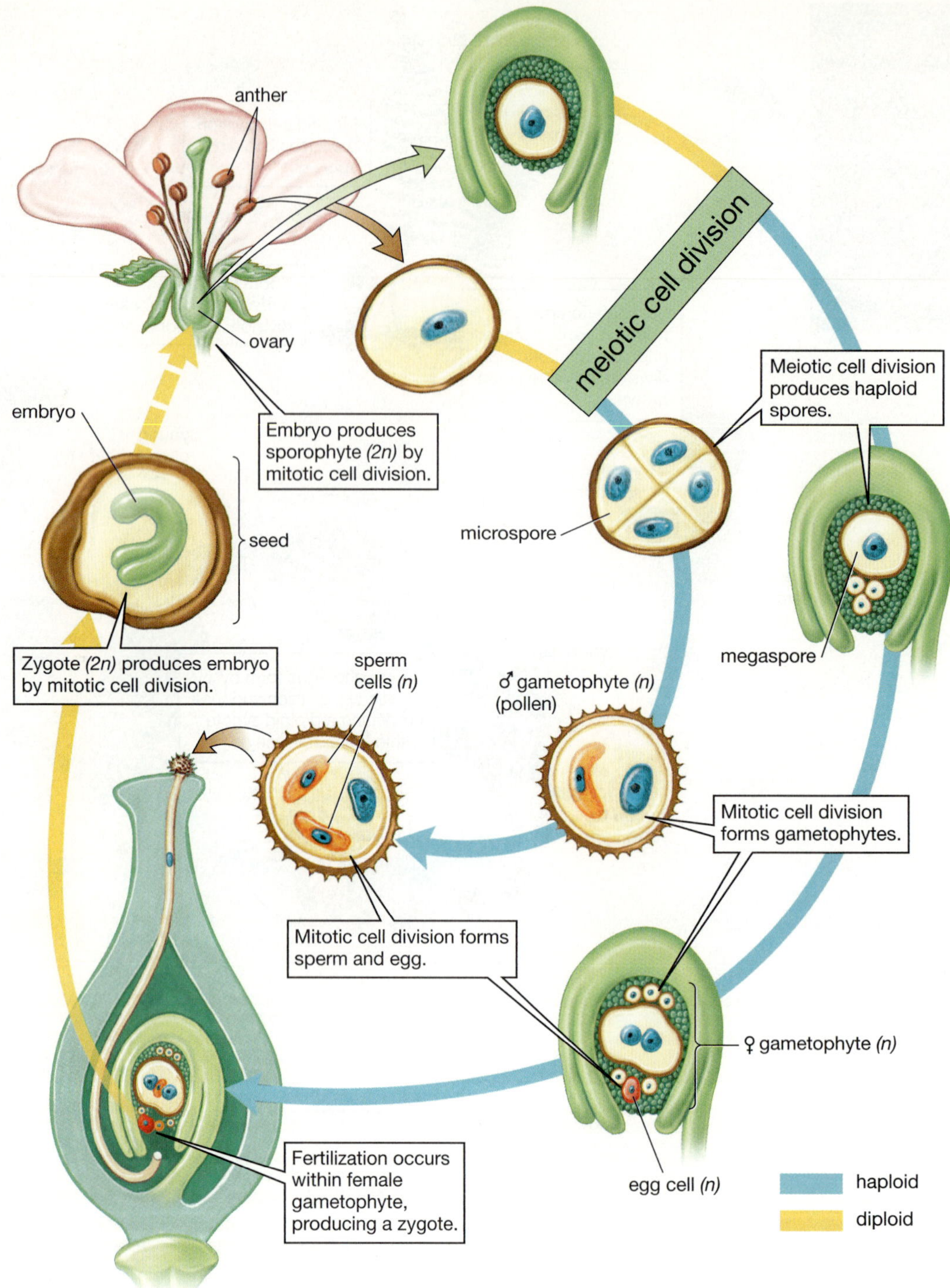

Figure 43-4 Flowering plant life cycle

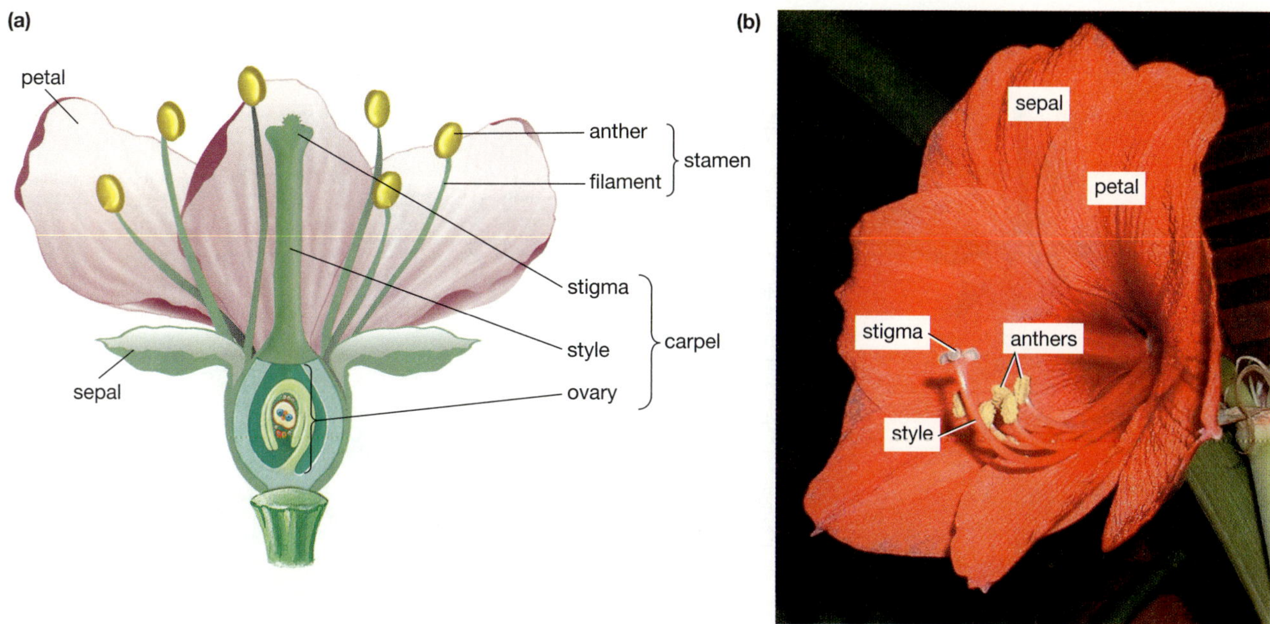

Figure 43-5 A complete flower

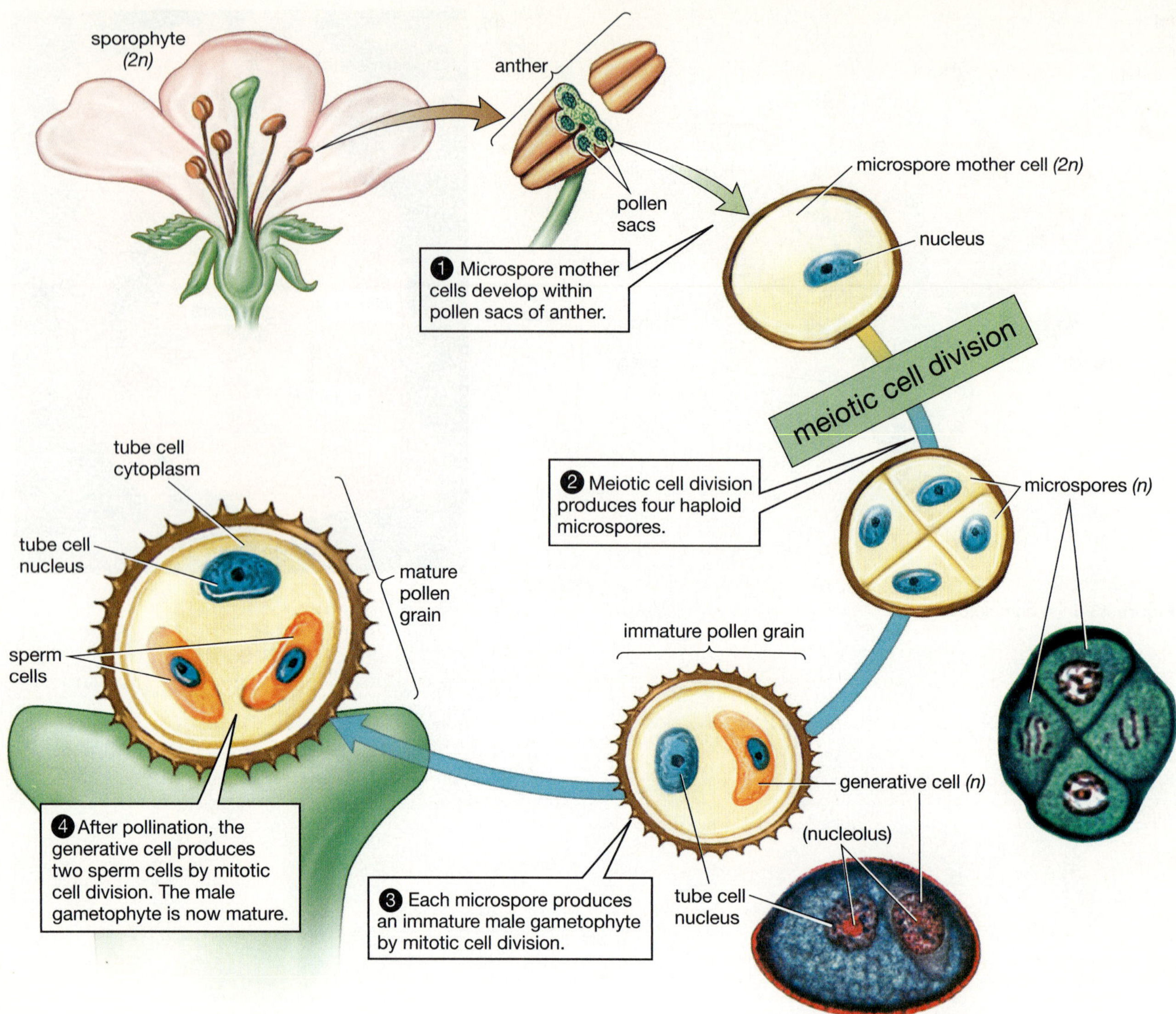

Figure 43-7 Male gametophyte development

Figure 43-9 Grass pollinating

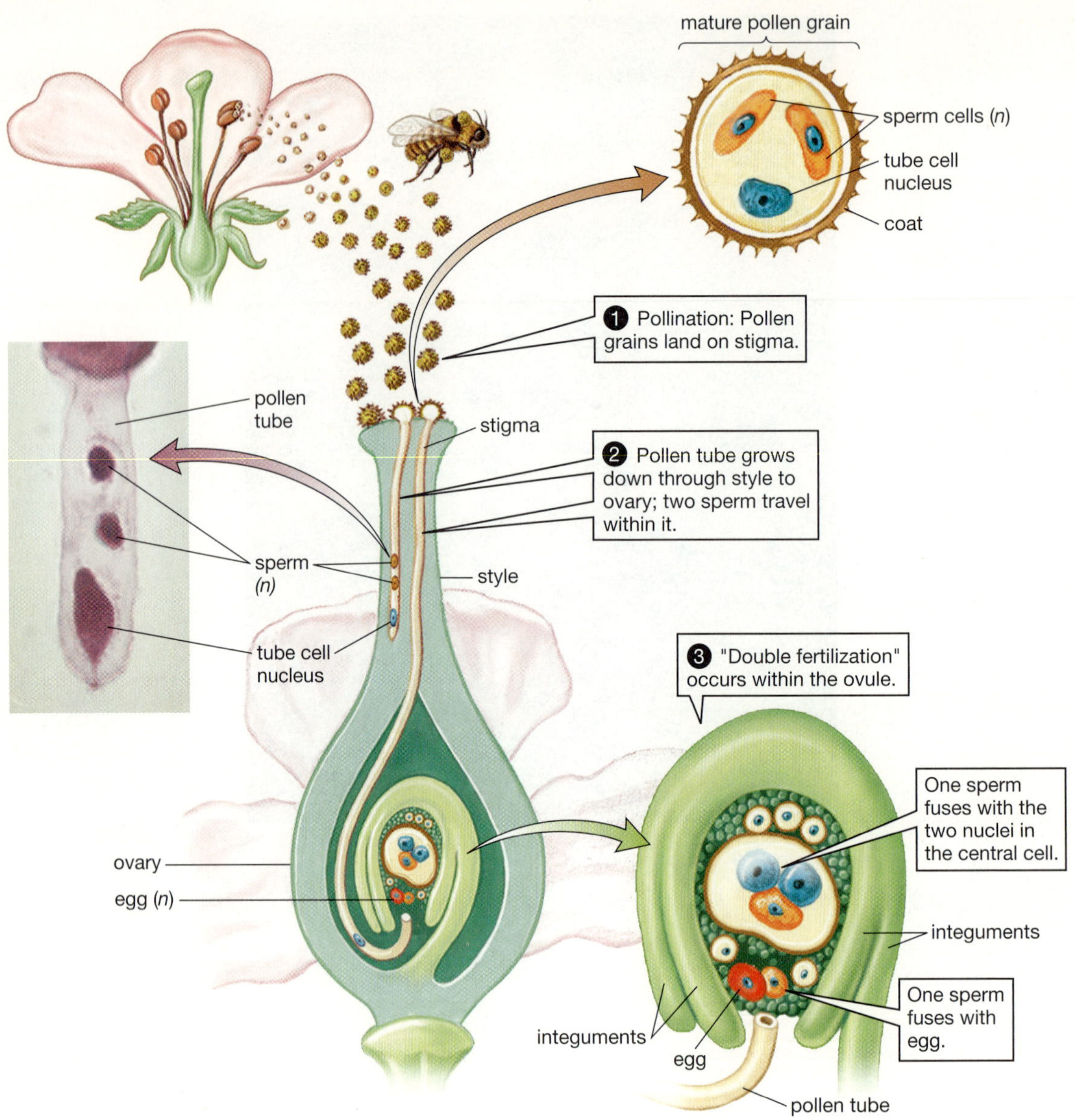

Figure 43-11 Pollination and fertilization

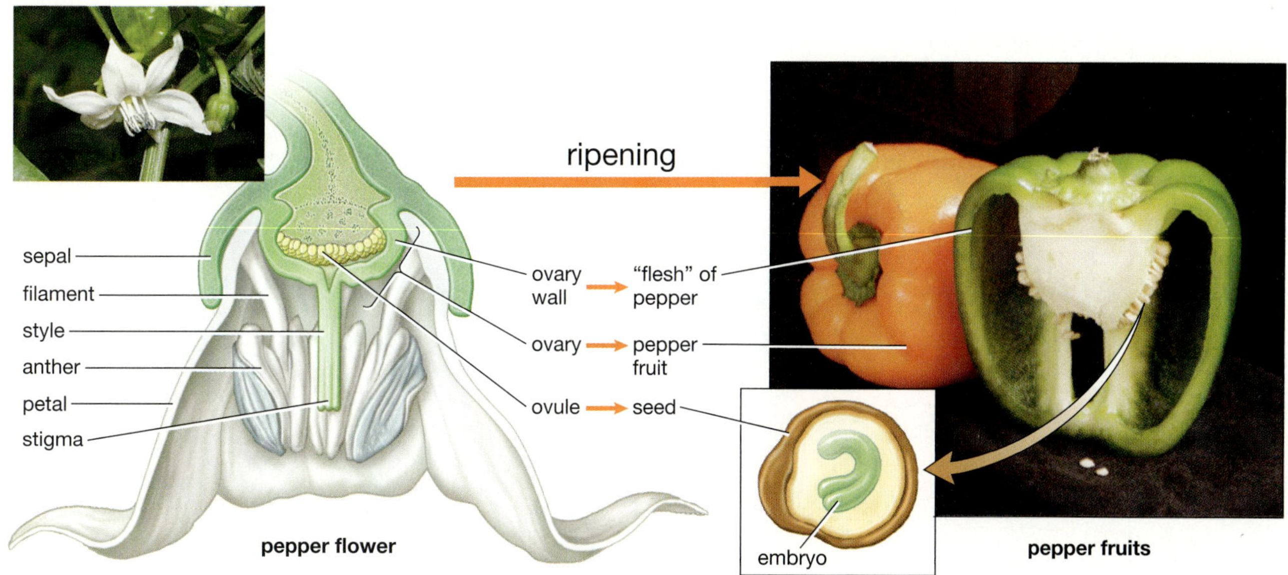

Figure 43-12 Pepper fruit and seed development

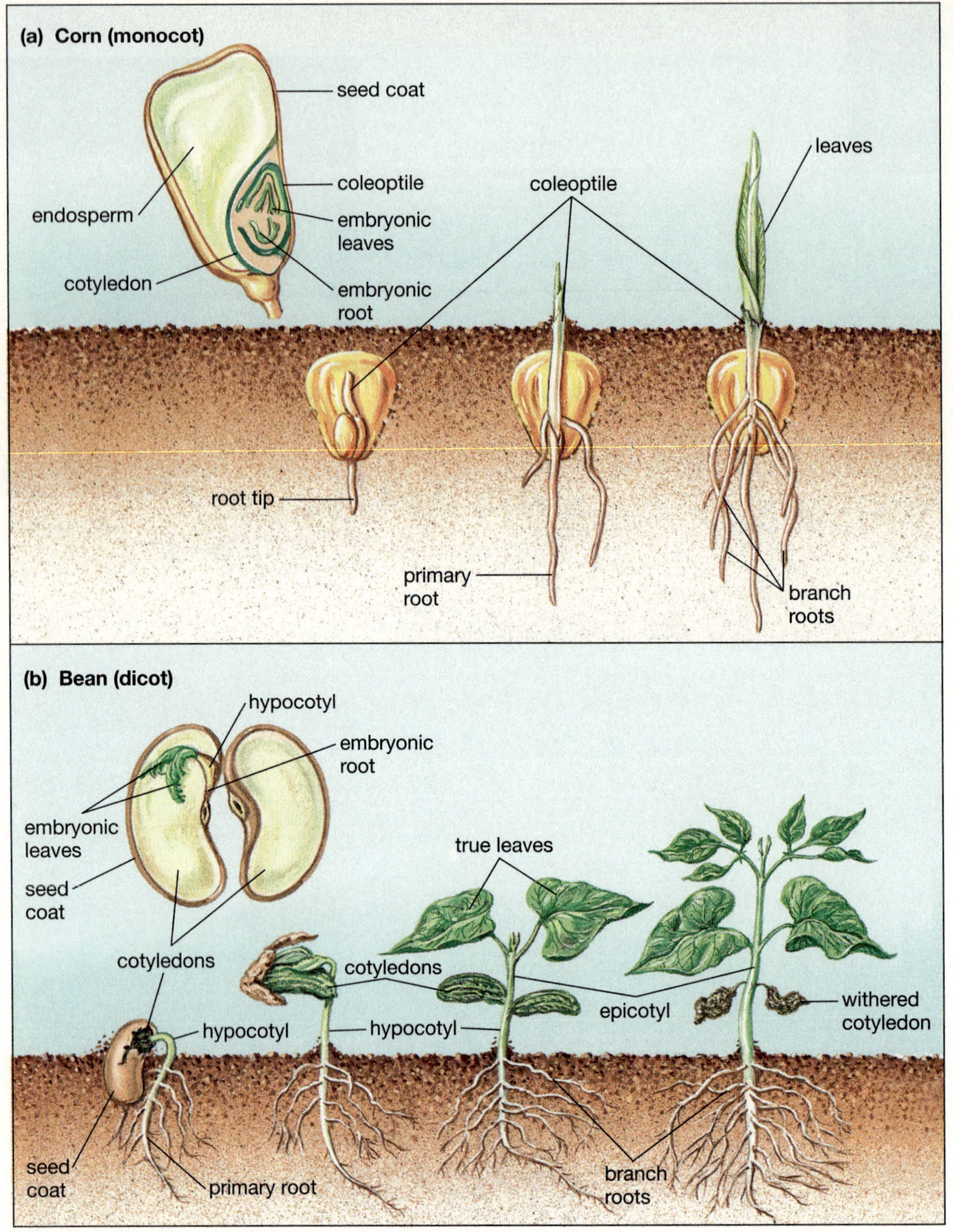

Figure 43-14 Seed germination

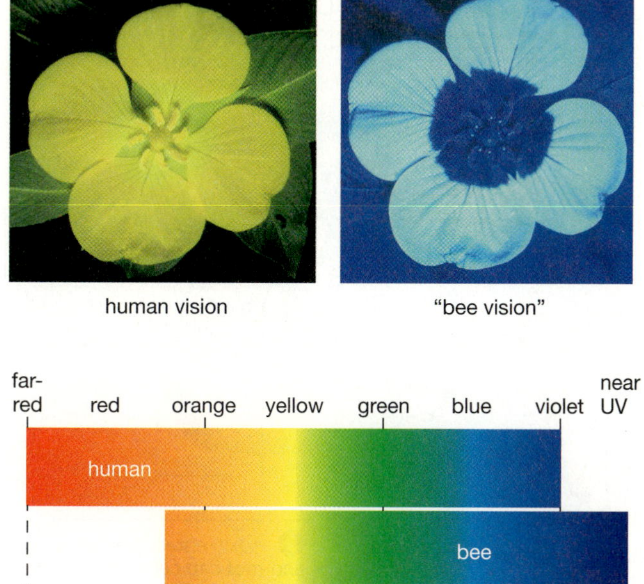

Figure 43-16 Human vs bee vision

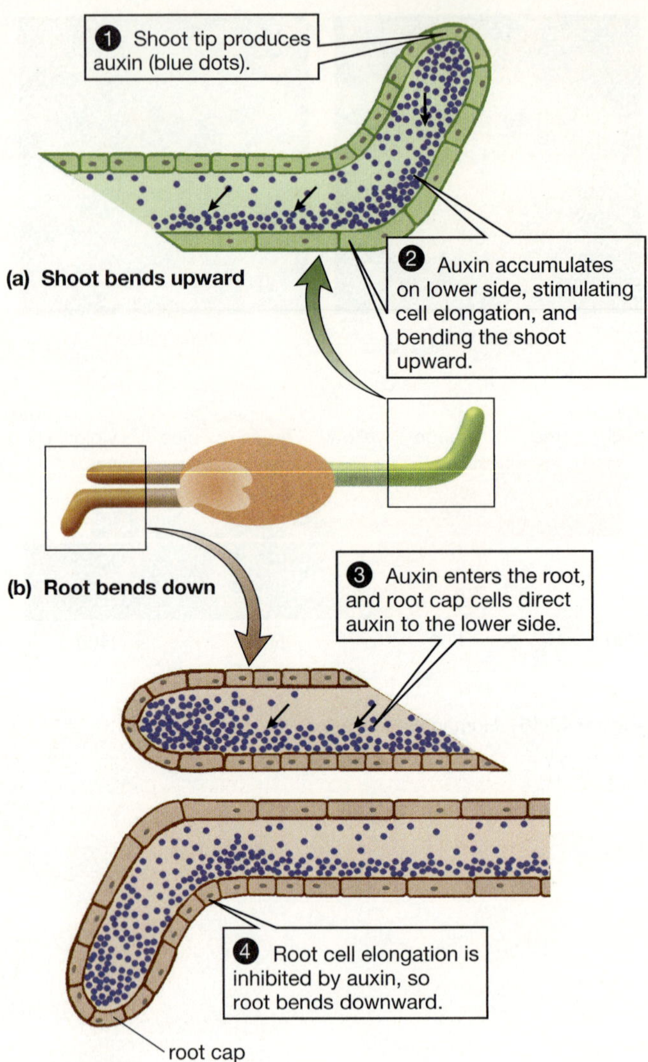

Figure 44-4 Gravitropism and phototropism

Negative gravitropism

Figure 44-4c Tomato plant bending

Positive phototopism

Figure 44-4d Seedlings bending

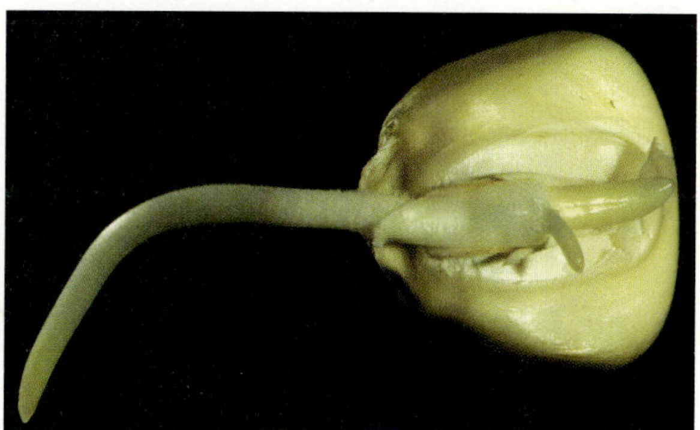

Positive gravitropism

Figure 44-4e Corn seed sprout

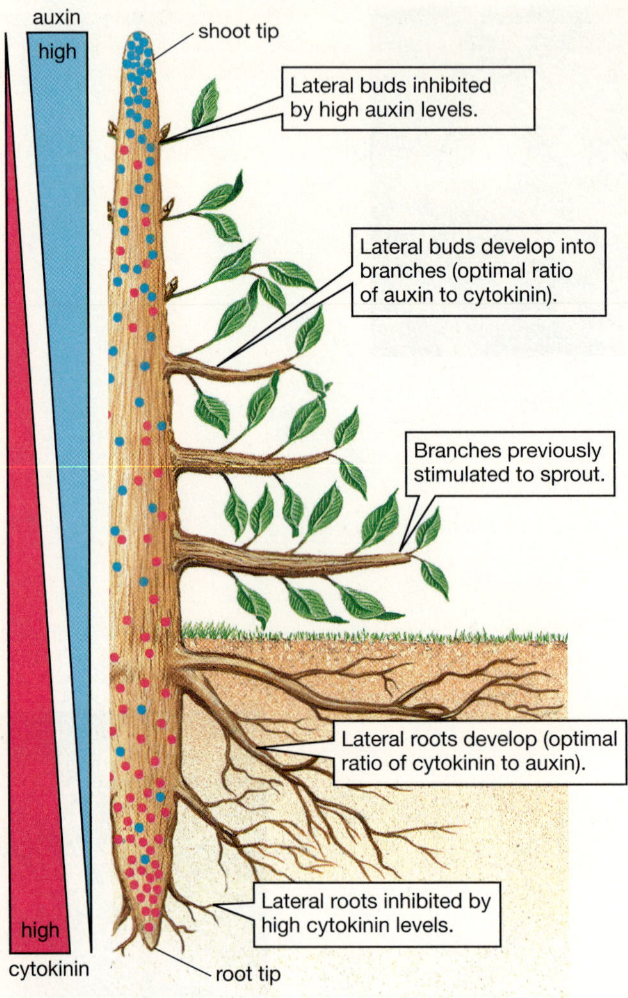

Figure 44-7 Cytokinin/auxin

Figure 44-8 Nightlength and flowering

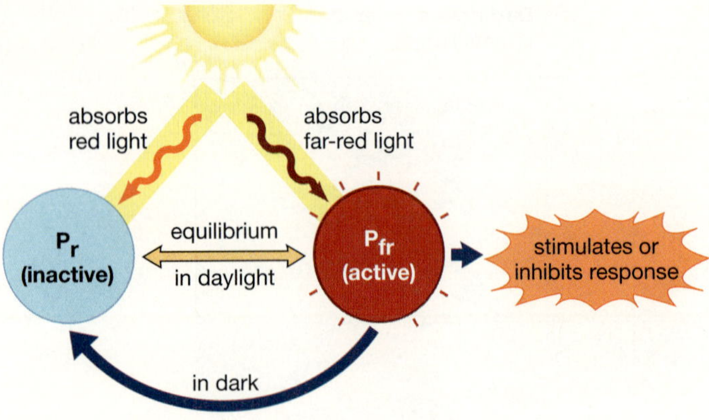

Figure 44-9 Phytochrome

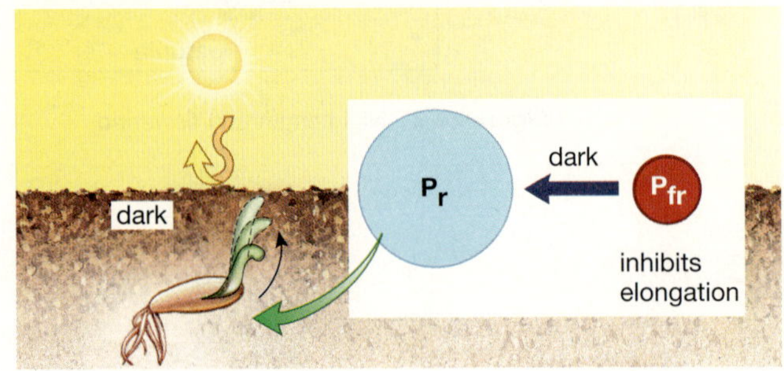

Figure 44-10 Reduced P_{fr}

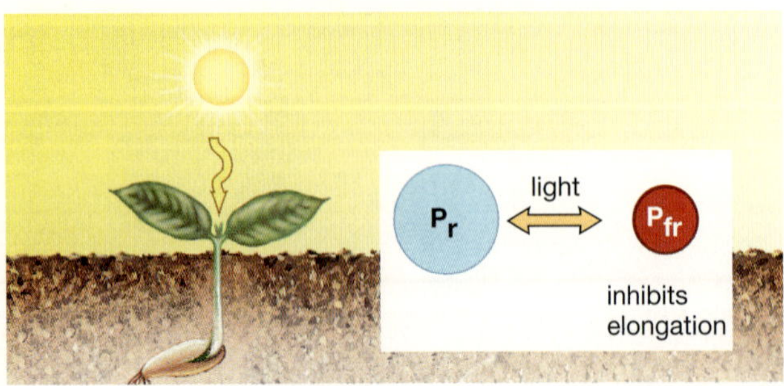

Figure 44-11 Increased P_{fr}

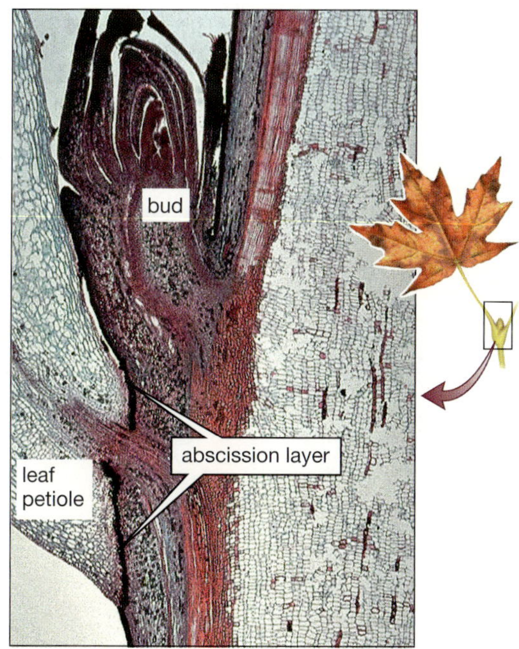

Figure 44-14 Abscission layer

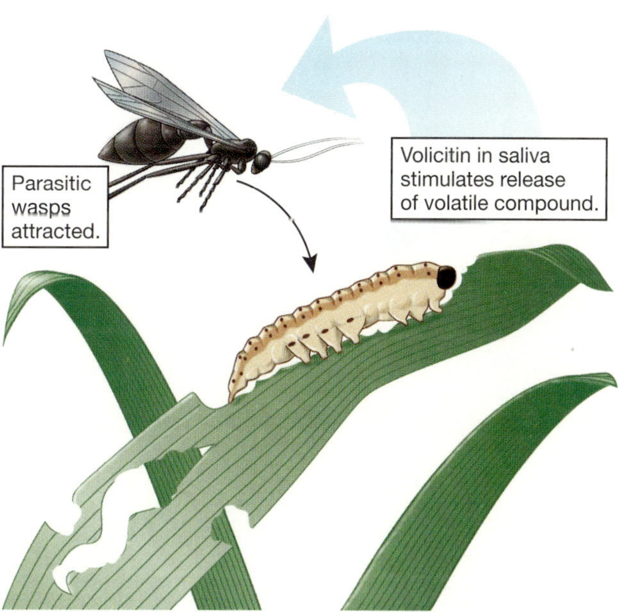

Figure 44-15 Chemical cry for help

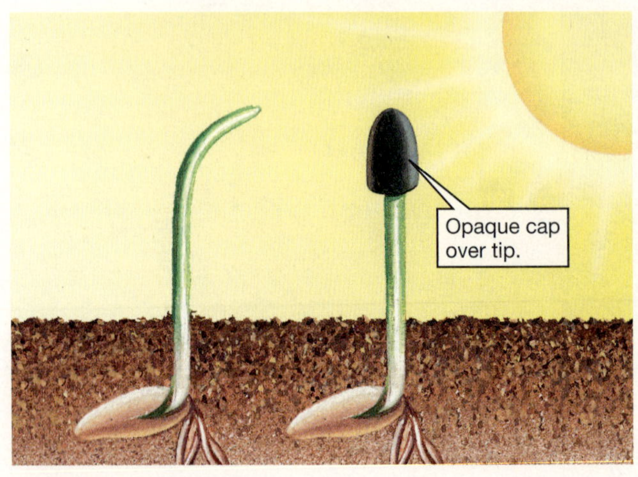

Figure E44-1 Seedlings

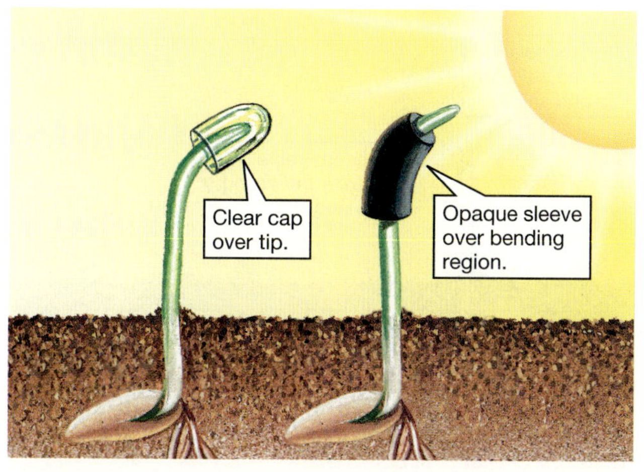

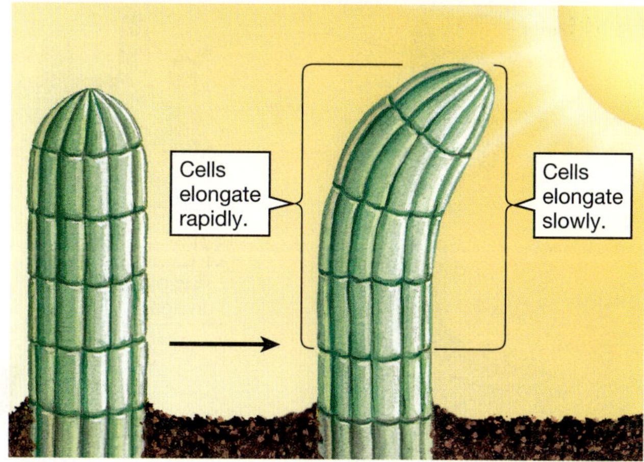

Figure E44-2, E44-3 Seedlings, tips

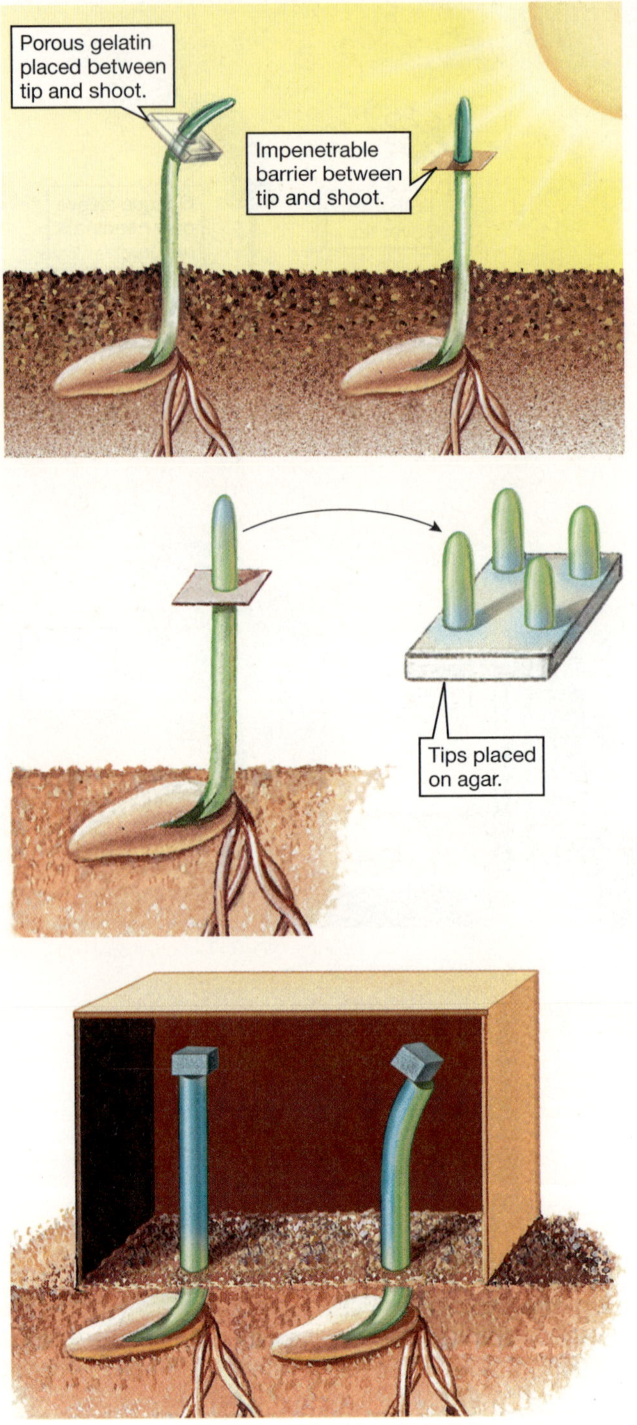

Figures E44-4, E44-5, E44-6 Seedlings